Nanotechnology Applications for Industry 4.0

This book explores the applications of nanotechnology in Industry 4.0, including how nanotechnology can be used to enhance various manufacturing processes. It discusses the use of nanotechnology in areas such as materials science, energy storage, electronics, biomedical and biotechnology, advanced computing and signal processing, and communication systems. Overall, it highlights the potential of these technologies to transform the manufacturing and production processes of the future.

Key Features:

- Explores nanotechnology applications within Industry 4.0.
- Built on a multidisciplinary approach, it offers a robust exploration of nanotechnology applications across various domains in science and engineering.
- Includes detailed case studies and real-world examples reflecting the integration of nanotechnology in Industry 4.0.
- Discusses communication protocols and networks.
- Reviews the development of targeted drug delivery systems, tissue engineering, medical imaging, and diagnostic tools.

This book is aimed at graduate students and researchers in nanotechnology, materials science, and industrial engineering.

Nanotechnology Applications for Industry 4.0

Edited by

Ratneshwar Kumar Ratnesh, Ramesh Kumar,
Jay Singh, and Ramesh Chandra

CRC Press
Taylor & Francis Group
Boca Raton London New York

CRC Press is an imprint of the
Taylor & Francis Group, an **informa** business

Designed cover image: Shutterstock

First edition published 2026
by CRC Press
2385 NW Executive Center Drive, Suite 320, Boca Raton FL 33431

and by CRC Press
4 Park Square, Milton Park, Abingdon, Oxon, OX14 4RN

CRC Press is an imprint of Taylor & Francis Group, LLC

ISBN: 9781032816906 (hbk)
ISBN: 9781041062554 (pbk)
ISBN: 9781003634553 (ebk)

DOI: 10.1201/9781003634553

Typeset in Times
by codeMantra

Dedication

May this book honor the beauty and wisdom of nature, serving as a bridge between the wonders of the natural world and the advancements of Industry 4.0, guiding us toward a future where technology enhances and preserves the beauty of our planet.

Contents

Preface

The fourth industrial revolution, commonly referred to as Industry 4.0, is transforming the manufacturing industry by integrating advanced technologies like artificial intelligence, the Internet of Things, and robotics into the production process. One of the most promising technologies in this field is nanotechnology, which involves manipulating and engineering materials at the atomic and molecular scale.

Nanotechnology has the potential to revolutionize manufacturing processes, making them more efficient, cost-effective, and environmentally sustainable. With the ability to create new materials, enhance existing ones, and improve the performance of devices, nanotechnology is at the forefront of the next wave of technological innovation.

This book explores the latest developments in the application of nanotechnology in Industry 4.0. It covers a range of topics, including the use of nanomaterials in advanced manufacturing, nanosensors for quality control, nanorobotics for precision manufacturing, and nanoelectronics for energy-efficient devices.

The contributors of this book are leading experts from academia, industry, and research institutions who have extensive experience and knowledge in the field of nanotechnology and Industry 4.0. They provide insights and perspectives on the current state and future directions of nanotechnology in Industry 4.0. They share their knowledge and expertise on the challenges and opportunities presented by this rapidly evolving field, as well as the potential implications for industry, society, and the environment.

We hope that this book will be a valuable resource for researchers, engineers, and professionals working in the field of nanotechnology and Industry 4.0. We also hope that it will inspire new ideas and collaborations that will accelerate the development and implementation of nanotechnology in Industry 4.0 and ultimately contribute to the advancement of science and technology for the benefit of society.

Acknowledgments

The authors would like to express sincere gratitude to the countless researchers, scientists, industry experts, and engineers whose groundbreaking work in nanotechnology has made this book possible. Their invaluable contributions have been instrumental in unraveling the vast potential of nanotechnology and its transformative role in Industry 4.0. Special thanks go to the academic and industrial collaborators whose insights and feedback were crucial in shaping the content and direction of this work and to the academic and industrial experts who provided insights and feedback during the writing process, enhancing the book's comprehensiveness and relevance. Additionally, the authors acknowledge the support and encouragement of MIET, Meerut, Uttar Pradesh; Chitkara University, Punjab; BHU, Varanasi; Maharaja Surajmal Brij University, Bharatpur, Rajasthan; and Delhi University, Delhi in bringing this book to fruition.

About the Editors

Ratneshwar Kumar Ratnesh is an associate professor in the Department of Electronics and Communication Engineering at MIET, Meerut, India. He received a B.Tech degree in Electronics and Communication Engineering from Uttar Pradesh Technical University (UPTU), Uttar Pradesh, India; an M.Tech degree in Nanotechnology from Jamia Millia Islamia Central University, Delhi, India; and a Ph.D. from Delhi Technological University, Delhi, India, in 2017. He serves as a visiting faculty at many universities and institutes in India and abroad (NGCEF, Australia). He is an editorial board member of more than five international journals and a reviewer for more than 15 international (SCI) journals. He has published more than 40 international research papers, with total citations of about 500 and an h-index of 12. He has been deeply involved in electronics research in both private and academic institutions. He has received a SERB research grant from the Government of India (ongoing until 2025) and three consultancy projects, one from the Government of India and two from private limited organizations. His current research interests include the development of very large-scale integrated circuits using nanotechnologies such as carbon nanotube field-effect transistors, flexible electronics, quantum dot-based sensors, nano-antennas, robotics with AI, qubits, metamaterials, and thin-film technology. Furthermore, he is also associated with the synthesis, characterization, modeling, and simulation of smart semiconductor materials (carbon nanotubes and quantum dots) at the nanoscale.

ORCID: https://orcid.org/0000-0002-3438-6541
Google Scholar: https://scholar.google.com/citations?user=W1IhIZEAAAAJ&hl=en

Ramesh Kumar was born in Jaipur, Rajasthan, India. He received a B.Tech degree in Electronics and Instrumentation and Control Engineering from the University of Rajasthan, Jaipur, Rajasthan, India, and M.Tech and Ph.D. degrees in Instrumentation and Control Engineering from Dr. B. R. Ambedkar National Institute of Technology, Jalandhar, Punjab, India. His research area of interest is Instrumentation and Control Engineering. He has more than 11 years of teaching and research experience and has published more than 20 papers in reputed journals, book chapters, and conferences.

ORCID: https://orcid.org/0000-0001-9822-8246
Google Scholar: https://scholar.google.com/citations?user=icvXPN8AAAAJ&hl=en

Jay Singh is currently working as an assistant professor in the Department of Chemistry at the Institute of Sciences, Banaras Hindu University, Varanasi, Uttar Pradesh, since 2017. He completed his M.Sc. and B.Sc. degrees at Allahabad University, Uttar Pradesh, India. He earned his Ph.D. degree in Polymer Science from Motilal Nehru National Institute of Technology in 2010. He has been a postdoctoral fellow at the National Physical Laboratory, New Delhi, Chonbuk National University, South Korea, and Delhi Technological University, Delhi. He has received many prestigious fellowships, such as CSIR (RA), DST-Young scientist fellowship, and the DST-INSPIRE faculty award. He is actively engaged in the development of various chemically and biologically synthesized nanomaterials and their nano-biocomposites, conducting polymers, and self-assembled monolayers for clinically important biosensors and sensors for the estimation of various bioanalytes based on enzymes, antibodies, DNA, and various toxic chemicals and gases. He has published more than 100 international research papers, with total citations of more than 3,900 and an h-index of 36. He has completed and is currently running various research projects with different funding agencies. He has more than 13 edited/authored books and has authored over 40 book chapters for internationally reputed press publications. He has also successfully handled many special issues in various reputed journals. He is currently engaged in fabricating sustainable metal oxide-based biosensors for clinical diagnosis, food packaging applications, drug delivery, and tissue engineering applications. His research has contributed significantly to the fundamental understanding of interfacial charge transfer processes and the sensing aspects of metal nanoparticles.

ORCID: https://orcid.org/0000-0002-3793-0450
Google Scholar: https://scholar.google.com/citations?user=ymd1ZksAAAAJ&hl=en

Ramesh Chandra is a distinguished scientist, a fellow of the Royal Society of Chemistry, London, and an outstanding researcher in the field of Biomedical Sciences. Currently, he is the Vice-Chancellor at Maharaja Surajmal Brij University, Bharatpur, Rajasthan, and the Founder-Director of Dr. B. R. Ambedkar Center for Biomedical Research (ACBR), University of Delhi. He is an official resource person to the Minister and Department of Science and Technology in the Kingdom of Lesotho, South Africa.

Professor Chandra served as the Vice-Chancellor of Bundelkhand University, Jhansi, for six years (1999–2005), as well as the President of the Indian Chemical Society (2004–2006); a member of the Planning Commission, Government of U.P., India; Secretary of the Zaheer Science Foundation, New Delhi; and Non-Official Director of Rashtriya Ispat Nigam Ltd (RINL), Government of India. He was also the Head of the Department of Chemistry at the University of Delhi for three years (August 2017–August 2020).

Professor Chandra shows a deep commitment to the cause of higher education and research and possesses a substantial quality of dynamic leadership and vision required for building academic institutions. He started his research career at the University of Delhi; thereafter, he pursued his research at The New York Hospital-Cornell University Medical College, New York, jointly with The Rockefeller University, New York, and the State University of New York at Stony Brook as assistant research professor. He conducted advanced research at Harvard University Medical School-Massachusetts General Hospital (MGH), jointly at MIT, Cambridge, USA.

Professor Chandra has excelled in his academic pursuits. To illustrate, he holds several international patents, nine books, 28 book chapters, and more than 375 original scientific research papers, including review articles and monographs in reputed international journals. He has mentored more than 100 Ph.D. and M.Phil. students, who are now occupying high positions globally. He is a member of several international scientific societies, an associate editor of *Artificial Cells, Nanomedicine and Biotechnology* (Taylor & Francis Publishers), and a member of editorial boards of several high-quality journals.

ORCID: https://orcid.org/0000-0002-7150-3604

Contributors

Mohsin Afroz
Department of Electronics and Communication Engineering
Meerut Institute of Engineering and Technology
Meerut, Uttar Pradesh, India

Ishteyaque Ahmad
Department of Electronics and Communication Engineering
Meerut Institute of Engineering and Technology
Meerut, Uttar Pradesh, India

Mohammad Aljaidi
Department of Computer Science
Zarqa University
Zarqa, Jordan

Mohammed H. Alsharif
Department of Electrical Engineering
College of Electronics and Information Engineering, Sejong University
Gwangjin-gu, Seoul, Korea

Muhammad Adeeb Ansari
Department of Electronics and Communication Engineering
Meerut Institute of Engineering and Technology
Meerut, Uttar Pradesh, India

Kamal Arora
Department of Chemistry
University of Missouri
Columbia, Missouri

Shonak Bansal
Department of Electronics and Communication Engineering
University Institute of Engineering, Chandigarh University
Gharuan, Mohali, India

Anamika Brahma
Department of Chemistry
University of Delhi
Delhi, India

Ramesh Chandra
Department of Chemistry
Maharaja Surajmal Brij University
Bharatpur, Rajasthan, India

Shivam Chaudhary
School of Computer Science Engineering and Technology
Bennett University, Greater Noida
Greater Noida, Uttar Pradesh, India

Prakhar Consul
School of Computer Science Engineering and Technology
Bennett University, Greater Noida
Greater Noida, Uttar Pradesh, India

Tanya Goel
Dr. B. R. Ambedkar Center for Biomedical Research
University of Delhi
Delhi, India

Anupma Gupta
Department of Electronics and Communication Engineering,
University Institute of Engineering, Chandigarh University,
Gharuan, Mohali, India,

Himanshu
Department of Chemistry
University of Delhi
Delhi, India

Neeraj Joshi
School of Computer Science Engineering and Technology
Bennett University, Greater Noida
Greater Noida, Uttar Pradesh, India

Gaurav Kumar
Department of Mechanical Engineering
Delhi Technological University
New Delhi, Delhi, India

Ramesh Kumar
Jadara University Research Center
Jadara University
Irbid, Jordan
and
Chitkara University Institute of Engineering & Technology
Chitkara University
Punjab, India

Vibhor Kumar
Department of Electrical and Computer Engineering
Texas A&M University
College Station, Texas

Vipan Kumar
Department of Electronics and Communication Engineering
Sri Sai College of Engineering and Technology
Pathankot, Punjab, India

Mohd Maaz
Department of Electronics and Communication Engineering
Meerut Institute of Engineering and Technology
Meerut, Uttar Pradesh, India

Jitender Madan
Department of Pharmaceutics
National Institute of Pharmaceutical Education and Research
Hyderabad, Telangana, India

Bhawna Mallick
Department of Computer Science and Engineering
SSET, Sharda University, Knowledge Park-III
Greater Noida, Uttar Pradesh, India

Ishani Mishra
Department of Biosciences and Bioengineering
Indian Institute of Technology
Roorkee, Uttarakhand, India

Neha Mittal
Department of Electronics and Communication Engineering
Meerut Institute of Engineering and Technology
Meerut, Uttar Pradesh, India

Nisha
Department of Chemistry
University of Delhi
Delhi, India

Pawan Kumar Pathak
School of Automation
Banasthali Vidyapith
Toda, Rajasthan, India

Abdullah Bin Queyam
School of Automation
Banasthali Vidyapith
Toda, Rajasthan, India

Rama Nand Rai
Department of Chemistry, Institute of Science
Banaras Hindu University
Varanasi, Uttar Pradesh, India

Priyanshu Rajput
Department of Chemistry
University of Delhi
Delhi, India

Sangita Ranjan
Department of Computer Science and Information Technology
Meerut Institute of Engineering and Technology
Meerut, Uttar Pradesh, India

Shubhanshu Rao
Department of Mechanical Engineering
Delhi Technological University
New Delhi, Delhi, India

Ratneshwar Kumar Ratnesh
Department of Electronics & Communication Engineering
Meerut Institute of Engineering & Technology
Meerut, Uttar Pradesh, India

Satyendra Kumar Satyarthi
Department of Ceramic Engineering
Indian Institute of Technology (Banaras Hindu University)
Varanasi, Uttar Pradesh, India

Yashaswi Sharma
Department of Chemistry
University of Delhi
Delhi, India

Akhilesh Kumar Singh
School of Materials Science and Technology
Indian Institute of Technology (Banaras Hindu University)
Varanasi, Uttar Pradesh, India

Aarushi Singh
Ramjas College
University of Delhi
Delhi, India

Jagbeer Singh
Department of Information Technology
Meerut Institute of Engineering and Technology
Meerut, Uttar Pradesh, India

Jay Singh
Department of Chemistry, Institute of Science
Banaras Hindu University
Varanasi, Uttar Pradesh, India

Raj Kumar Singh
Department of Mechanical Engineering
Delhi Technological University
Rohini, New Delhi, Delhi, India

Rohit Kumar Singh
Department of Information Technology
Meerut Institute of Engineering and Technology
Meerut, Uttar Pradesh, India

Sanjay Kumar Singh
Department of Electronics and Communication Engineering
Meerut Institute of Engineering and Technology
Meerut, Uttar Pradesh, India

Snigdha Singh
Department of Chemistry
University of Delhi
Delhi, India

Vishwa Pratap Singh
UNAM-Materials Science and Nanotechnology
Bilkent University
Ankara, Turkey

Aditi Singhal
Department of Pharmaceutical Technology
Meerut Institute of Engineering and Technology
Meerut, Uttar Pradesh, India

Manish Kumar Singla
Applied Science Research Center
Applied Science Private University
Amman, Jordan
and
Department of Biosciences, Saveetha School of Engineering, Saveetha Institute of Medical and Technical Sciences, Chennai, India

Rajesh Kumar Thakur
Department of Computer Science and Information Technology
Meerut Institute of Engineering and Technology
Meerut, Uttar Pradesh, India

Manisha Tiwari
Dr. B. R. Ambedkar Center for Biomedical Research
University of Delhi
Delhi, India

Subodh Kumar Tripathi
Department of Electronics and Communication Engineering
Meerut Institute of Engineering and Technology
Meerut, Uttar Pradesh, India

Mansi Verma
Department of Pharmaceutical Technology
Meerut Institute of Engineering and Technology
Meerut, Uttar Pradesh, India

Vineet Vishnoi
Department of Computer Science
Shobit Institute of Engineering & Technology (Deemed to be University)
Meerut, Uttar Pradesh, India

1 An Overview of Industry 4.0 Applications in the Area of Nanotechnology

Ramesh Kumar, Ratneshwar Kumar Ratnesh, Jay Singh, and Ramesh Chandra

1.1 INTRODUCTION

Nanotechnology has gained traction across a diverse range of industries worldwide, with rapid adoption in the developing world over the past decade [1]. As a general technology, nanotechnology is not a new concept. Four generations of nanomaterials, namely, active and inactive nanocomposites, multi-system nanosystems, and small molecular nanosystems, have emerged and are being used in scientific fields, demonstrating deep and widespread technological impact [1]. The rapid progress in nanoscience indicates that nanotechnology will soon be integrated into various scientific and technological fields [2]. This review article discusses recent advanced applications of nanotechnology in major industries, such as agriculture, food, cosmetics, pharmaceuticals, healthcare, automotive, oil and gas, chemicals, and machinery. In addition, it covers the drawbacks of nanotechnology present in each industry, raising awareness about its risks and benefits [3–4]. Nanotechnology is a multidisciplinary field that combines principles from biological, physical, and chemical sciences at the nanometer scale. It enables growth reduction, new interactions with chemicals, and biological processes such as targeted drug delivery [5] Nanotechnology bridges classical and quantum mechanics through mesoscopic systems, enabling natural nanoassemblies designed for agricultural purposes, nanomedicine, and nano tools for therapeutic treatment and diagnosis [6]. It has allowed the treatment of previously incurable diseases through nano-based drugs and diagnostic tools. In addition, nanotechnology has revolutionized manufacturing in many industries by using the principle of reverse engineering inspired by nature [7]. Instead of cutting through materials, it enables bottom–up manufacturing at the atomic level, facilitating the creation of materials from a deeper level. Worldwide, significant investments are being made in nanotechnology, especially in developing countries like Europe, China, and the United States, to advance this new knowledge [8].

However, developing countries are lagging behind, finding it difficult to keep pace with past industrial progress [9], mainly due to economic constraints. Nevertheless, both developed and developing scientific communities acknowledge nanotechnology

DOI: 10.1201/9781003634553-1

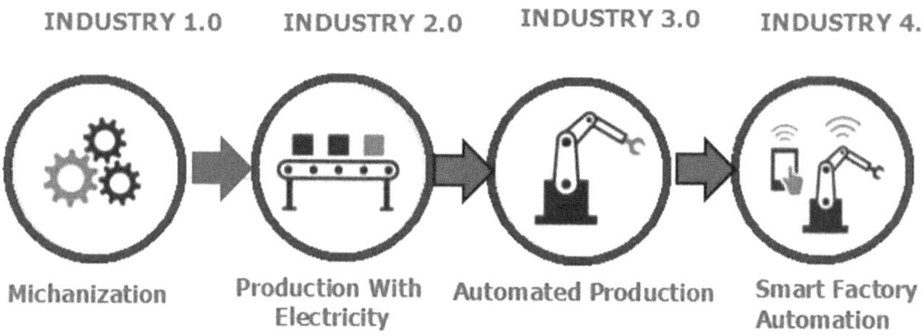

FIGURE 1.1 Industry revolution.

as the next technological frontier [10], which necessitates further industrial development and investment in this field in the coming years. As environmental resources dwindle [11], the scientific community seeks cost-effective, safe, and clean solutions. Nanotechnology offers an effective solution, providing affordable and environmentally friendly alternatives to conventional mass production and heavy machinery [13]. With potential applications in nanomaterials, nanoelectronics, nanomedicine, chemistry, physics, and biology [12], nanotechnology is poised to be a very important field of study for future generations. This review explores key applications of nanotechnology in important global industries and their implications for future industrial development [14].

1.1.1 Definition of Industry 4.0

Industry 4.0, additionally referred to as the Fourth Industrial Revolution, refers to the ongoing transformation of traditional production and commercial practices through the combination of advanced digital technologies, as shown in Figure 1.1. It represents a paradigm shift in how industries operate, leveraging current technologies such as Internet of Things (IoT), cyber-physical systems, cloud computing, artificial intelligence (AI), and big data analytics [1–5].

The core idea of Industry 4.0 is the digitization and interconnection of numerous processes and systems within the commercial environment. This interconnectivity allows for the seamless exchange of information and data among machines, systems, and people, leading to improved automation, enhanced efficiency, and better decision-making capabilities. By harnessing the power of real-time data analysis and intelligent systems, Industry 4.0 aims to optimize manufacturing processes, enhance product quality, reduce downtime, and enable mass customization to meet consumer needs more effectively [11–14]. Furthermore, Industry 4.0 promotes the integration of smart factories, where machines and systems can communicate and collaborate autonomously, enabling self-optimization, predictive maintenance, and efficient resource allocation, as shown in Figure 1.2. This paradigm shift also encourages the development of new business models, fostering innovation and creating opportunities for new value chains and services within the industrial landscape [1–7].

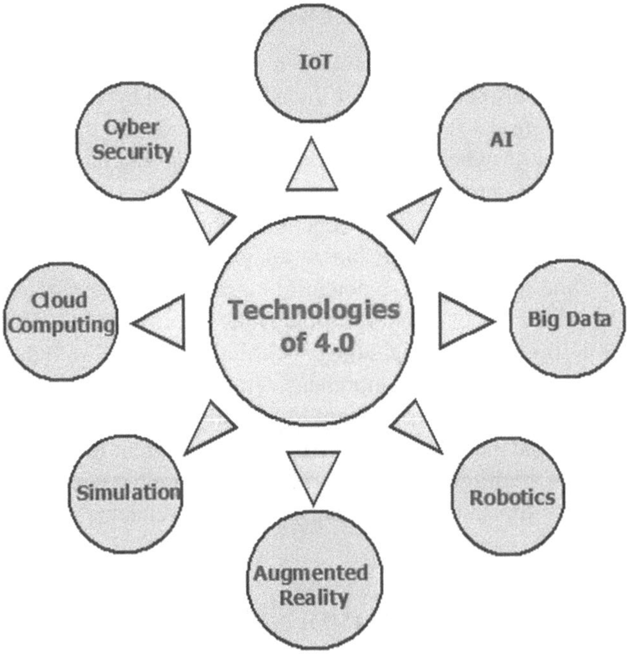

FIGURE 1.2 Industry revolution 4.0.

1.1.2 Significance of Nanotechnology in the Fourth Industrial Revolution

Nanotechnology plays a key role in the Fourth Industrial Revolution, enabling major advances and breakthroughs across various sectors. The ability to manipulate and control materials at the nanoscale has opened up a world of possibilities, transforming industries and driving innovation. In the context of Industry 4.0, nanotechnology is crucial as it facilitates the development of materials, sensors, and devices that are essential for the connected and intelligent systems envisioned in this revolution. Nanostructured materials exhibit unique properties, such as high strength, conductivity, and reactivity, enabling lighter, stronger, and more efficient industrial products [1–10]. Nano sensors, capable of detecting minute changes in their environment, enable precise monitoring and real-time data collection, enabling predictive maintenance and efficient resource allocation. In addition, nanotechnology supports the development of advanced computing and communication technologies, which form the backbone of Industry 4.0. Nanoscale transistors and integrated circuits allow for the miniaturization of electronic devices, resulting in increased computing power and energy efficiency [7–10]. Nanostructured materials in energy storage devices, such as batteries and supercapacitors, facilitate the integration of energy sources and enable the development of sustainable industrial processes. In the nanotechnology sector, Industry 4.0 has the potential to achieve unprecedented levels of automation, efficiency, and sustainability, driving innovation and transforming traditional manufacturing processes into intelligent, connected, and adaptable systems [11–13].

1.1.3 Objectives and Scope of the Chapter

The objective of this chapter is to provide a comprehensive overview of the synergistic relationship between the Fourth Industrial Revolution and nanotechnology. Its goals are to explore the fundamental concepts of Industry 4.0 and its core technologies, including IoT, cyber-physical systems, AI, and big data analytics, as well as advanced materials, sensors, electronics, energy storage, and biotechnology, in enabling smart and interconnected production techniques. This leads to increased performance, sustainability, and customization, along with scalability, safety, regulatory concerns, and forecasting the potential impact on business practices, societal changes, and their relevance to nanotechnology. It will also discuss emerging trends and future possibilities of nanotechnology within the context of Industry 4.0. The scope of this chapter includes a comprehensive review of the applications of nanotechnology across diverse sectors relevant to Industry 4.0, including manufacturing, energy, healthcare, and electronics. It delves into the current research and development (R&D) efforts, highlighting real-world examples and case studies that demonstrate the successful implementation of nanotechnology-enabled solutions within the realm of Industry 4.0 [1–14].

1.2 NANOTECHNOLOGY IN INDUSTRY 4.0

Nanotechnology plays a key role in the Fourth Industrial Revolution, or Industry 4.0, with the potential to enhance advanced materials, equipment, and processes with other important capabilities. Nanomaterials, whose unique properties depend on their nanoscale dimensions, are being incorporated into various industrial materials, increasing their strength, durability, conductivity, and efficiency [15]. Nanoelectronics and nanophotonics are driving breakthroughs in computing, communication, and sensing technologies, enabling faster data processing, higher bandwidth, and more sophisticated detection systems. Nanobiotechnology is revolutionizing healthcare through targeted drug delivery, tissue engineering, and advanced diagnostics. In addition, non-manufacturing technologies, such as 3D printing and self-assembly, are enabling the fabrication of complex products with precise control of properties and functionality, paving the way for more efficient manufacturing processes for stable solutions [16]. Therefore, we discuss individually all the technological advancements in nanotechnology in relation to Industry Revolution 4.0, presented below [17].

1.2.1 Convergence of Nanotechnology in Industry 4.0

The convergence of nanotechnology and other innovative technologies is the driving force behind Industry 4.0, enabling the seamless integration of physical and digital environments. The integration of nanotechnology with biotechnology, information technology, and cognitive science has led to the development of advanced nanobiosensors, nanoelectronics, and nanorobotics. These innovations facilitate real-time monitoring, data analysis, and autonomous decision-making in industrial processes. The combination of nanotechnology and additive manufacturing technologies, such as 3D printing, has opened up new opportunities to create complex, customizable nanomaterials and devices [18]. In addition, the convergence of nanotechnology with AI and machine learning algorithms contributes to the design and management of nanomaterials, as well as the development of intelligent nanorobots for various applications [19]. This convergence of nanotechnology and other innovative technologies is transforming manufacturing, enabling smarter factories, mass production, and customized products, ultimately paving the way for the Industry 4.0 revolution [20].

1.2.2 Enabling Technologies for Nano-enabled Industry 4.0 Applications

Technologies that enable nanotechnology in Industry 4.0 have emerged as a key driving force for the Fourth Industrial Revolution [3]. Nanotechnology, with its ability to manipulate materials at the atomic and molecular levels, has opened up many ways to improve industrial processes, manufacturing, and operational efficiency [21]. The integration of nanomaterials, nano devices, and nanoscale sensors with advanced technologies such as IoT, AI, and big data analysis has paved the way for new technologies in various sectors [22,23]. For example, nanosensors provide unprecedented sensitivity and accuracy in monitoring and controlling industrial processes, enabling real-time data collection and analysis, as shown in Figure 1.3. Nanomaterials with unique properties, such as high strength, ductility, and performance, are being incorporated into the manufacturing process, resulting in lighter, stronger products that perform well. Furthermore, the convergence of nanotechnology with AI and machine learning algorithms enables predictive maintenance, more efficient use of resources, and automated decision-making, thereby improving overall efficiency and sustainability [1–2].

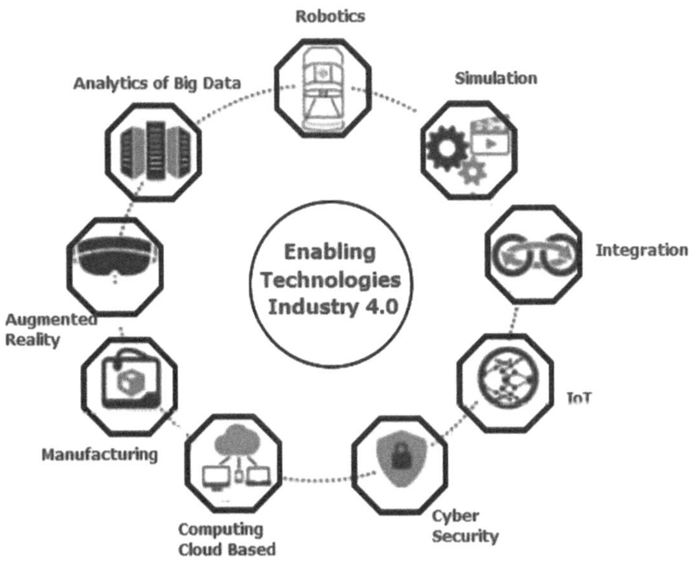

FIGURE 1.3 Enabling technologies of Industry 4.0.

1.2.3 Internet of Nano-Things (IoNT)

IoNT represents a visionary concept that extends the reach of IoT to the nanoscale realm. While the IoT has revolutionized the way we connect and interact with ordinary devices, the IoNT takes this idea to an extraordinary level, envisioning a network of interconnected nanoscale devices capable of sensing, computing, and communicating on the molecular scale. The origins of IoNT can be traced back to the rapid advancements in nanotechnology, a field that has been exploring the manipulation of matter at the atomic and molecular scales. As early as the Nineties, researchers began exploring the capability of nanoscale devices for numerous applications, ranging from healthcare to environmental monitoring. However, the idea of interconnecting these devices and establishing communication networks at the nanoscale level emerged more recently, driven by the convergence of nanotechnology with fields such as wireless communications, networking, and advanced computing, as shown in Figure 1.4. [24].

The IoNT envisions a future where nanoscale sensors, actuators, and computing devices are seamlessly integrated, forming a vast network that can monitor and interact with the surrounding at an unprecedented level of detail. These nanodevices, frequently made of advanced nanomaterials, possess unique properties that enable them to operate on the nanoscale, overcoming the limitations of traditional devices. One of the key enabling technologies for IoNT is the development of nanonetworks, which facilitate communication and data transfer among nanoscale devices. These nanonetworks use unconventional communication paradigms, such as molecular communication, in which data is encoded and transmitted using molecules, or electromagnetic communication at terahertz frequencies [25].

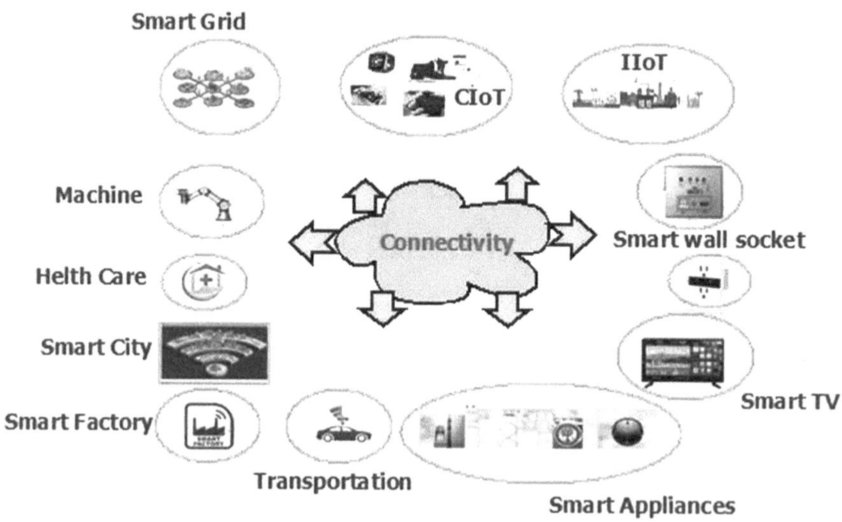

FIGURE 1.4 Industry internet of nano-things revolution.

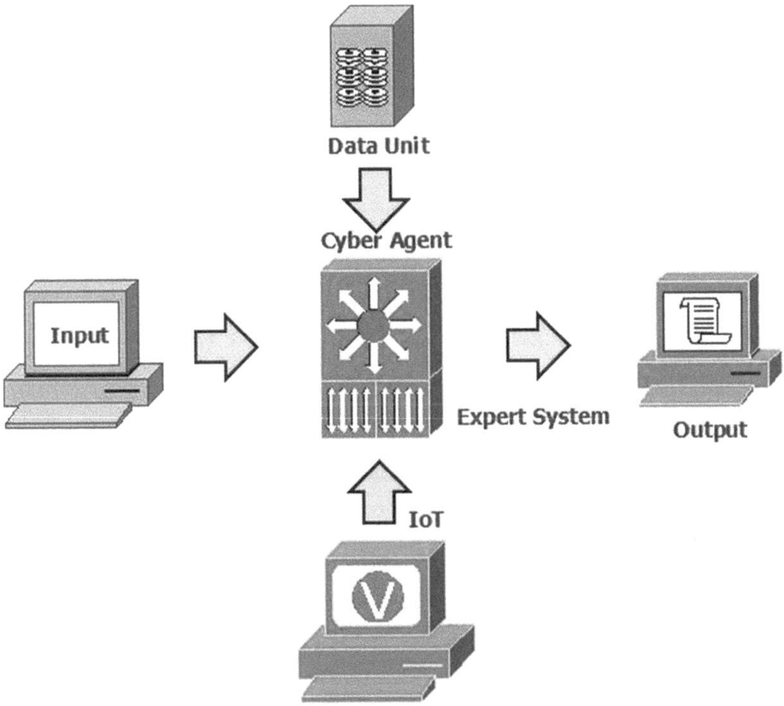

FIGURE 1.5 Cyber-physical nano systems in industry.

1.2.4 Cyber-Physical Nano Systems

Cyber-physical nano systems are an emerging field at the intersection of nanotechnology, computer science, and engineering, as shown in Figure 1.5. These systems involve the integration of mathematical and physical processes at the nanoscale, allowing materials to be controlled and manipulated at the atomic and molecular levels [26]. These tightly connected and interconnected systems have the potential to revolutionize a wide range of applications, from pharmaceuticals and healthcare to energy and manufacturing. One of the key applications of cyber-physical nano systems is in the field of personalized medicine. These techniques can be used to design and develop targeted drug delivery systems, where nanoparticles are programmed to release drugs at specific sites within the body [27]. This results in more effective treatment, fewer side effects, and improved patient outcomes. In addition, cyber-physical nanosystems have the potential to transform manufacturing processes. The combination of nanotechnology and cyber-physical approaches allows for precise and automated control at the nanoscale, leading to improvements in product design, assembly, and quality control [28].

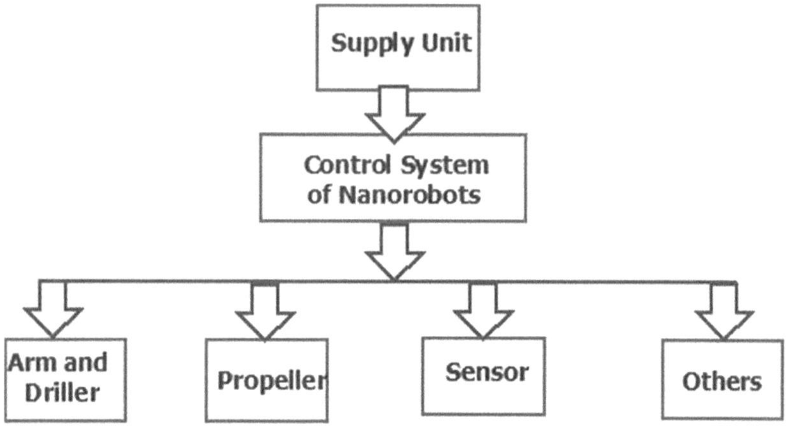

FIGURE 1.6 Nanorobotic systems in industry.

1.2.5 Nanorobotics

Nanorobotics is a discipline that focuses on the design, development, and application of small robotic devices at the nanoscale. These nanorobots are typically a few nanometers or less and can be programmed to perform specific tasks and functions, such as drug delivery, targeted therapy, and nanoscale assembly. One of the key challenges of nanorobots is to create such small nanorobot, as shown in Figure 1.6 [29]. They must be mobile, scalable, and efficient. Researchers are exploring the use of various propulsion techniques such as magnetic fields, chemical reactions, and biological motors to drive and control these miniature robots, along with advances in materials science and nanomanufacturing techniques that enable complex Nano robots [30]. As this field continues to evolve, the nanoscale holds great promise for revolutionizing industries such as pharmaceuticals, electronics, and manufacturing. Nanorobots have the potential to significantly impact these industries [31]. In medicine, they can be used for targeted delivery of drugs, allowing for precise and localized treatment of diseases while minimizing side effects. In addition, nanorobots can revolutionize surgical techniques by enabling unprecedented precision and capability at the cellular and molecular levels. In electronics, nanorobots can be used to assemble complex nanoscale components, resulting in more advanced and efficient electronic devices. Additionally, nanorobots are being developed to facilitate the creation of nanoscale systems and materials, opening new possibilities for materials and technologies that could further the development of nanorobotics [32].

1.2.6 Nanosensors and Nano-Actuators

Nanosensors and nano-actuators are emerging technologies that have revolutionized sensing and processing at the nanoscale. These tiny devices can sense and respond to small changes in their environment, making them extremely useful for a variety of applications. Nanosensors detect and measure incredibly small changes, such

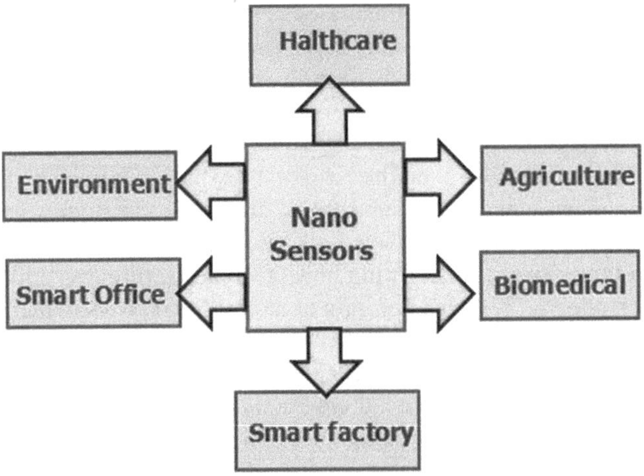

FIGURE 1.7 Nanosensors and nano-actuators in industry.

as changes in temperature, pressure, or the presence of specific substances [33]. Nanoactuators, on the other hand, take that information and put it into action, creating controlled motion or manipulating the environment at the nanoscale, as shown in Figure 1.7. This combination of small-scale phenomena holds tremendous potential in fields such as medicine, environmental monitoring, and even miniaturized robot design [34]. They represent the future of precise monitoring and ultra-sensitive detection. One of the most exciting properties of nanosensors is their potential in medicine. They can be used for targeted drug delivery, monitoring biological processes at the cellular level, and even detecting and diagnosing diseases in their early stages. Conversely, nano-actuators can manipulate and control nanoscale objects with unprecedented accuracy, opening up new opportunities such as nanorobotics and nanomanufacturing [35,36].

1.3 INDUSTRY 4.0 APPLICATIONS OF NANOTECHNOLOGY

Industry 4.0, characterized by smart factories and interconnected machines, receives a significant boost from nanotechnology. Nanoparticles unlock a new degree of control. Imagine ultra-sensitive sensors embedded in machines, feeding actual-time data into the network, or advanced materials with self-cleaning properties [1–5]. Nanotechnology enhances Industry 4.0, pushing performance, precision, and sustainability to unseen heights.

1.3.1 Manufacturing

Nanotechnology is transforming manufacturing from the ground up. By operating on the atomic level, it allows unique control over material properties. This means creating extremely robust, lightweight components, self-repairing surfaces, or even materials with entirely new functionalities. This revolutionizes everything from stronger airplane wings to self-cleaning paints, making manufacturing more efficient and creating products with previously unattainable competencies [1–10].

1.3.1.1 Nanomaterials for Advanced Manufacturing Processes

The world of manufacturing is on the cusp of a revolution, driven by the power of nanomaterials. These engineered substances, crafted at the atomic and molecular stage, have homes in contrast to their bulk counterparts. This newfound control over material behavior is transforming how we manufacture everything from airplanes to smartphones [37]. One key gain of nanomaterials lies in their capacity to enhance current materials. By incorporating nanoparticles into conventional substances like metals or polymers, producers can create composites with superior strength, lighter weight, or improved conductivity. Imagine airplane wings that are lighter yet sturdier, or sports equipment that is both long-lasting and flexible, as shown in Figure 1.8. Nanomaterials for advanced manufacturing processes in Industry 4.0. Nanotechnology also allows for the development of self-healing coatings, extending product lifespan and reducing maintenance needs [37–39].

1.3.1.2 Nano-Enabled Additive Manufacturing

The manufacturing of hardware, also known as 3D printing, has made great strides with the introduction of nanotechnology. Nanoparticles are used as components in this arena. They can be embedded in printers or filaments, allowing for the creation of complex products with different features and functions. Imagine designing a lightweight yet incredibly strong wristband, or a microfluidic device with built-in sensors; it all depends on the unique properties of these nanomaterials [40–44]. This nanoscale technology, 3D printing, opens the door to entirely new possibilities for product design and functionality, as shown in Figure 1.9 [16–18].

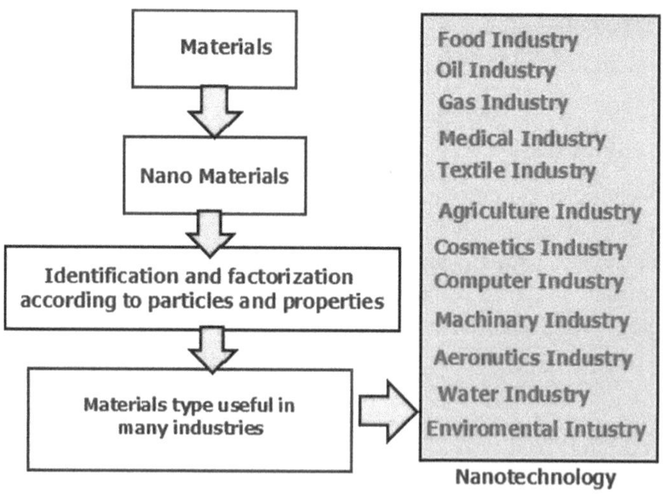

FIGURE 1.8 Nanomaterials for advanced manufacturing processes in Industry 4.0.

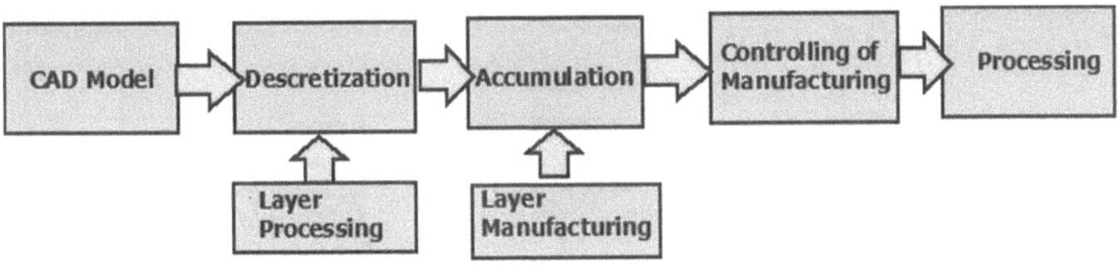

FIGURE 1.9 Nano-enabled additive manufacturing in Industry 4.0.

1.3.1.3 Nano-Enabled Smart Tools and Equipment

Ordinary tools and devices are experiencing a major breakthrough, thanks to nanotechnology. With the incorporation of nanomaterials, these tools are evolving into smart tools and nanoscale smart devices. Imagine a filter that could not only tighten bolts, but also detect when they have reached the right weight, preventing overtightening and damage [41–44]. These smart tools can be embedded with sensors that monitor various parameters such as temperature, pressure, or even wear and tear. This information can be used in real-time to perform predictive maintenance, alerting operators when a piece of equipment needs attention before it affects proper operation. In addition, nanomaterials can enable tools that are self-cleaning or corrosion-resistant, reducing maintenance needs and extending lifespan. Overall, nano-powered intelligent tools offer increased efficiency, improved safety, and new controls in a wide variety of industries [41–44].

1.3.2 Energy

In the area of energy, nanotechnology emerges as a game-changer. It offers a suite of tools to revolutionize how we address electricity, from harvesting to storage and utilization. Imagine solar panels that capture sunlight with extraordinary efficiency, thanks to light-manipulating nanoparticles, as shown in Figure 1.10. Waste heat, a cutting-edge energy drain, could be transformed into electricity using nanomaterials with exceptional thermoelectric properties [45]. Nanotechnology can also usher in a new generation of batteries, paving the way for high-capability options that charge faster and last longer – perfect for electric vehicles and a suture powered by renewables. The easy power ability of hydrogen fuel cells gets a boost too, with nanoparticles acting as catalysts to enhance performance. Even the way we manage energy becomes smarter. Nanosensors enabled by nanotechnology can be embedded in buildings and power grids, providing real-time data for optimized energy use and reduced waste. Nanotechnology is poised to fundamentally transform our energy landscape, creating a more sustainable and green future [45,46].

1.3.2.1 Nano-Enabled Energy Storage Systems

Nanotechnology is revolutionizing energy storage systems, providing a much-needed boost to our ability to capture and store electricity. One key area of focus is batteries. By incorporating nanoparticles into electrodes, scientists are developing batteries with significantly higher capacities, allowing them to store more energy

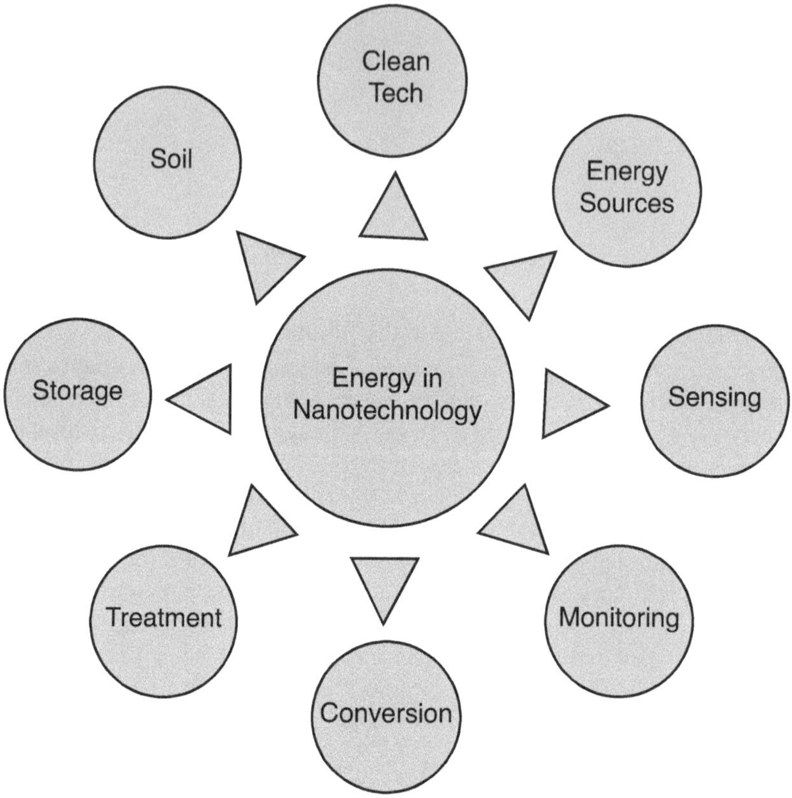

FIGURE 1.10 Energy storage manufacturing in Industry 4.0.

in a smaller space. These nanoparticles also improve the rate at which batteries charge and discharge, making them ideal for applications like electric vehicles and grid-scale energy storage [46]. Beyond batteries, nanomaterials are being used to develop new thermal energy storage systems. Here, nanoparticles can be used to create phase-change materials that absorb and release heat more efficiently, smoothing out fluctuations in energy production from renewable sources like solar and wind. Overall, nano-enabled energy storage systems promise a more efficient and sustainable energy future [45–47].

1.3.2.2 Nano-Enabled Energy Harvesting and Conversion

The conventional methods of electricity generation have received a nanoscale upgrade. Nano-enabled energy harvesting and conversion techniques are revolutionizing how we capture ambient energy. Imagine solar cells embedded with nanoparticles that can absorb a broader spectrum of sunlight, significantly boosting their efficiency. Waste heat, a pervasive energy drain, can be harnessed using novel thermoelectric materials engineered with nanostructures [45]. These materials effectively convert heat into electricity, providing a clean energy source from industrial processes or even body heat. Nanotechnology extends its reach to other forms of energy harvesting as well. Specialized nanomaterials can be used to create highly sensitive vibration energy harvesters, capturing energy from machinery or even human motion. By manipulating materials at the atomic level, nano-enabled energy harvesting and conversion offer a pathway toward a future powered by clean and readily available energy sources from our surroundings [47].

1.3.2.3 Nano-Catalysts for Efficient Energy Production

Nano-catalysts are emerging as tiny superstars in the realm of efficient energy production. These engineered nanoparticles can significantly improve the efficiency of chemical reactions vital for generating clean energy. Imagine hydrogen fuel cells, a promising clean energy source, receiving a boost thanks to nano-catalysts that accelerate hydrogen production and improve fuel cell performance. Even conventional methods like fossil fuel combustion can benefit from nano-catalysts that optimize reaction rates and reduce harmful emissions. In the realm of renewable energy, nano-catalysts can play an important role in enhancing the efficiency of solar fuel production, where sunlight is converted into storable fuels like hydrogen gas. By manipulating reactions at the atomic level, nano-catalysts provide a promising path toward purifying and manufacturing greener energy across numerous technologies [45–47].

1.3.3 Healthcare

Healthcare is making a revolutionary leap forward with the advent of nanotechnology. Imagine small machines swimming through your bloodstream, delivering remedies directly to diseased cells or performing targeted maintenance [48]. This futuristic vision holds great potential, thanks to the development of nanomedicine. Nanoparticles can be engineered to interact with specific biological targets, allowing for early disease detection, targeted drug delivery, and even gene therapy. Additionally, nanomaterials may be used to create biosensors that monitor vital signs or track the progress of treatment in real-time. Overall, nanotechnology offers exciting opportunities for personalized medicine, enhanced treatment efficacy, and a deeper understanding of diseases at the cellular and molecular level, as shown in Figure 1.11. [49].

1.3.3.1 Nano-Biosensors for Disease Diagnosis and Monitoring

Early detection and precise monitoring are critical for effective disease management. Nano-biosensors are emerging as powerful tools in this area. These nanosensors, constructed with nanoparticles, are capable of detecting specific biomarker molecules indicative of disease with remarkable sensitivity and accuracy. Imagine a simple blood test that can identify cancerous cells at their earliest stages or a wearable device that continuously monitors blood sugar levels in real-time for diabetic patients [50,51]. Nano-biosensors offer the potential for non-invasive, rapid, and personalized diagnostics, enabling earlier intervention and improved treatment outcomes. Their ability to continuously monitor disease progression allows for adjustments to treatment plans and personalized medicine processes. The future of disease diagnosis and monitoring is indeed shrinking in size but growing in power, thanks to nano-biosensors [52,53].

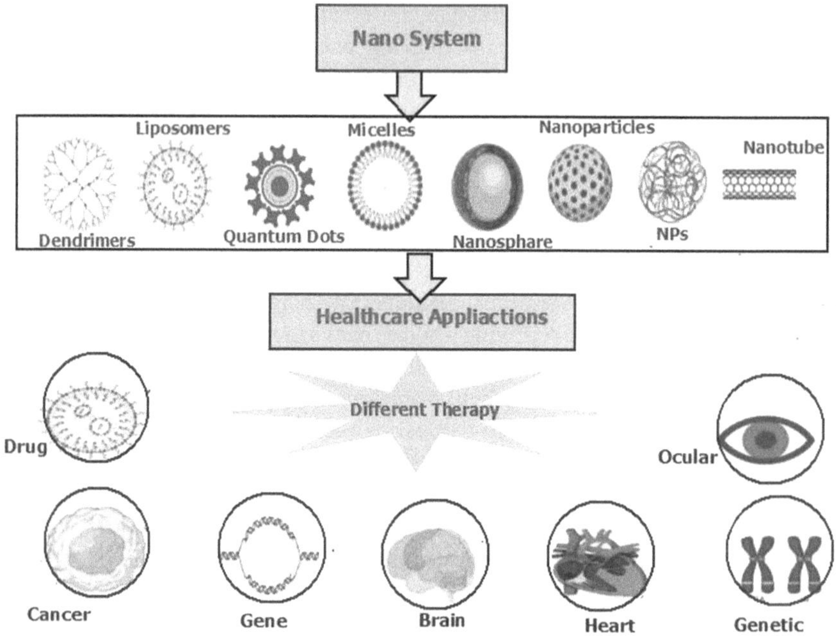

FIGURE 1.11 Nano-enabled healthcare in Industry 4.0.

1.3.3.2 Nano-Drug Delivery Systems

The world of medicine is witnessing a revolution in drug delivery thanks to nano-drug delivery systems. These microscopic carriers, made from nanoparticles, provide a much more targeted and efficient approach to treatment [54]. Unlike conventional drugs that can have widespread effects throughout the body, nano-drug delivery systems can be designed to target specific cells or tissues. This allows for higher drug concentrations at the disease web, reducing side effects on healthy tissues. Additionally, these systems can be engineered to release their load in a controlled manner, ensuring sustained drug delivery and improved treatment efficacy. Imagine a nanoparticle loaded with a cancer drug that homes in on tumors, delivering an effective dose directly to diseased cells while minimizing harm to healthy ones. Nano-drug delivery systems hold significant promise for revolutionizing the treatment of various diseases, from cancer and infections to chronic conditions, as shown in Figure 1.12 [55,56].

1.3.3.3 Nano-Enabled Tissue Engineering and Regenerative Medicine

In the area of healing and restoration, nanotechnology offers a glimmer of hope via nano-enabled tissue engineering and regenerative medicine. This exciting discipline focuses on the use of nanoparticles to create scaffolds and other tools that can guide the body's natural healing processes. Imagine biocompatible scaffolds, meticulously crafted with nanomaterials, that mimic the structure of natural tissues [57]. These scaffolds can provide a supportive environment for damaged tissues to regenerate, promoting cell growth and repair. Nanoparticles can also be used to deliver growth factors or other therapeutic agents directly to injured areas, stimulating the body's

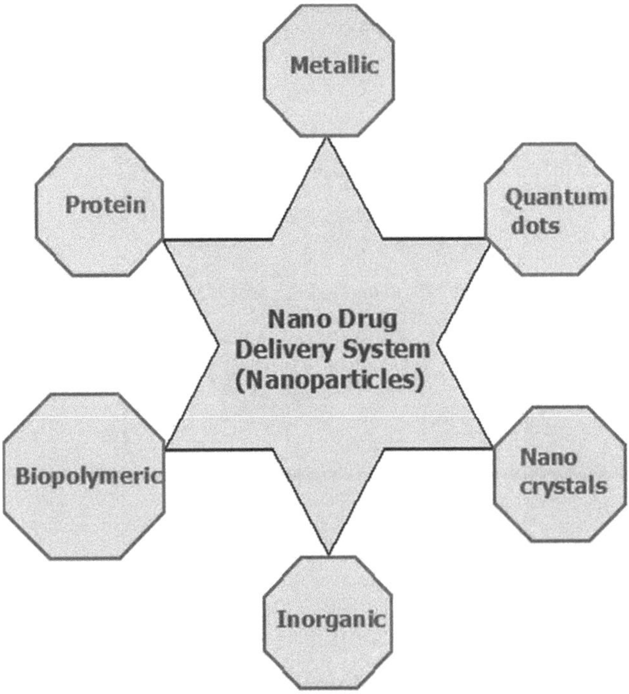

FIGURE 1.12 Nano-drug delivery systems in Industry 4.0.

own healing mechanisms [58]. This targeted approach holds enormous potential for treating a wide range of conditions, from bone fractures and burns to damaged nerves and organs. Overall, nano-enabled tissue engineering and regenerative medicine offer a revolutionary approach to healing, promoting the body's own ability to repair and regenerate damaged tissues [3].

1.3.4 Environment

Nanotechnology offers a double-edged sword when it comes to the environment. On the positive side, it provides exciting opportunities for cleaning up existing pollutants and developing more sustainable technologies. Imagine nanoparticles that act like tiny sponges, soaking up contaminants from polluted water or soil [59]. These nanoparticles could revolutionize remediation efforts for industrial spills and waste sites. Additionally, nanotechnology can be used to develop more efficient solar panels, wind turbines, and other renewable energy sources, reducing our dependence on fossil fuels and their associated environmental impact, as shown in Figure 1.13. However, concerns exist regarding the potential risks of nanomaterials themselves. Because they are so small, they can easily enter the environment, and their long-term effects on ecosystems and human health are still being studied. Responsible development and rigorous testing are crucial to ensure that the environmental benefits of nanotechnology outweigh the potential risks [60–62].

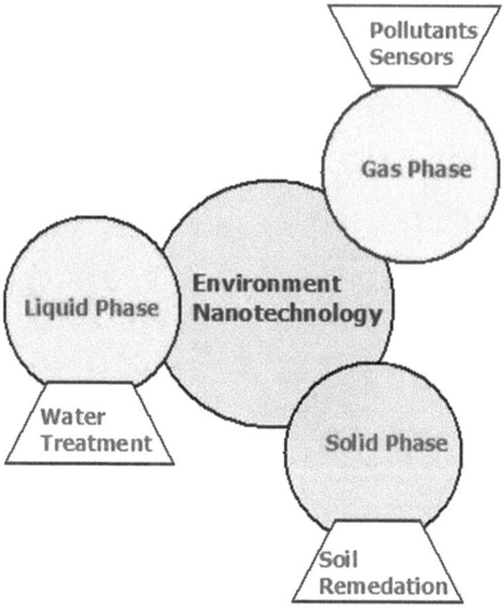

FIGURE 1.13 Environment nanotechnology in Industry 4.0.

1.3.4.1 Nano-Enabled Water Treatment and Purification

Traditional water treatment methods are receiving support from nanotechnology. Nano-enabled water treatment and purification use the unique properties of nanoparticles to remove contaminants and purify water more effectively. Imagine nanoparticles acting like tiny magnets, attracting and capturing pollutants like heavy metals and bacteria from water [59]. These nanoparticles can be integrated into filters or membranes, significantly enhancing their ability to remove contaminants. Additionally, nano-catalysts can be used to break down harmful substances or even kill bacteria through light-activated processes. Overall, nano-enabled water treatment offers a promising approach for providing clean drinking water, especially in areas with limited access to safe water sources. However, careful consideration needs to be given to the potential environmental impact of these nanoparticles themselves [62].

1.3.4.2 Nano-Based Air Pollution Control

Our fight for clean air receives a boost from nanotechnology with the emergence of nano-based air pollution control solutions. These modern approaches target air pollution at the atomic stage. Imagine catalysts embedded with nanoparticles that can break down dangerous gases like nitrogen oxides or volatile organic compounds into harmless byproducts at room temperature. This substantially reduces the energy needed for air purification compared to standard techniques. Nanotechnology can also be harnessed to create highly efficient air filters. Imagine filters infused with nanoparticles that could capture even the tiniest pollutants, including particulate matter like PM2.5, with tremendous efficiency. These advanced filters promise cleaner air for homes, industries, or even entire towns. However, research is ongoing to evaluate the environmental impact of these nanoparticles themselves, ensuring a sustainable approach to purifying air [63].

1.3.4.3 Nano-Enabled Environmental Remediation

The fight for a cleaner planet receives an effective ally with nano-enabled environmental remediation. This exciting field harnesses the unique properties of nanoparticles to tackle existing environmental challenges. Imagine polluted soil or water being treated with nanoparticles that act like microscopic sponges, absorbing contaminants like heavy metals or organic pollutants. These nanoparticles can be tailor-made to target specific pollutants, leading to more efficient and focused remediation strategies. Additionally, nanocatalysts can be used to break down harmful chemical compounds into less toxic substances or even render them harmless altogether. This approach holds significant potential for cleaning up industrial waste sites, oil spills, and polluted water resources. However, responsible development and deployment are essential. The long-term environmental impact of these nanoparticles needs careful study to ensure a sustainable path towards a cleaner planet [63,64].

1.3.5 Electronics and Communication

In the field of electronics and communications, nanotechnology is ushering in a new era of miniaturization and functionality, as shown in Figure 1.14. Imagine transistors, the building blocks of electronic devices, shrunk to sizes thousands of times smaller than a human hair. This remarkable achievement, made possible by nanotechnology, paves the way for much smaller and more powerful devices [65]. In addition, nanomaterials with unique electrical properties can be used to fabricate new electronic components with improved performance. Nanotechnology also plays an important role in transforming information technology. For example, the development of nanoantennas with significantly enlarged surface areas can lead to smaller and more

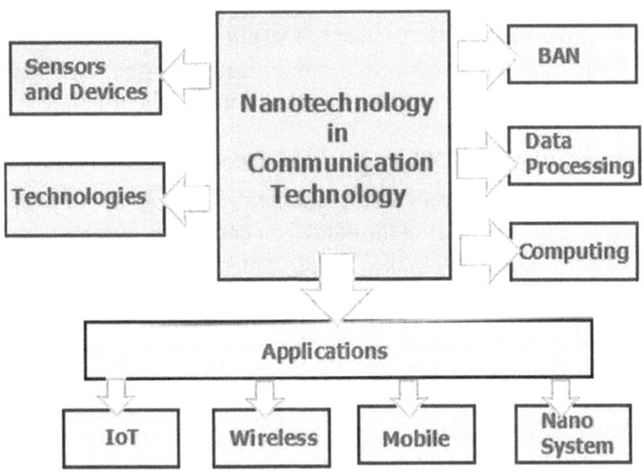

FIGURE 1.14 Communication nanotechnology in Industry 4.0.

efficient antennas, allowing for faster data transmission and greater coverage, which can be used with a telephone. Overall, nanotechnology is transforming electronics and communication, promising a future of miniaturized devices with unparalleled processing power and communication capabilities [66].

1.3.5.1 Nano-Enabled Electronics and Optoelectronics

The world of electronics and optics is shrinking while dramatically expanding its reach, thanks to nano-enabled electronics and optics. Imagine compressing the power of a computer chip into a container smaller than a grain of sand. Nanotechnology does this by enabling the creation of transistors, the basic building blocks of electronics, at the atomic scale [67]. This requires very limited equipment with incredible processing power. Beyond mineralization, nanomaterials with unique electrical and optical properties are employed to fabricate new electronic components with high performance. In the field of optoelectronics, which deals with the interaction of light and electronic devices, nanotechnology allows for the development of next-generation solar cells, LEDs, and lasers with high performance and efficiency. This combination of nanotechnology with electronics and optoelectronics paves the way for a future filled with powerful, compact devices and significant advances in optical technology [67,68].

1.3.5.2 Nano-Antennas for Communication Systems

Communication systems are experiencing a small revolution thanks to nano-antennas. These are antennas crafted from carefully engineered nanomaterials, thousands of times smaller than their conventional counterparts. Despite their size, nano-antennas offer significant advantages. Their unique properties allow them to interact with a wide range of frequencies, potentially enabling communication systems that operate across multiple spectrums. Furthermore, their compact size opens the door to the development of highly integrated devices with built-in communication capabilities [68,69]. Imagine a world where sensors, medical implants, or even everyday clothing seamlessly communicate with each other and with their surroundings – all thanks to the power of nano-antennas. However, challenges remain in terms of optimizing their signal strength and overall efficiency. Despite these obstacles, nano-antennas hold tremendous potential to transform communication systems, paving the way for a future of ubiquitous connectivity with small and multifunctional devices [70].

1.3.5.3 Nano-Enabled Data Storage and Processing

The future of data storage and processing drastically shifts with nanoscale data storage and processing. Imagine storing the data of an entire library on a chip smaller than your fingertips. Nanotechnology makes this dream a reality by enabling high-density storage systems. This is achieved by manipulating materials at the atomic level to create storage devices that can hold large amounts of data in a very small space. Beyond storage capabilities, nanotechnology is revolutionizing the data industry. By creating nanoscale transistors that switch extremely fast, nanoelectronics paves the way for processors with unparalleled processing power. This allows for faster computers, better AI, and the ability to handle ever-increasing amounts of data. However, improving heat dissipation and maintaining data transmission integrity at these small scales are challenges that need to be addressed. Despite these challenges, nanoscale data storage and processing promise a future in which information is stored efficiently and immediately available for dynamic analysis like never before [71,72].

1.4 CHALLENGES AND FUTURE PROSPECTS

Nanotechnology, regardless of its modern capability, faces a double-edged sword. While it offers thrilling opportunities for improvements in diverse fields, demanding situations remain that need to be addressed. A foremost concern lies within the potential environmental impact of nanoparticles. Their tiny size raises questions about their interplay with ecosystems and potential long-term fitness effects. Responsible development and rigorous testing are critical to ensure the environmental benefits outweigh the risks [15].

Another challenge includes the safe and environmentally friendly manufacturing of nanomaterials at scale. Current methods may be costly and energy-intensive. Developing cost-effective and sustainable manufacturing techniques is essential for the broader adoption of nanotechnology. The future of nanotechnology, however, remains bright. As research progresses, scientists are finding ways to mitigate environmental risks and discover greener methods for nanoparticle manufacturing. Additionally, ongoing exploration of novel nanomaterials with unique properties opens doors to even more groundbreaking applications. With careful attention and continued research, nanotechnology has the potential to revolutionize our world, providing solutions for clean energy, sustainable production, personalized medicine, and a future brimming with technological marvels [6].

1.4.1 Regulatory and Safety Concerns

The promise of nanotechnology is undeniable, but its rapid development is accompanied by a growing shadow of uncertainty: regulatory and safety concerns. A fundamental challenge lies in the very nature of these engineered products. Their small size allows them to bypass biological barriers and can interact with organs and systems in unpredictable ways. Long-term health effects from exposure to nanoparticles are still under investigation, raising concerns about potential toxicity or unintended consequences. Regulatory frameworks struggle to keep pace with this rapidly growing industry, making it difficult to ensure the safe manufacture and use of nanomaterials. Additionally, ethical considerations arise over potential misuses of nanotechnology, such as the development of undetectable weapons or surveillance equipment. Addressing these regulatory and safety issues is of utmost importance. Rigorous testing, open communication, and strong international cooperation are needed to ensure the development and use of quality nanotechnology for the benefit of humanity [4].

1.4.2 Scalability and Cost-effectiveness

Despite the fact that nanotechnology boasts significant potential, there are two key challenges: scalability and cost. Exciting discoveries in research often face challenges when transitioning to larger projects. The development of methods to efficiently synthesize and produce nanomaterials in large quantities is crucial for their maximum applications. Currently, many nanotechnology processes are expensive and energy-intensive, limiting their commercialization. Cost-effective technologies and new manufacturing techniques are needed to make nanotechnology accessible to a wide range of industries. Overcoming these challenges is vital to harness the power of nanotechnology from the realm of scientific research into real-world applications that benefit all societies [7].

1.4.3 Integration with Existing Systems and Infrastructures

Bridging the gap between the promise of nanotechnology and its real-world application hinges on successful integration with current systems and infrastructure. This presents a captivating puzzle with two distinct facets [6].

On the positive side, nanotechnology offers exciting opportunities. Imagine fortifying existing bridges or homes with nano-improved substances, leading to lighter systems with superior energy and prolonged lifespans. Nanosensors seamlessly woven into modern infrastructure can provide real-time information on essential parameters, allowing for predictive maintenance and preventing costly downtime. Nanotechnology even holds the potential to breathe new life into aging infrastructure. Self-cleaning coatings or self-healing materials could significantly reduce maintenance needs and extend the lifespan of existing structures. Additionally, ensuring interoperability is vital. Ideally, these nanosolutions could communicate seamlessly with existing communication protocols, integrating results into contemporary monitoring and control systems [15].

However, challenges remain. Compatibility is a major hurdle. New nanomaterials must mesh seamlessly with current infrastructure to avoid unintended consequences. Upgrading existing structures with nanotech may be expensive, demanding careful cost–benefit analysis. Standardization across nanomaterials and functionalities is crucial to avoid compatibility issues during integration. Finally, the workforce must adapt. Integrating nanotechnology may require specialized knowledge for safety and operation, necessitating upskilling the existing staff or training new personnel [15].

1.4.4 Workforce Development and Skill Requirements

As nanotechnology rapidly evolves, the future workforce requires a significant upgrade to translate its potential into reality. Developing a skilled staff hinges on addressing specific needs. A strong foundation in core sciences like physics, chemistry, and biology is essential. This equips workers to understand the behavior of materials on the atomic level and their interactions with biological systems. In-depth knowledge of various nanomaterials, their properties, and how to create them is crucial. Engineering skills for designing, building, and analyzing nanostructures and devices are necessary. The ability to use advanced computer modeling tools allows for predicting the behavior of nanomaterials before physical introduction. Data analysis skills are important to interpret the vast quantities of data generated from simulations and experiments. Finally, a deep understanding of safety protocols, handling procedures, and relevant regulations is paramount for the responsible development of nanotechnology [6].

To bridge this skills gap, revamping STEM education to include nanotechnology concepts can spark early interest and inspire future generations to pursue careers in this field. Specialized training programs and workshops can equip current professionals with the necessary knowledge to transition into nanotech roles. Collaboration between academia and industry is critical. By working together, universities and research institutions can tailor their programs to the specific needs of the industry, ensuring a job-ready workforce with the skills to propel nanotechnology forward. Additionally, a culture of lifelong learning is crucial to keep the workforce adaptable in the face of rapid advancements in this ever-evolving field. By prioritizing workforce development and addressing these skill requirements.

1.4.5 Future Trends and Emerging Applications

Nanotechnology is a rapidly growing field that has the potential to transform various aspects of our lives. Here are a few examples of exciting future trends and emerging applications: Biomimetic Nanomaterials: Nature continues to inspire scientists. The future holds promise for biomimetic nanomaterials that mimic the structures and functions found in nature. Imagine artificial tissues inspired by biological systems, leading to advances in robotics and prosthetics. Nanorobotics: While robotic microscopy is still in its infancy, advances in this field could lead to intelligent nanobots capable of performing tasks within the human body [1–10]. Imagine nanorobots programmed to deliver drugs directly to diseased cells or perform delicate surgical procedures with unparalleled precision. Brain–computer interfaces: Nanotechnology could play an important role in the development of next-generation brain–computer interfaces. Imagine a neural implant with ultra-fine nanowires that communicate seamlessly with the brain, enabling a more natural and efficient exchange of information between the human brain and computer. This could have profound effects on communication, rehabilitation, and even cognitive development. Personalized Medicine: Nanotechnology offers exciting possibilities for tailoring treatments to individual patients. Imagine diagnostic tools that can detect diseases at an early stage or customized nanoparticles that deliver targeted therapies directly to diseased cells, minimizing side effects. Sustainable Materials and Processes: Nanotechnology can pave the way for new sustainable products and processes. Think of self-healing products for businesses that can self-clean, reducing waste and maintenance needs. Furthermore, nano-catalysts hold promise as clean and efficient methods for energy production. Compatibility with Other Technologies: The future of nanotechnology depends on how it integrates with other rapidly growing industries such as AI, big data, and genetic engineering. This convergence will unlock even greater potential for innovation and discovery. Ethical Considerations: As nanotechnology evolves, the ethical implications of its use must be carefully considered. Open dialogue and strong regulation are critical to ensuring responsible access to and use of this powerful technology. Overall, the future of nanotechnology is full of possibilities. By addressing challenges, advancing responsible growth, and harnessing the power of collaboration, we can leverage nanotechnology to create a better future for all [6,15].

1.5 CONCLUSION

This review provides an overview of cost-effective and innovative nanotechnology in materials and engineering sciences. It will enable researchers and practitioners in various disciplines to delve deeper into the application of nanotechnology in their fields. While the application of nanotechnology is vast, the risks associated with its unlimited use remain elusive and understudied. Further research is needed to address the issues of nanotoxicity and develop appropriate solutions. It is recommended that researchers, engineers, and technologists collaborate both practically and academically to explore and optimize nanotechnology through field trials. Furthermore, careful R&D of new nanoscale materials is needed to prepare for a future dominated by these revolutionary technologies. The widespread application of nanotechnology in the industrial sector means that it will eventually be incorporated into every possible category. However, caution is needed to address the environmental, pollution, and health impacts that may result from its misuse, given the increasing emphasis on sustainability in the world for various reasons. By aligning nanotechnology with sustainability principles, future prosperity can be ensured for this industry.

1.6 RECOMMENDATIONS FOR FUTURE R&D AND APPLICATIONS

R&D is very important. Here are some basic suggestions to guide us toward this goal. First, a focus on environmental protection is of utmost importance. Intensive research is needed to understand the long-term environmental impact of nanomaterials, encompassing environmentally friendly manufacturing processes and methods for safe disposal. Second, prioritizing scalability and cost reduction is essential for widespread adoption. Research efforts should focus on developing cost-effective and scalable methods for synthesizing high-quality nanomaterials. Additionally, differentiation between lab and industry through collaboration is essential, ensuring that fundamental research translates into commercially viable real-world applications. Investing in workforce development by equipping workers with the necessary skills and competencies through restructuring core education and training programs is essential. Finally, promoting responsible development and ethical considerations through open dialogue and strong international law is critical. By prioritizing these recommendations, we can navigate the exciting future of nanotechnology, paving the way for a world that benefits all societies, promotes sustainability, improves human health, and provides a realm of technological wonders.

REFERENCES

1. Oláh J., Aburumman N., Popp J., Khan M. A., Haddad H., Kitukutha N., Impact of Industry 4.0 on environmental sustainability. *Sustainability*, 2020, 12(11), p. 4674. https://doi.org/10.3390/su12114674
2. Franciosi C., Iung B., Miranda S., Riemma S., Maintenance for sustainability in the Industry 4.0 context: a scoping literature review. *IFAC-PapersOnLine*, 2018, 51(11), pp. 903–908. ISSN 2405-8963. https://doi.org/10.1016/j.ifacol.2018.08.459

3. Bai C., Dallasega P., Orzes G., Sarkis J., Industry 4.0 technologies assessment: a sustainability perspective. *International Journal of Production Economics*, 2020, 229, p. 107776, ISSN 0925-5273. https://doi.org/10.1016/j.ijpe.2020.107776
4. Shabur M. A., A comprehensive review on the impact of Industry 4.0 on the development of a sustainable environment. *Discover Sustainability*, 2024, 5, p. 97. https://doi.org/10.1007/s43621-024-00290-7
5. Dilberoglu U. M., Gharehpapagh B., Yaman U., Dolen M., The role of additive manufacturing in the era of Industry 4.0. *Procedia Manufacturing*, 2017, 11, pp. 545–554. ISSN 2351-9789. https://doi.org/10.1016/j.promfg.2017.07.148
6. Bhushan B., *Handbook of Nanotechnology*. Springer Handbook of Nanotechnology; Berlin/Heidelberg, Germany: 2017. Introduction to Nanotechnology; pp. 1–19.
7. Singh N. A., Nanotechnology innovations, industrial applications and patents. *Environmental Chemistry Letters*, 2017, 15, pp. 185–191. https://doi:10.1007/s10311-017-0612-8
8. Waldron A., Spencer D., Batt C., The current state of public understanding of nanotechnology. *Journal of Nanoparticle Research*, 2006, 8, pp. 569–575. https://doi:10.1007/s11051-006-9112-7
9. Waldron A., Spencer D., Batt C., The current state of public understanding of nanotechnology. *Journal of Nanoparticle Research*, 2006, 8, pp. 569–575. https://doi:10.1007/s11051-006-9112-7
10. Hulla J., Sahu S., Hayes A., Nanotechnology: history and future. *Human and Experimental Toxicology*, 2015, 34, pp. 1318–1321. https://doi:10.1177/0960327115603588
11. Tsaramirsis G., Kantaros A., Al-Darraji I., Apostolopoulos D. P. C., Pavlopoulou A., Alrammal M., Ismail Z., Buhari S. M., Stojmenovic M., Tamimi H., A modern approach towards an Industry 4.0 model: from driving technologies to management. *Journal of Sensors*, 2022, 2022, p. 5023011. https://doi:10.1155/2022/5023011
12. Ratnesh R. K., Singh M. K., Kumar V., Singh S., Chandra R., Singh M., Singh J., Mango leaves (Mangifera Indica) derived highly florescent green graphene quantum dot nanoprobes for enhanced on-off dual detection of cholesterol and Fe^{2+} ions based on molecular logic operation. *Biomaterials Science & Engineering, American Chemical Society*, 2024, 7, pp. 4417–4426. https://doi.org/10.1021/acsabm.4c00292
13. Talebian S., Rodrigues T., Neves J., Sarmento B., Langer R., Conde J., Facts and figures on materials science and nanotechnology progress and investment. *ACS Nano*, 2021, 15, pp. 15940–15952. https://doi:10.1021/acsnano.1c03992
14. Bongomin O., Ocen G. G., Nganyi E. O., Musinguzi A., Omara T., Exponential disruptive technologies and the required skills of Industry 4.0. *Journal of Engineering*, 2020, 2020, p. 4280156. https://doi:10.1155/2020/4280156
15. Kumar N., Dixit A., *Nanotechnology for Defence Applications*. Springer; Berlin/Heidelberg, Germany: 2019. Role of Nanotechnology in Futuristic Warfare; pp. 301–329.
16. Santos J., de Oliveira R. S., de Oliveira T. V., Velho M. C., Konrad M. V., da Silva G. S., Deon M., Beck R. C. R., 3D printing and nanotechnology: a multiscale alliance in personalized medicine. *Advanced Functional Materials*, 2021, 31, p. 2009691. https://doi:10.1002/adfm.202009691
17. Opait G., Nanotechnology and drones, boom-art in the advanced tech. *Economics Applies Informa*, 2020, 1, pp. 93–104. https://doi:10.35219/eai1584040986
18. Yamamoto B. E., Trimble A. Z., Minei B., Ghasemi Nejhad M. N., Development of multifunctional nanocomposites with 3-D printing additive manufacturing and low graphene loading. *Journal of Thermoplastic Composite Materials*, 2019, 32, pp. 383–408. https://doi:10.1177/0892705718759390
19. Goyal R. K., *Nanomaterials and Nanocomposites*. CRC Press; Boca Raton, FL: 2021. The Role and Applications of Nanomaterials in the Automotive Industry; pp. 51–59

20. Becker S., Nanotechnology in the marketplace: how the nanotechnology industry views risk. *Journal of Nanoparticle Research,* 2013, 15, p. 1426. https://doi:10.1007/s11051-013-1426-7
21. Zhang P., Wang L., Wei H., Wang J., A critical review on effect of nanomaterials on workability and mechanical properties of high-performance concrete. *Advances in Civil Engineering,* 2021, 2021, pp. 1–24. https://doi:10.1155/2021/8827124
22. Kumar R., Ratnesh R. K., Singh J., Kumar A., Chandra R., IoT-driven experimental framework for advancing electrical impedance tomography. *ECS Journal of Solid State Science and Technology*, 2024, 13(2), p. 027002. https://doi:10.1149/2162-8777/ad2331
23. Usman M., Farooq M., Wakeel A., Nanotechnology in agriculture: current status, challenges and future opportunities. *Science of the Total Environment*, 2020, 2020, p. 137778. https://doi:10.1016/j.scitotenv.2020.137778
24. F. Akyildiz, J. M. Jornet, The internet of nano-things. *IEEE Wireless Communications*, 2010, 17(6), pp. 58–63. https://doi:10.1109/MWC.2010.5675779
25. Civas M., Akan O. B., Molecular communication transmitter architectures for the internet of bio-nano things. In: *2022 International Balkan Conference on Communications and Networking (BalkanCom)*, Sarajevo, Bosnia and Herzegovina, 2022, pp. 132–136. https://doi:10.1109/BalkanCom55633.2022.9900784
26. Sharma P. K., Dorlikar S., Rawat P., Malik V., Vats N., Sharma M., Rhyee J. S., Kaushik A. K. Nanotechnology and its- application: a review. *Nanotechnology in Cancer Management*, 2021, pp. 1–33
27. Fino M. H., Nanoelectronic challenges and opportunities for cyber-physical systems. In: *2022 29th International Conference on Mixed Design of Integrated Circuits and System (MIXDES)*, Wrocław, Poland, 2022, pp. 15–21. https://doi:10.23919/MIXDES55591.2022.9837959
28. Shi C., Kojima M., Tercero C., et al., A cyber-physical system for strain measurements in the cerebral aneurysm models. In: *2012 IEEE/RSJ International Conference on Intelligent Robots and Systems*, Vilamoura-Algarve, Portugal, 2012, pp. 4137–4142. https://doi:10.1109/IROS.2012.6385754
29. Opait G., Nanotechnology and drones, boom-art in the advanced tech. *Economics Applies Informa*, 2020, 1:93–104. https://doi:10.35219/eai1584040986
30. Gupta A., Kumari M., Sharma M., Alsharif M. H., Uthansakul P., Uthansakul M., Bansal S., 8-port MIMO antenna at 27 GHz for n261 band and exploring for body centric communication. *PLoS One*, 2024, 19(6), p. e0305524. https://doi.org/10.1371/journal.pone.0305524
31. Pisarenko Z., Ivanov L., Wang Q., Nanotechnology in construction: state of the art and future trends. *Nanotechnology in Construction: A Science Internet-Journal*, 2020, 12, pp. 223–231. https://doi.org/10.15828/2075-8545-2020-12-4-223-231
32. Tang C., Liu J., Chen X., Huang T., Micro/nano-robotics in biomedical applications and its progresses. In: *2015 Chinese Automation Congress (CAC)*, Wuhan, 2015, pp. 376–380. https://doi.org/10.1109/CAC.2015.7382528
33. Ghosh T., Raj G. B., Dash K. K., A comprehensive review on nanotechnology based sensors for monitoring quality and shelf life of food products. *Measurements in Food,* 2022, p. 100049. https://doi:10.1016/j.meafoo.2022.100049
34. Çelik-Butler Z., Hamid H. M. A., ZnO nano-sensors and nano-energy harvesters. In: *2017 IEEE Sensors*, Glasgow, UK, 2017, pp. 1–3. https://doi:10.1109/ICSENS.2017.8233872
35. Ratnesh R. K., Singh M. K., Singh J., Enhancing ZnO/Si heterojunction photodetector performance for ultra high responsivity across wide spectral range. *Journal of Material Science: Materials in Electronics*, 2024, 35, p. 756. https://doi.org/10.1007/s10854-024-12516-5

36. Kumar R., Tripathi S., A novel GUI-based image reconstruction algorithm of EIT imaging technique. *International Journal of Cognitive Informatics and Natural Intelligence (IJCINI)*, 2021, 15(3), pp. 31–46. https://doi:10.4018/IJCINI.20210701.oa3
37. Yadav, M. K., Kumar, R., Ratnesh, R. K., Singh, J., Chandra, R., Kumar, A., & Singh, A. K., Revolutionizing technology with spintronics: devices and their transformative applications. *Materials Science and Engineering: B*, 2024, 303, p. 117293. https://doi.org/10.1016/j.mseb.2024.117293
38. Hosnedlova B., Sochor J., Baron M., Bjorklund G., Kizek R., Application of nanotechnology based-biosensors in analysis of wine compounds and control of wine quality and safety: a critical review. *Critical Reviews in Food Science and Nutrition*, 2020, 60, pp. 3271–3289. https://doi:10.1080/10408398.2019.1682965
39. Teizer J., Venugopal M., Teizer W., Felkl J., Nanotechnology and its impact on construction: bridging the gap between researchers and industry professionals. *Journal of Construction Engineering and Management*, 2012, 138, pp. 594–604. https://doi:10.1061/(ASCE)CO.1943-7862.0000467
40. Ren H., Zou G., Zhao Z. et al., High-reliability wireless packaging for high-temperature SiC power device sintered by novel organic-free nanomaterial. *IEEE Transactions on Components, Packaging and Manufacturing Technology*, 2020, 10(12), pp. 1953–1959. https://doi:10.1109/TCPMT.2020.3038430
41. Kumar Singla M., Gupta J., Safaraliev M., Zeinoddini-Meymand H., Ghanizadeh A. J., Mathematical modeling for solar cell optimization: evaluating sustainability with different diode configurations. *IEEE Access*, 2024, 12, pp. 93802–93822. https://doi:10.1109/ACCESS.2024.3424416
42. Vamsi Krishna P., Srikant R. R., Padmini R., Viswaditya J., Application of nanomaterials as coolants/ lubricants in machining. In: *International Conference on Advanced Nanomaterials & Emerging Engineering Technologies*, Chennai, India, 2013, pp. 674–682. https://doi:10.1109/ICANMEET.2013.6609389
43. Tamiya E., Nanomaterials based optical and electrochemical biosensors. In: *2006 IEEE Nanotechnology Materials and Devices Conference*, Gyeongju, Korea (South), 2006, pp. 288–289. https://doi:10.1109/NMDC.2006.4388869
44. Bupesh Raja V. K., Prabhu S., Processing of cellular materials – an overview. In: *International Conference on Advanced Nanomaterials & Emerging Engineering Technologies*, Chennai, India, 2013, pp. 631–633. https://doi:10.1109/ICANMEET.2013.6609371
45. Stelmakh V., Chan W. R., Ghebrebrhan M., et al., Sputtered tantalum photonic crystal coatings for high-temperature energy conversion applications. *IEEE Transactions on Nanotechnology*, 2016, 15(2), pp. 303–309. https://doi:10.1109/TNANO.2016.2522423
46. Chau R., Datta S., Doczy M., et al., Benchmarking nanotechnology for high-performance and low-power logic transistor applications. *IEEE Transactions on Nanotechnology*, 2005, 4(2), pp. 153–158. https://doi:10.1109/TNANO.2004.842073
47. Costello B., Davis J. A., Breakdown field strength variations and energy density limits of nanoparticle composite materials. *IEEE Transactions on Nanotechnology*, 2020, 19, pp. 811–819. https://doi:10.1109/TNANO.2020.3037897
48. Kumar R., Mahadeva R., An experimental measurement and control of human body stomach using electrical impedance tomography. *Journal of Circuits, Systems and Computers*, 2021, 30(6), p. 2150103. https://doi.org/10.1142/S0218126621501036
49. Pramanik P. K. D., Solanki A., Debnath A., Nayyar A., El-Sappagh S., Kwak K.-S., Advancing modern healthcare with nanotechnology, nanobiosensors, and internet of nano things: taxonomies, applications, architecture, and challenges. *IEEE Access*, 2020, 8, pp. 65230–65266. https://doi:10.1109/ACCESS.2020.2984269

50. Queyam A. B., Kumar R., Ratnesh R. K., Chauhan R. K., LabVIEW-enabled synthetic signal for empowering fetal-maternal healthcare. *ECS Journal of Solid State Science and Technology*, 2024, 13(5). https://doi:10.1149/2162-8777/ad4dde
51. Kumar R., Kumar S., Sengupta, A., Optimization of bio-impedance techniques-based monitoring system for medical & industrial applications. *IETE Journal of Research*, 2022, 68(5), pp. 3843–3854. https://doi.org/10.1080/03772063.2020.1780957
52. Rizwan A., Zoha, A., Zhang, R. et al., A Review on the role of nano-communication in future healthcare systems: a big data analytics perspective. *IEEE Access*, 2018, 6, pp. 41903–41920. https://doi:10.1109/ACCESS.2018.2859340
53. Kumar R., Kumar S., Sengupta A., An experimental analysis and validation of electrical impedance tomography technique for medical or industrial application. *Biomedical Engineering: Applications, Basis and Communications*, 2019, 31(2), p. 1950010. https://doi.org/10.4015/S1016237219500108
54. Hornsby T., Kashkooli F. M., Jakhmola A., Kolios M. C., Tavakkoli J., Measuring drug release induced by thermal and non-thermal effects of ultrasound in a nanodrug delivery system, In *2022 IEEE International Ultrasonics Symposium (IUS)*, Venice, Italy, 2022, pp. 1–4. https://doi:10.1109/IUS54386.2022.9958697
55. Jakhmola, T. H., Rod K., Tavakkoli J., A novel gold nanoparticles drug delivery system: design and ex vivo tissue testing. In: *2020 IEEE International Ultrasonics Symposium (IUS)*, Las Vegas, NV, USA, 2020, pp. 1–3. https://doi:10.1109/IUS46767.2020.9251430
56. Asemani D., Haemmerich D., A unified mathematical model for nano-liposomal drug delivery to solid tumors. *IEEE Transactions on NanoBioscience*, 2018, 17(1), pp. 3–11. https://doi:10.1109/TNB.2017.2783889
57. Bajpai V. K., Kamle M., Shukla S., Mahato D. K., Chandra P., KyuHwang S., Kumar P., Huh Y. S., Han Y.-K., Prospects of using nanotechnology for food preservation, safety, and security. *Journal of Food and Drug Analysis,* 2018, 26, pp. 1201–1214. https://doi:10.1016/j.jfda.2018.06.011
58. Huang H., Wang D., Zhu J., Seeger S., Chu Z., Sand-based economical micro/nanocomposite materials for diverse applications. *ACS Applied Materials & Interfaces*, 2022, 14, pp. 43656–43665. https://doi:10.1021/acsami.2c08778
59. Abdirad N., Kamran-Pirzaman A., Investigating green chemistry and the effect of nanotechnology on the environment In: *2023 8th International Conference on Technology and Energy Management (ICTEM)*, Mazandaran, Babol, Iran, Islamic Republic of, 2023, pp. 1–4. https://doi:10.1109/ICTEM56862.2023.10084268
60. McNally H. A., Maximizing nanotechnology education at purdue university: its integration into the electrical engineering technology curriculum. *IEEE Nanotechnology Magazine*, 2013, 7(3), pp. 19–22. https://doi:10.1109/MNANO.2013.2275026
61. Pathak S., Kishore N., Upadhyay G., Ratnesh R. K., Mishra R., A compact size planar microstrip-fed patch antenna with hexagonal DGS slot for WLAN application, recent trends in electronics and communication. In: *LNEE*, Springer, 2021, 777, pp. 263–271. https://doi.org/10.1007/978-981-16-2761-3_25
62. Inshyna N., Chorna I., Ethical and societal aspects of nanotechnology applications in medicine. In: *2022 IEEE 12th International Conference Nanomaterials: Applications & Properties (NAP)*, Krakow, Poland, 2022, pp. 1–5. https://doi:10.1109/NAP55339.2022.9934298
63. Hebbar S., Karuppasam V., Kiran G. K., et al., System engineering and deployment of envirobat An urban air pollution monitoring device. In: *2014 IEEE International Conference on Electronics, Computing and Communication Technologies (CONECCT)*, Bangalore, India, 2014, pp. 1–6. https://doi:10.1109/CONECCT.2014.6740321
64. Stone V., Johnston H., Clift M. J. D., Air pollution, ultrafine and nanoparticle toxicology: cellular and molecular interactions. *IEEE Transactions on NanoBioscience*, 2007, 6(4), pp. 331–340. https://doi:10.1109/TNB.2007.909005

65. Veletić M., Floor P. A., Babić Z., Balasingham I., Peer-to-peer communication in neuronal nano-network. *IEEE Transactions on Communications*, 2016, 64(3), pp. 1153–1166. https://doi:10.1109/TCOMM.2016.2526657
66. Atakan B., Galmes S., Akan O. B., Nanoscale communication with molecular arrays in nanonetworks. *IEEE Transactions on NanoBioscience*, 2012, 11(2), pp. 149–160. https://doi:10.1109/TNB.2011.2181862
67. Kumar R., Aljaidi M., Singla M. K., Gupta A., Alhomoud A. M., Alsuwaylimi A. A., Alenezi S. M., Development of a prototype global positioning system based stick for blind patients. *International Journal of Online & Biomedical Engineering*, 2024, 20(8), p. 21. https://doi:10.3991/ijoe.v20i08.49343
68. Kothiyal S. R., Ratnesh R. K., Shukla A. K., Enhancing sensing with surface plasmons resonance: theoretical insights and design strategies in photonic crystal fiber. In: *International Conference Artificial Intelligence and Information Technologies, (ICAIIT)-2024*, Taylor & Francis, 2023, pp. 279–285. https://doi.org/10.1201/9781003510833-46
69. Kumar R., Ratnesh R. K., Singh J., Chandra R., Singh G., Vishnoi V., Recent prospects of medical imaging and sensing technologies based on electrical impedance data acquisition system. *Journal of The Electrochemical Society*, 2023, 170(11), p. 117507. https://doi.org/10.1149/1945-7111/ad050f
70. Petrov V., Moltchanov D., Balasubramaniam S., Koucheryavy Y., Incorporating bacterial properties for plasmid delivery in nano sensor networks. *IEEE Transactions on Nanotechnology*, 2015, 14(4), pp. 751–760. https://doi:10.1109/TNANO.2015.2440296
71. Pandey P. K., Kumar R., Gupta V. et al. An innovative hybrid controller-based combined grid-connected hybrid renewable energy system. *Electrical Engineering*, 2024. https://doi.org/10.1007/s00202-024-02363-2
72. Wender B. A., Seager T. P., Towards prospective life cycle assessment: single wall carbon nanotubes for lithium-ion batteries. In: *Proceedings of the 2011 IEEE International Symposium on Sustainable Systems and Technology*, Chicago, IL, USA, 2011, pp. 1–4. https://doi:10.1109/ISSST.2011.5936889

2 Nanotechnology's Utility in Renewable Energy

Satyendra Kumar Satyarthi, Ishani Mishra, Vishwa Pratap Singh, Kamal Arora, Rama Nand Rai, and Akhilesh Kumar Singh

2.1 INTRODUCTION

Rapid growth in the global population, sub-urbanization, and large-scale industrial development are responsible for high energy demand. Due to excessive energy needs and limited energy supply, an energy shortage occurs when available sources are insufficient to meet the energy demand for all types of applications and usage [1]. The gap between energy demand and supply indicates that fossil fuels alone are cannot provide for our needs; rather, there is a significant negative impact on the environment. Living creatures are also affected by the increased reliance on conventional energy sources such as oil, natural gas, and coal, which are known as fossil fuels. This happens because of the combustion of fossil fuels, which emits large amounts of greenhouse gases (GHGs), such as nitrous oxide (N_2O), chlorofluorocarbons, carbon dioxide (CO_2), and water vapor. These GHGs are mainly responsible for climate change on the planet due to global warming, as agreed upon by researchers, scientists, and environmentalists in several research studies and reports [2]. Several decades ago, various reports predicted that GHG emissions would increase over time because of growth in civilization, industrial expansion, and transportation. Health issues for living creatures, as well as the impact of global warming – such as changes in weather, rising sea levels, increased temperatures on Earth, and disturbances in the ecosystem of the planet – will also increase if concrete steps are not taken. In a single line, we can say that humanity is at risk due to the negative impacts of fossil fuels on the climate as well as human health [3,4]. Based on all these facts and future estimations, almost every country has started various initiatives at their level, led by their governments and some non-government organizations. To improve conditions and minimize these issues, governments have also begun examining their energy strategies and determining if there is a need for regulation. Several approaches and plans have been proposed to reduce GHG emissions and related issues while promoting more efficient and environmentally friendly energy technologies. Environmentally friendly energy generation sources or alternatives are still limited due to their high costs (manufacturing versus efficiency) and ecological consequences (wildlife costs, hazardous wastes, cooling water, etc.). Therefore, several reports have stated that even

DOI: 10.1201/9781003634553-2

first-world countries cannot produce and utilize energy through alternative energy generation methods [5,6]. For the first time, nanotechnology has paved the way for developing new energy generation sectors with cost-effectiveness and enhanced efficiency to promote sustainable economic growth.

2.2 NANOTECHNOLOGY

Nanotechnology is the area of science, technology, and engineering that focuses on the properties of materials at the nanoscale level. It is well known and proven that the properties of any material at the nanoscale are entirely different from those at the bulk level of the same material. For example, the behavior of fullerenes, such as superconductors at the nanoscale, differs significantly from that of carbon nanotubes, whose optical, electrical, and mechanical properties at the nanoscale are entirely different from those at the bulk level [7].

2.2.1 Nano Technology in Renewable Energy Source Generation

With great potential to improve the quality of renewable energy, nanotechnology can play a significant role. Nanotechnology can enhance capability, reduce economic factors, and improve the efficiency of several renewable energy resources by altering the material's properties at the atomic and molecular levels. One of the most promising examples is solar photovoltaic (PV) cells, enhanced through nanotechnology applications. The recombination of electron–hole pairs and reflection losses are considered the two main issues that restrict the efficiency of traditional silicon-based solar cells; however, the performance is now enhanced by using nanostructured perovskite materials, quantum dots, and nanowires. For more efficient energy production from solar cells, nanostructured materials can be very helpful in absorbing different wavelengths of sunlight [8]. One area for renewable energy is energy storage, where the performance of batteries can be significantly enhanced. Regarding battery applications, nanostructured materials like graphene and carbon nanotube (CNTs) are under consideration for the development of next-generation battery materials with improved density, more effective charging and discharging times, and an increased number of cycles. Researchers are also considering these nanostructured materials for multilayered capacitors, supercapacitors, and Li-ion batteries. Nanoscaled electrochemical battery materials also exhibit reduced degradation and improved life cycles. In the case of hydrogen production with higher efficiency, nanotechnology plays a vital role by focusing on economic factors [4,9]. The large surface area-to-volume ratio of nanocatalysts enhances reactivity, significantly increasing the water splitting for hydrogen production. Nanostructured turbine blades are more robust and lighter, improving efficiency and durability. Other parts are also being developed with nanomaterials, resulting in low maintenance costs by increasing lifespan. This chapter provides a detailed discussion of the employment of nanotechnology in the production of renewable energy through various sources, considering all possible factors that affect it [7].

2.2.2 Significance of Materials at the Nanoscale

In the 1950s, physicist Richard Feynman envisioned a future in which we might manipulate things on a tiny scale, but it was not until the 1980s that scientists began research in nanotechnology. In 1981, the discovery of the scanning tunneling microscope marked a milestone in the field; it enabled researchers and scientists to observe and control properties at the atomic level, opening new paths for the world. One Feynman = 10^{-9} m = 10^{-3} μm = 10 Å. "Nano" is a Greek prefix meaning "dwarf," which denotes a size that is 10^{-9} meters and one-billionth of a meter. It can contain 3–5 atoms of 10 Å or be smaller than 1/10th of 1 μm in at least one direction. The nanometer scale is often used for particles smaller than 1 μm. A science-based approach has been developed in the field of nanomaterials, mainly applied to materials with unique morphology-based characteristics [10,11]. Figure 2.1 illustrates the size ranges of various micrometer and nanometer objects, demonstrating that nanotechnology applies to tiny objects as well as systems.

The development and study of nanotechnology always concentrate on materials with the following significant characteristics:

i. The objects must have at least one dimension smaller than 10 nm.
ii. The physio-chemical characteristics of molecular-scale structures fundamentally control the processes.
iii. The presence of building block properties indicates that creating a larger structure must be possible.

The reduction in the size of materials or particles to less than 100 nm can significantly change the associated characteristics of the same materials; at this scale, desirable properties depend on size confinement, particle size distribution, the morphology of the materials, associated interfacial phenomena, as well as quantum effects [11].

There are many important reasons why the nanoscaling of materials is important for the new generation of technological innovations with efficiency, which are discussed below.

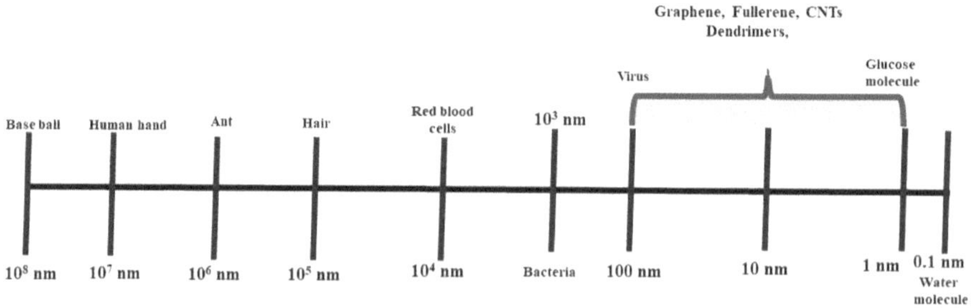

FIGURE 2.1 Various size ranges of objects in micrometers and nanometers.

i. The increased volume-to-surface ratio in nanoscale materials is one of the key characteristics, opening up new possibilities for quantum mechanical (wave-like) effects. Thus, the quantum size effect ensures changes and improvements in the electronic properties of the materials with a significant reduction in size. It is also possible to change the melting temperature and magnetic properties without altering their chemical composition [11].
ii. At the nanoscale, the chemical characteristics enhance the catalytic properties of materials due to the increased surface-to-volume ratio, which is beneficial for interfacial and surface chemistry that is important for a wide range of technological applications in catalysis, composite materials, electrical thermal energy storage, pharmacology, and drug delivery systems [11].
iii. In nanostructured mechanical systems, highly improved mechanical characteristics exist, such as enhanced strength and hardness. Lightweight nanocomposites exhibit improved compression characteristics along with enhanced malleability. Nanoscale materials have a higher density than bulk or microscale mechanical systems, which improves their electrical conductivity, benefiting the electronics industry by enabling the development of smaller, faster, and more sophisticated devices with less power consumption by controlling nanostructure and the interactions and complexity of nanostructures [11].

2.2.3 Synthesis Techniques and Development of Nano Materials

There are many synthesis methodologies that have been successfully implemented in research labs as well as in industries for novel application-based materials for technological advancement. The concept of nanotechnology has been widely accepted and recognized by everyone. It has different meanings in different areas of scientific research. In fact, we can say that it is a multidisciplinary area of research that encompasses almost all branches of science, engineering and technology, medicine, biology and pharma, and environmental and social sciences. The rearrangement of molecules and atoms for the synthesis and development of different kinds of nanomaterials and nanodevices raises two different categories for the synthesis and development of nanomaterials and nanodevices: the first is the top–down approach and the second is the bottom–up approach, both of which are based on atomic-level fabrication of materials, as shown in Figure 2.2 [12,13].

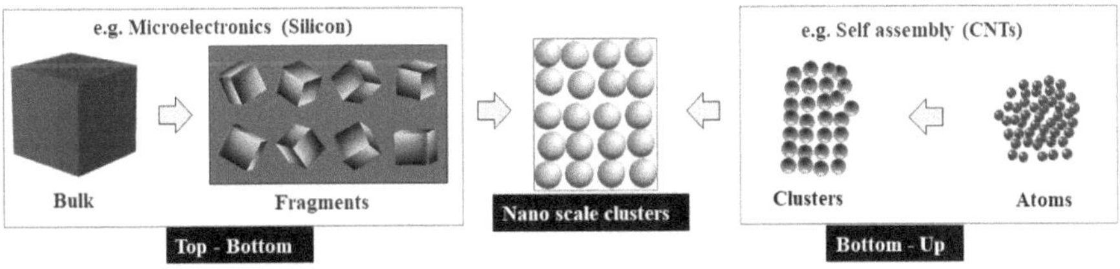

FIGURE 2.2 Nanosynthesis approach.

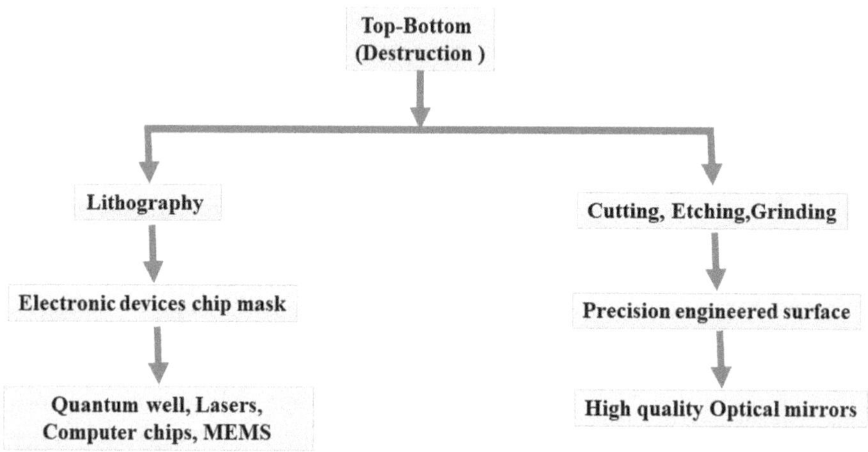

FIGURE 2.3 Subcategories of top–down synthesis of nanomaterials.

2.2.3.1 Top–Down Synthesis Method for Nanomaterials and Devices

One of the most accepted and valuable techniques for nanotechnology is top–down, which is used to miniaturize materials for the synthesis and development of highly efficient machines and gadgets. The top–down methodology begins with breaking down substances using mechanical energy, chemical energy, or other forms of energy. The electronics, optoelectronics, and mechanical industries are the most successful sectors utilizing this method. Chemical vapor deposition technique, physical vapor deposition technique, various types of lithography such as photo, X-ray, and electron beam lithography, wet and plasma etching techniques, and sometimes high-energy ball milling are also used for the synthesis of nanoscale materials and devices. Figure 2.3 shows the subcategories of top–down synthesis of nanomaterials [14,15].

2.2.3.2 Bottom–Up Synthesis Method for Nano Materials and Devices

The bottom–up method of nanomaterials synthesis is also a promising alternative technique for scientific and technological development. This method is the opposite of the top–down approach. It involves the self-assembly of atoms and molecules, as well as the chemical building blocks used for material and device fabrication (Figure 2.4). This method is particularly interesting for researchers across all scientific domains, especially those working on the development of nanoscience and nanotechnology [15]. It is cheaper than lithography techniques and very useful for developing nanostructures around 2–10 nm by controlling atoms with the use of chemical reactors. Drexler hypothesized that bottom–up protein design could be used for the fabrication of nanodevices. Nanodevices that mimic biological processes can perform specific functions, replicate themselves, and self-destruct. In 1991, S. Iijima discovered that carbon nanotubes have long, thin, cylindrical structures. The bottom–up strategy was utilized to develop CNTs and quantum dots, which are crystals with a few hundred atoms that emit light at different wavelengths based on their size [11].

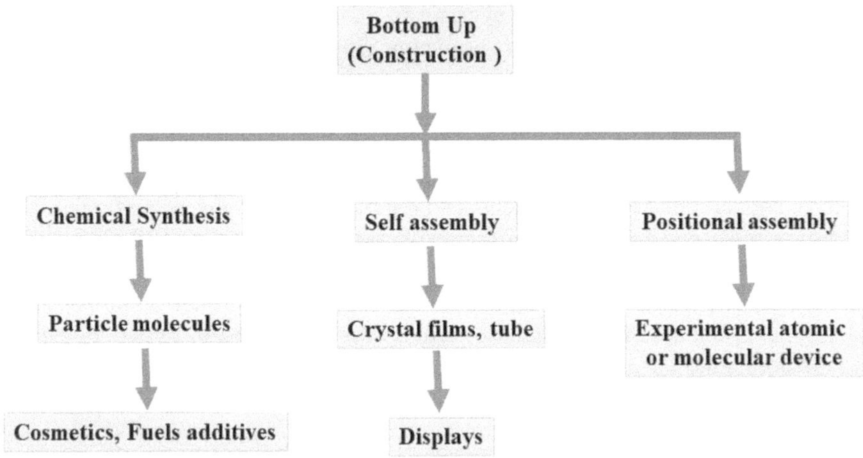

FIGURE 2.4 Subcategories of bottom–up synthesis of nanomaterials.

2.3 INTRODUCTION OF RENEWABLE ENERGY

Max Planck defined energy as the ability of any system to perform external work; it can be classified based on energy sources such as mechanical, thermal, electric, chemical, nuclear, or solar. Renewable energy is different from finite fossil fuels like coal, oil, and natural gas, which also pollute the environment, whereas renewable energy sources consistently provide a sustainable and cleaner option for powering various locations, whether on the planet or in space [16].

Renewable energy is used in a variety of ways, including devices that capture energy from natural resources and convert it into useful power. A few examples include solar panels, which use photovoltaic (PV) cells to collect sunlight and convert it to power; wind turbines, which harness the kinetic energy of wind to generate electrical energy; hydropower, which uses flowing water to produce electricity; and biomass energy, which is generated by burning or converting organic materials such as wood, agricultural waste, and municipal solid waste into biofuels for heat or power generation. There are various reasons why renewable energy is necessary [17]. First of all, after being utilized as the main energy source for a long time, fossil fuels are limited resources that will eventually run out. The burning of fossil fuels emits GHGs along with other dangerous chemicals into the atmosphere, which enhances climate change, air and water pollution, and affects people's health. Therefore, to minimize these negative effects on the environment and guarantee a sustainable energy future, switching to renewable energy sources is very important. Renewable energy has been used for thousands of years. Irrigation, milling, and heating are just a few examples of how early civilizations utilized energy from the sun, wind, and water. However, the development of renewable energy technology for large-scale electricity generation did not start until the late 19th and early 20th centuries. For example, hydroelectric power plants first came into use in the late 19th century, and the first modern wind turbine was built in the 1940s [18]. Interest in renewable energy increased during the 1970s oil crises as countries started considering ways to reduce their dependency on foreign petroleum. Since then, significant technological advancements have been made in the renewable energy sector, leading to increased efficiency, affordability, and widespread adoption around the world. Renewable energy holds enormous promise to meet the world's energy demands while reducing climate change in the future. It is expected that sustainable technological development, in cooperation with investments and regulations from governing bodies, will encourage this trend and inspire further expansion and innovation in the renewable energy sector [19]. For example, solar power is becoming more competitive with traditional energy sources due to advancements in energy storage technology and solar panel efficiency. The capacity of wind power continues to grow and evolve, and offshore wind farms are becoming more common and useful. In the bioenergy sector, new technologies for biogas and advanced biofuels are being developed as a result of advancements in bioenergy research. Even new technologies like wave and tidal energy present attractive opportunities to harness the vast energy of the seas. It is possible for renewable energy to cover all other power sources globally with continuous effort, providing generations to come with inexpensive, sustainable, and clean energy.

The adoption of renewable energy provides various environmental benefits, which are discussed briefly as follows:

- **Reduced GHG emissions:** The adoption and use of renewable energy sources emit fewer GHGs than fossil fuels, contributing to a reduced carbon footprint of energy production and helping to mitigate climate change.
- **Improved air quality:** Harmful air pollutants, including sulfur dioxide, nitrogen oxides, and particulate matter, are significantly decreased when fossil fuel burning is replaced with cleaner renewable energy sources. This also improves air quality and public health as a consequence.
- **Conservation of natural resources:** Renewable energy sources are abundant and inexhaustible; they put less strain on finite energy resources like coal, oil, and natural gas. In the absence of this, resource extraction would have an adverse effect on natural ecosystems and habitats.
- **Water conservation:** Most renewable energy sources, like solar and wind power, require very little water, which helps to save freshwater resources. This contrasts with traditional power plants, which need vast amounts of water for cooling.
- **Reduction of environmental degradation:** The extraction, transportation, and combustion of fossil fuels frequently cause environmental degradation, such as habitat destruction, water pollution, and soil contamination. By switching to renewable energy, we can mitigate these negative effects and promote environmental conservation.

2.4 TYPES OF RENEWABLE ENERGY

There are several types of renewable energy, as shown in Figure 2.5, categorized based on energy sources [20].

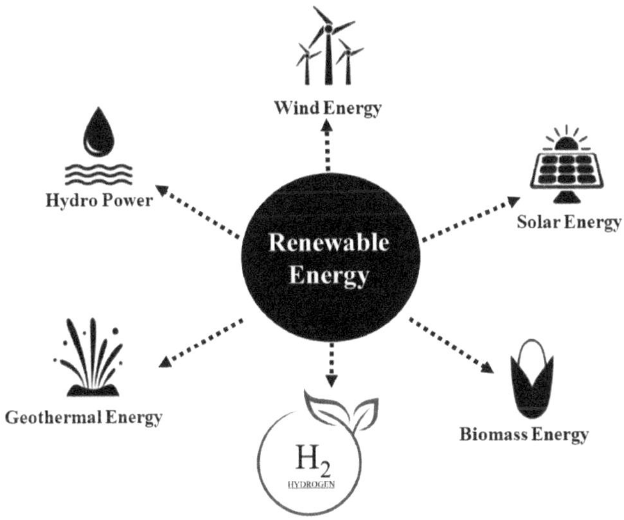

FIGURE 2.5 Different types of renewable energies.

2.4.1 Solar Energy

Solar energy is one of the most effective sources of renewable energy. It has numerous practical applications, including solar power plants, solar cells, seawater desalination, and solar collectors. In fact, sunlight on Earth provides a solution because the incident solar flux on the Earth's surface in an hour is greater than the energy consumed by humans in a year. Therefore, the sun is considered an unmatched source of energy for the Earth, and the amount of solar energy collected in a year varies seasonally, with the maximum obtained during the summer [21]. One of the major difficulties in using solar energy systems is the deficiency in the typical fluids' absorption qualities, which lowers efficiency, but such problems can be reduced by using nanotechnology [22]. The use of nanoparticles, which have a higher surface-to-volume ratio, can effectively enhance solar energy efficiency by providing more conductive surfaces for sunlight. Nanotechnology is also used to improve solar cell efficiency by using nanomaterials like lead selenide, as these materials can release more electrons when photons strike the substrate. The cost-effectiveness of any energy device is a major concern, so the use of nanomaterials in solar cells is also beneficial for enhancing efficiency, making it superior to fossil fuels [23]. Three main types of solar cells are widely used: dye-sensitized solar cells, where sensitizers are used with a nanocrystalline structure as a photoanode to absorb light with charge separation and injection. This method uses interpenetrating networks of mesoscopic semiconductor materials and electrolytes as an alternative to inorganic solid-state semiconductor p–n junctions in traditional solar cells [24]. Organic-polymer-based PV solar cells separate excitons into free electron–hole pairs using an effective field formed by the heterojunction of two different organic materials, known as the donor and acceptor molecules. This type of solar cell is a low-cost solution for producing energy from sunlight in response to the growing demand for renewable energy sources [24].

In hot carrier solar cells, free electrons are pushed into the conduction band by highly energetic photons, causing the electron temperature to reach up to 3000 K. The hot electron relaxes to the bottom of the conductivity band within a few hundred femtoseconds, transferring heat to the lattice [23].

2.4.1.1 Use of Nanotechnology in the Solar Cells

Conventional PV solar cells are not very efficient, and in terms of cost-effectiveness, they are not suitable for large-scale electric power generation. One significant issue associated with efficiency is that when solar panels are heated, they lose almost half the energy they receive. Using nanoscience and nanotechnology techniques, we can now build cost-effective and efficient solar panels [9]. Through advancements in nanotechnology for solar cells, the following advantages can be achieved: their small size allows them to operate without any adverse effects. Nanofluids can absorb energy directly by bypassing intermediate heat transmission steps. Nanofluids can be optically selective. The solar collector allows for a more uniform receiver temperature by minimizing material constraints. Improved heat transfer through increased convection and thermal conductivity can also enhance receiver performance. Tuning the size and shape of nanoparticles can improve absorption efficiency for specific applications.

2.4.2 Biomass Energy

All organic material comes from plants, including algae, trees, and various types of crops, which are referred to as biomass. Biomass results from photosynthesis, which collects and stores solar energy. The conversion of biomass into usable forms of energy is known as biomass energy or bioenergy; these energies can take the form of liquid fuels, heat, and electricity. Biomass for bioenergy comes from two sources: waste left over after crops are processed for food or other goods, such as pulp and paper from the wood industry, or directly from the soil as energy crops. Post-consumer waste streams, which include the clean portion of municipal solid waste, pallets used for transportation, and wood used for construction and demolition, all make significant contributions. In modern civilization, the management of the flow of solar-generated materials, food, and fiber can be equated with the biomass-to-bioenergy system [25,26].

Figure 2.6 displays the various types of resources and their uses.

Throughout history, human civilization has relied heavily on biomass as its primary energy source. Due to advancements in extraction and combustion technology, coal, oil, and natural gas have supplanted wood as the principal fuel for the industrialized world in the last century. Currently, biomass is used for energy at a rate of 40–55 exajoules (EJ = 1018 joules) each year, accounting for 10%–14% of worldwide energy consumption. This positions biomass as the fourth main source of energy, with the other sources being oil (33%), coal (21%), and natural gas (19%) [27].

2.4.2.1 Role of Nanotechnology for the Biomass Renewable Energy

Nanotechnology plays an important role in increasing the efficiency and effectiveness of biomass renewable energy resources. Organic resources that can be transformed into energy are referred to as biomass, including plant and animal waste. The

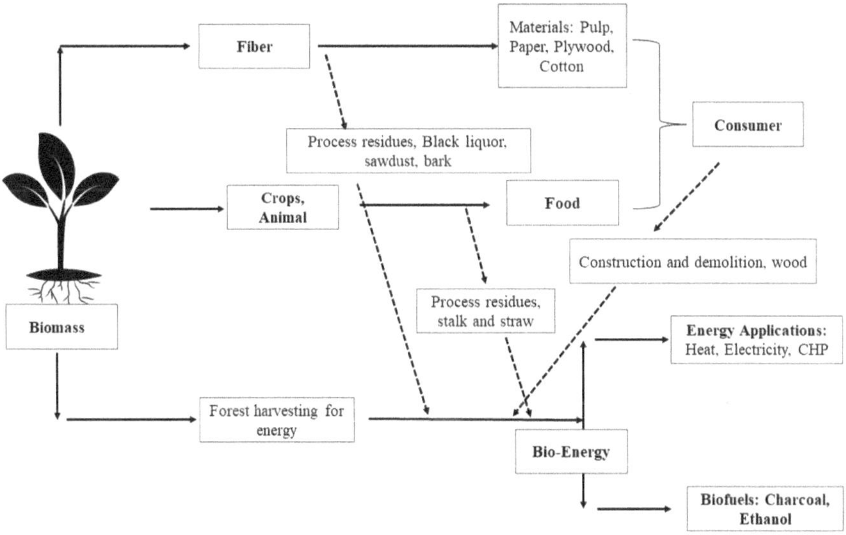

FIGURE 2.6 Various kinds of resources and their uses.

use of nanotechnology in this field improves the processes involved in converting biomass into usable energy sources such as biofuels, biogas, and bioelectricity. The following is a brief discussion of how nanotechnology is used in biomass renewable energy sources.

2.4.2.1.1 Improving Biomass Conversion

One of the most important applications of nanotechnology in biomass energy is the conversion process. To convert biomass into energy, organic resources are broken down into simpler molecules that can be utilized as fuel. Nanocatalysts, which are microscopic particles composed of metals or metal oxides, can considerably improve this process. Nanocatalysts provide a higher surface area for chemical reactions to occur and accelerate the conversion of biomass into fuels such as bioethanol and biodiesel [28]. For example, in the synthesis of biodiesel from vegetable oils, nanocatalysts can improve reaction efficiency, shorten conversion time, and reduce production costs.

2.4.2.1.2 Enhancing Biogas Production

Methane and carbon dioxide are the main constituents of biogas, produced during the anaerobic digestion of organic waste materials. The utility of nanotechnology can enhance biogas production by improving the efficiency of microorganisms involved in the digestion process of organic waste materials. Nanomaterials can be added to anaerobic digesters to increase the metabolic activity of bacteria, resulting in enhanced methane gas production [28].

2.4.2.1.3 Advanced Biofuel Production

Nanotechnology plays an important role in the more efficient production of biofuels such as bioethanol and biodiesel as renewable energy sources. During bioethanol production, plant cellulose is broken down into sugars, which are subsequently fermented to yield ethanol. By utilizing nanoscale enzymes, the breakdown of cellulose into sugar can be made more efficient. The application of nano-based technology is also very useful for enhancing the stability and performance of biofuels. Nanoscaled particles can improve the combustion properties of biofuels by providing a high surface-to-volume ratio, resulting in more efficient energy release and reduced pollutant emissions [29].

2.4.2.1.4 Improving Biomass-Derived Electricity

Nanotechnology can also enhance the efficiency of converting biomass to power. One approach is to produce improved materials for fuel cells. Fuel cells use chemical energy to generate electricity and can be powered by biomass-derived hydrogen. Nanomaterials such as catalysts and membranes can be utilized to improve the efficiency and durability of fuel cell components. These advancements can result in higher electrical generation and longer-lasting fuel cells, making biomass a more viable choice for producing clean electricity [30].

2.4.2.1.5 Carbon Capture and Utilization

Carbon capture and the high potential use of biomass energy through nanotechnology are possible. When biomass is burned, CO_2 is emitted, but with the application of nanotechnology, it can be converted for further use. Nanoscale materials have also been developed to absorb CO_2 very effectively. For example, nanostructured materials with large surface areas and a high surface-to-volume ratio can absorb CO_2 from biomass power plant exhaust gases. Captured CO_2 can be transformed into fuel or chemicals for further utility; this can help reduce the carbon loop and minimize the GHG emission effect [31].

2.4.3 Geothermal Energy

Geothermal energy is another valuable form of renewable energy, which involves the extraction of thermal energy from the Earth's crust. The range extends almost 5–10 km, where the observed temperatures are very high. Therefore, at greater depths, where very high temperatures are present, it is necessary to use nanofluids as cooling fluids to prevent damage to the pipes used in these high-temperature zones. Connecting electronic components, such as sensors, is also necessary for the drilling machine to ensure the smooth extraction of thermal energy from the Earth's crust, as the drilling machine operates under high friction and high temperatures [32]. Geothermal energy sources are categorized based on the measured temperature range: low (<100°C), medium (100°C –150°C), and high temperature (>150°C). The geothermal gradient is defined as the temperature difference between the Earth's core and its surface, which allows for continuous thermal energy conduction in the form of heat from the Earth's core to its surface. Nanofluids are used for the extraction of energy from the Earth's core and are subsequently transferred to the energy power plant system for the production of energy for various applications. Geothermal energy is utilized in various ways; for example, in colder regions, it is used as a central heating system for entire cities by using piped hot water to heat multiple buildings [33]. More than 72 countries are officially documented as using geothermal energy, with Iceland being the leading country in its consumption. In Iceland, almost 93% of houses use geothermal energy, saving the country approximately 100 million dollars each year. Additionally, the consumption of this form of energy has made Iceland one of the cleanest countries in the world. Geothermal-based power plants can operate 24 hours a day by providing load-based capacity; in the upcoming 30 years, geothermal power generation could reach 85 GW [34].

2.4.3.1 Utility of Nanotechnology for the Geothermal Energy

Nanotechnology can revolutionize the geothermal energy sector by improving efficiency and sustainability, as explained briefly below.

2.4.3.1.1 Improved Drilling and Exploration

Sophisticated materials and sensors can be developed utilizing nanotechnology to endure the harsh conditions found deep underground. For smooth and efficient operation, the drilling equipment must be stronger and heat-resistant, properties that can be achieved through nanotechnology. To accurately locate geothermal energy resources, it is essential to detect parameters such as temperature, pressure, and chemical composition with precision, which can be accomplished using nanoscale sensors [35].

2.4.3.1.2 Enhanced Heat Transfer

Compared to conventional fluids, nanofluids have significantly higher thermal conductivity. Therefore, the use of nanofluids is more efficient in transferring heat from the Earth's core to its surface. This increased efficiency benefits power plants by generating more electricity from the same amount of geothermal energy compared to conventional fluids [36].

2.4.3.1.3 Improved Materials for Energy Systems

The development of new materials using nanotechnology can better resist corrosion and scaling. With the use of these materials, the lifespan of geothermal equipment extends several years and reduces maintenance costs. For example, the nanoscale material coating on the pipes can protect them from the harsh geothermal environment [37].

2.4.3.1.4 Enhanced Geothermal Systems (EGS)

To enhance heat extraction and create cracks in hot rock formations, EGS can be used in fluid injection. The use of nanoparticles improves this process by making the rocks more penetrable and facilitating heat extraction. This makes EGS more feasible and effective, increasing the potential for geothermal energy in regions without naturally occurring geothermal reservoirs [38].

2.4.3.1.5 Environmental Benefits

More efficient geothermal systems using nanotechnology reduce the environmental impact of geothermal energy production. Because of the high efficiency of these systems, there is less wastage and more sustainable energy solutions.

2.4.4 Wind Energy

Wind energy is another source of renewable energy where the blowing wind is harnessed for power. Wind energy is captured by wind turbines or other energy devices that transform the kinetic energy of wind into mechanical or electrical energy for further use. In this case, no fuel needs to be burned to generate energy, making it highly environmentally friendly; wind and solar energy are also interconnected, as wind occurs whenever the sun shines. Compared to other conventional energy sources, wind energy has fewer adverse effects. Currently, 23 billion kWh of electric energy is generated every year worldwide through wind energy. Germany alone generated almost 4,400 MW of electricity from wind energy, whereas India generated 1,000 MW annually [39]. In 2010, electrical energy production from wind energy accounted for about 2.5% of global electricity demand, and it is increasing at a rapid rate of about 25% per year. During power generation, wind power plants rely on wind speeds of at least 13–15 m/s to operate optimally, while small power plants should have a minimum speed of 10 m/s. An average efficiency of 30%–40% has been observed from these power plants due to variable conditions. The generated energy depends on the square of the blade length. Nanocomposite materials exhibit high strength-to-weight and stiffness-to-weight ratios, making nanotechnology very useful for developing different materials for longer and stronger turbine blades [40].

The use of different materials for different parts of the wind turbine is discussed below [41–43].

i. **Blades:** Graphene, CNTs, silica, and a hybrid of graphene and CNTs can be used for blade construction. Nanotechnology can also be used for coatings, structural adhesives, and sealants to improve blade performance.
ii. **Nacelle:** The nacelle, a significant component, can be produced using graphene, CNTs, silica-based nanomaterials, and nanotechnology for coating purposes.
iii. **Bearings:** This can be made using silica, copper, tungsten disulfide, molybdenum disulfide, boron nitride, graphene-based nanomaterials, and nanotechnology, which can be used for the lubricants, coatings, and sealants of the bearing.
iv. **Main shaft:** This part can be made using CNTs, silica, a hybrid of graphene and CNT-based nanomaterials, and nanotechnology, which can also be used for the coating.
v. **Gearbox:** This part can be made using silica, copper, tungsten disulfide, molybdenum disulfide, boron nitride, graphene, CNTs, and nanotechnology, which can also be used for the lubricants and coatings
vi. **Generator:** This part can be made from alumina material, copper oxide (CuO), silica (Si), and a metal matrix composite made of Cu–CNT/graphene-based nanomaterials, and nanotechnology can also be used for the sensors, coolants, and windings.
vii. **Tower:** This part can be made using a hybrid form of graphene and CNTs; graphene, CNTs, and nanotechnology can also be applied to cement and concrete.

viii. **Energy storage and transmission:** Graphene, along with a hybrid of graphene and CNTs, can also be used to manufacture this part. The use of CNTs and nanotechnology can be beneficial for battery and ultracapacitor applications. The emergence of new highly efficient materials through nanotechnology development also decreases the weight of devices and enhances their performance.

2.4.5 Hydrogen Energy

Since hydrogen is the most abundant element in the universe (making up more than 90% of all atoms), is the lightest element (molecular weight ¼ 2.016), and has the highest known energy content (in terms of calorific or heating value) of any available energy fuel, it is widely considered a potential cost-effective clean fuel for future use. Additionally, hydrogen is sustainable and non-toxic, and unlike coal, natural gas, or petroleum, it is an energy carrier that is good for the environment and produces water as the only byproduct when converted to energy [44].

It can be produced using renewable energy resources as well as non-renewable energy sources. Examples of renewable energy resources include hydro, wind, solar, biomass, and geothermal, while coal and natural gas are examples of non-renewable energy sources that can be used for hydrogen fuel generation. The storage and subsequent use of hydrogen as a fuel is also possible in internal combustion engines and turbines for electricity generation. It can also be used for fuel cells and transportation vehicles with greater efficiency [45].

Because hydrogen possesses a remarkable ability to store energy for later use, one kilogram of hydrogen has an energy content of around 120 MJ (¼ 33.33 kWh), which surpasses the energy content of most conventional fuels by two times. While some other energy sources like liquefied natural gas exist at around 48.62 MJ/kg and liquefied petroleum gas exists at around 46.60 MJ/kg, solid fuel from coal (wet basis) exists at around 22.73 MJ/kg and ethanol exists with around 26.395 MJ/kg of energy content [46]. The storage of hydrogen as fuel can pave the way for new technological development, which will be very beneficial for the future fuel economy. For a hydrogen and hydrogen storage-based fuel economy, it is necessary to develop both mobile and stationary energy storage systems. There are three different types: gaseous-based hydrogen fuel energy storage systems, liquid-based hydrogen fuel energy storage systems, and solid-state hydrogen fuel energy storage, which are used for hydrogen fuel storage to meet various needs.

Nanotechnology and nanostructured materials play very important roles in the generation of highly efficient hydrogen fuel as well as in hydrogen energy storage with greater capacity in different types of storage systems [47].

2.4.5.1 Application of Nanotechnology for the Generation of Hydrogen Energy

There are several types of hydrogen energy generation where fossil and non-fossil fuels are used as resources. Some of the primary sources of generation are electricity and chemicals, while the secondary sources of hydrogen energy generation are fossil and renewable energy sources.

The following classes of hydrogen production techniques can be distinguished based on the reactions using different reactants:

- Production of hydrogen fuel through the direct reaction of metals, acids, and bases with chemical reagents.
- Production of hydrogen using hydrocarbons, such as biomass and fossil fuels.
- Production of hydrogen through the direct splitting of water.

2.4.5.1.1 Production of Hydrogen Fuel through the Direct Reaction of Metals, Acids, and Bases with Chemical Reagents

The generation of hydrogen using chemical reactants has been in process for a long time and is widely studied by researchers and scientists. The associated reactions of hydrogen production through chemical reactions are further divided into three types, which are as follows:

2.4.5.1.1.1 Reactions in Which Acids Play Main Role The typical chemical reaction for this type of hydrogen production is given by the following reaction in which sulfuric acid plays a crucial role with zinc.

$$Zn + H_2SO_4 \rightarrow ZnSO_4 + H_2 \uparrow \tag{2.1}$$

It is a chemical reaction that occurs spontaneously and generates a lot of heat and hydrogen. This reaction can be considered highly viable and profitable if the produced heat and hydrogen are used simultaneously.

2.4.5.1.1.2 Reactions in Which Bases Play Main Role In this type of chemical reaction, bases take the place of strong acids for the reaction; in the following reaction, sodium hydroxide plays a crucial role with silicon:

$$2NaOH + Si + H_2O \rightarrow Na_2SiO_3 + 2H_2 \uparrow \tag{2.2}$$

2.4.5.1.1.3 Reactions in Which Water and Reactive Metals Reacts Each Other This type of hydrogen production reaction is carried out using water with a highly abundant metal; in the following reactions Li and Al are used for the reaction and production of hydrogen,

$$2Li + 2H_2O \rightarrow 2LiOH + H_2 \uparrow \tag{2.3}$$

$$Al + 3H_2O = Al(OH)_3 + 3/2H_2 \uparrow \tag{2.4}$$

2.4.5.1.1 Production of Hydrogen by Using Hydrocarbons

There are numerous forms of hydrogen-rich hydrocarbons available that can be used to generate hydrogen; these hydrocarbons can come from renewable biomass sources as well as non-renewable fossil fuels. The use of heat and catalysts can easily convert fossil fuels into hydrogen, particularly those in gaseous conditions, which can be easily converted into hydrogen [48].

2.4.5.1.1 Production of Hydrogen Generation from the Fossil Fuels

In the current scenario, almost 95% of useful hydrogen is generated from natural gases using steam methane reforming. There are other methods also used for fossil fuels as a source of hydrogen generation, including coal gasification, auto-thermal reforming, partial oxidation, and plasma reforming, but steam reforming is considered the major factor in the generation of hydrogen fuel due to its complete thermodynamic feasibility along with higher efficiency [48].

$$\begin{gathered}\mathrm{Cm} + \mathrm{Hn} + \mathrm{MH_2O} = \mathrm{mCO} + (\mathrm{m} + \mathrm{n}/2)\mathrm{H_2CH_3OH} + \mathrm{H_2O} = \mathrm{CO_2} + 3\mathrm{H_2}, \\ \Delta\mathrm{H} = 49\ \mathrm{kJ/mol}\end{gathered} \tag{2.5}$$

For methanol as an example, ΔH: hydrocarbon dependent, endothermic.

2.4.5.2 Use of Nanotechnology in Hydrogen Generation from Fossil Fuels

A $Pd@CeO_2$ core–shell nanostructured dispersible material was synthesized by a researcher at the University of Pennsylvania and found to be a highly effective catalyst precursor for methanol steam reforming processes. In this process, an Al_2O_3 substrate is required to support the catalytic function, meaning that the $Pd@CeO_2$ nanostructure alone cannot be used as a steam-reforming catalyst. Some other nanostructured materials are also used as catalysts for hydrogen production, such as nanopowder of zirconia, which is employed as a catalyst during the steam reforming of bioethanol for hydrogen production. Nanostructured In_2O_3 is also used as a catalyst for hydrogen production in the process [48].

2.4.5.3 Use of Nanotechnology in Hydrogen Generation from the Biomass

To improve the efficiency and quality of the hydrogen produced during biomass steam gasification, a nanostructured Ni-La-Fe/-Al_2 catalyst is used. With the evolution of different cases, the activity and effectiveness of the nanostructured tri-metallic catalyst in biomass gasification applications are assessed. When the nanostructured catalyst was used, tar production was significantly decreased, and gas yield noticeably improved. For catalytic gasification at 800°C, the tar removal efficiency reached up to 99%. After catalysis, the gaseous percentages of C_2 and CH_4 were clearly lower, but the valuable H2 content significantly increased to more than 10% vol.% as a result of the process. It has also been demonstrated that another nanostructured NiO/-Al_2O_3 catalyst with a similar composition is helpful for producing hydrogen from the gasification of biomass [48].

2.4.5.4 Producing Hydrogen by Directly Splitting Water

Water splitting for hydrogen generation is the most promising technique. There are several other kinds of water splitting methods: electrolysis of water and thermolysis of water. In electrolysis of water, the transformation of energy using electricity to produce hydrogen is not suitable because electric energy is not a self-generated energy source, and in thermolysis of water, a considerable amount of energy is required for high-temperature treatment. Photoelectrochemical water splitting is known to be a very efficient and almost continuous energy source for producing hydrogen from water splitting. There are three major techniques of water splitting for the production of hydrogen at lower temperatures [48].

2.4.5.4.1 Thermochemical Decomposition and Thermolysis Processes in Water Splitting

In this method, compounds break into two or more elements when subjected to sufficient heat. During this, chemical bonds get broken and the compound undergoes decomposition. It mainly indicates the decomposition of water, and the simplest chemical reaction is given as follows in the net chemical reaction, resulting hydrogen production [48]:

$$H_2O \rightarrow H_2 + 1/2\ O_2 \tag{2.6}$$

2.4.5.5 Applications of Nanotechnology for the Decomposition of Water

A nanostructured porous zirconia ceramic membrane is used to separate hydrogen and oxygen at a higher temperature (> 2500 K) to avoid an explosion, which can happen when the temperature is relatively lower. Other nanotechnology applications are also necessary during the heating and cooling cycle of thermochemical water decomposition, serving as nanostructured catalytic materials. For example, Zn nanoparticles are employed for the production of solar hydrogen through Zn/ZnO thermochemical water splitting [9].

2.4.5.5.1 Direct Electrolysis of Water

This technique is simpler and more efficient than others for hydrogen production. Almost 4% of the hydrogen produced worldwide uses this method, and from this method, heavy water is also generated, which is highly beneficial for nuclear applications [48].

The simplified form of the related reactions during electrolysis is as follows:

$$2H_2O\ (\text{liquid/vapor}) + \text{electrical energy} \rightarrow 2H_2(\text{gas}) + O_2(\text{gas}) \tag{2.7}$$

The half-reactions of this electrolysis process can be balanced through

With acid:

$$\text{Reaction at Cathode (reduction)} \rightarrow 2H + (aq) + \ 2e^- \rightarrow H_2(\text{gas}) \tag{2.8}$$

$$\text{Reaction at Anode (oxidation)} \rightarrow 2H_2O \rightarrow O_2(\text{gas}) + 4H^+(aq) + 4e^- \tag{2.9}$$

With base:

$$\text{Reaction at Cathode (reduction)} \rightarrow 2H_2O\ (1) + 2e^- \rightarrow H_2(\text{gas}) + 2OH^-(aq) \tag{2.10}$$

$$\text{Reaction at Anode (oxidation)} \rightarrow 4OH^-(aq) \rightarrow O_2(\text{gas}) + 2H_2O(a) + 4e \tag{2.11}$$

2.4.5.5.2 *Applications of Nanotechnology for the Direct Electrolysis of Water*

The well-controlled electrodeposition process into a nonporous alumina template is facilitated by the Ni nanoparticle structure. Using nanostructures as stable and efficient cathode catalysts for the hydrogen evolution reaction in alkaline water electrolysis, along with the application of Ni nanostructured pottery, provides a higher quantity and larger surface area as reactive sites, which leads to more hydrogen adsorption during the electrolysis of water. Thus, the activity at the cathode is significantly boosted compared to conventional cathode catalysts. Carbon nanotube-based materials are also used as novel materials in the electrolysis of water as cathode-related materials because their electrically conducting surfaces are very beneficial for (a) functioning as cathodes for hydrogen generation and adsorption and (b) improving the electrolysis of water. Nanostructured hydrophobic CNT-based electrochemical cells are also considered a good alternative for hydrogen storage due to their unique combined properties related to hydrophobic and conductive nature [48]. Nano-range Ni particles are also a good alternative as electrodes for H_2 generation in SOECs because they are highly stable in humidified hydrogen.

2.4.5.5.3 *Decomposition of Water through Photo-Electrolysis and Photo-Catalysis*

Photo-electrolysis is a form of electrolysis for hydrogen generation where light is used directly. In the photo-electrolysis method, the chemical conversion of H_2O into H_2, along with the conversion of light into electrical current, takes place in a single step. Water decomposition through photo-catalysis is quite different from water decomposition through the photo-electrochemical method [49]. The main differences are associated with the reaction site location within the whole system, but generally, their mechanisms are the same. In the photo-electrochemical method, two separate reactions take place at the photo-anode and the photo-cathode, where oxygen and hydrogen gas are produced, respectively. On the surface of the photo-catalyst, both types of reactions take place concurrently and produce a mixture of the two gases [50,51].

2.4.5.5.4 *Application of Nanotechnology in Water Decomposition through Photo-Electrolysis and Photo-Catalysis*

Nanostructured electrodes made of semiconductor materials are used in both photoelectrochemical processes and water splitting through photo-catalysis. Nanostructured photoelectrodes made of TiO_2 are very effective in minimizing the mismatch between the recombination of electrons and holes because their emission of electrons is rapid and may be several thousand times greater than that of charge recombination. Nanostructure-based CdS and CdSe materials are deposited on the surface of mesoporous TiO_2 films, which improves the charge transport characteristics of TiO_2 films and enhances performance [48,50].

2.4.6 Hydropower Renewable Energy and Nanotechnology

Nanotechnology Hydropower renewable energy is one of the oldest renewable energy sources for various applications; in terms of power generation, it is currently used to generate electricity using the power of moving large-scale water. It is called "renewable" because it depends on the natural water cycle, which is continuously replenished by rain and snowfall. Electricity generation occurs as water from rivers or dams flows through turbines, which consist of large wheels. As the water flows through these turbines, it causes them to spin at a desired speed. A power generator is connected to the turbine, which rotates with the help of this hydropower force, and the generator produces electricity. The process is very efficient and can generate a significant amount of power without producing pollution [52].

2.4.6.1 Application of Nanotechnology in the Power Generation through Hydro Energy

Applications of nanotechnology in power generation through hydro energy are extensive. Here we cover diverse areas as follows.

2.4.6.2 Turbine Energy Efficiency Improvement

In hydropower generation plants, turbines are used to convert the kinetic energy of flowing large-scale water into mechanical energy, which is then converted into electricity. The efficiency of these turbines is crucial for the overall performance of the plant. Nanotechnology is applied in various ways to improve the efficiency of the turbines, which are discussed below [53,54].

2.4.6.3 Nanostructured Materials Coatings for Turbines

Nanocoating on the turbine blades is an effective way to improve the overall efficiency of the generators; this nanostructured coating can help reduce friction and wear on turbine blades, enabling smooth and efficient spinning. This not only increases the efficiency of energy conversion but also extends the life of the turbines, reducing maintenance costs, which is made possible by nanostructure-based properties [55].

2.4.6.4 Use of Nanostructured Anti-Corrosion Materials

During power generation, turbines are constantly exposed to water, which can lead to corrosion over time. Nanotechnology can develop materials that are highly resistant to corrosion. The use of these materials in turbine components like blades and other parts can significantly extend the durability of the turbines, ensuring that hydropower plants operate more reliably [56].

2.4.6.5 Enhancing Water Quality Monitoring

The efficiency of hydropower plants depends on the quality of water. Turbines can be damaged, and efficiency can also be reduced due to poor water quality.

2.4.6.6 Nanosensors

Nanostructure-based sensors are used to improve water quality monitoring. These nanosensors efficiently detect pollutants and other harmful substances present at very low concentrations. They provide real-time data on water quality and also act effectively to maintain optimal conditions for hydropower generation.

2.4.6.7 Electrical Efficiency Improvement

Generated electricity needs to be transmitted and stored efficiently, and improvements have been made in these sectors with the help of nanotechnology.

Use of nano-enhanced conductors and storage: The use of nanostructured materials in electrical conductors can reduce energy loss during transmission. For example, graphene, which has excellent electrical conductivity, can be used to create more efficient transmission lines. This means that most of the generated electricity reaches the final point by enhancing the overall efficiency of the power system. Nanotechnology-based storage batteries can store more energy for a longer time.

2.4.6.8 Environmental Benefits

Nanotechnology not only improves efficiency but also has potential environmental benefits.

2.4.6.8.1 Reduced Material and Recycled Water Usage

The use of nanostructured materials enhances properties, so less material is needed to achieve better performance, and the extraction of materials is also reduced, which significantly impacts the environment. Nanostructured filters can ensure that wastewater from power generation can be cleaned and returned to rivers or lakes for further use, contributing to the preservation of aquatic ecosystems.

2.5 REQUIRED CHARACTERIZATION TECHNIQUES FOR RENEWABLE ENERGY SOURCES

Characterization techniques are essential for understanding and optimizing materials and devices used in renewable energy generation. These techniques help analyze the properties, performance, and stability of the materials and devices. Below are some of the key characterization techniques used across different renewable energy technologies (Figure 2.7).

2.5.1 X-RAY DIFFRACTION

X-ray diffraction is a useful technique in the field of renewable energy for better understanding the materials and systems that harness clean and sustainable energy. It offers an in-depth understanding of phase composition, crystallinity, and crystal structure, among other structural characteristics of the materials involved. This information aids in optimizing the performance of the materials and the efficiency of the device. It is essential for understanding mechanical stability and potential defects that could impact efficiency by providing insights into the stress and strain of the material. It is used to identify the correct stoichiometry of the compounds used in thin-film technologies, such as cadmium telluride (CdTe) and copper indium gallium selenide (CIGS) solar cells, by assessing their different phases. It helps to determine the size of crystalline grains and their degree of crystallinity, which can be further used to reduce recombination losses and improve charge carrier mobility, thereby increasing the efficiency of thin-film solar cells. The crystal structure of perovskite materials such as methyl ammonium lead iodide ($MAPbI_3$) and composite materials used for wind turbine blades, such as polymer matrices reinforced with nanomaterials like CNTs or graphene, aids in determining crystallinity, phase stability,

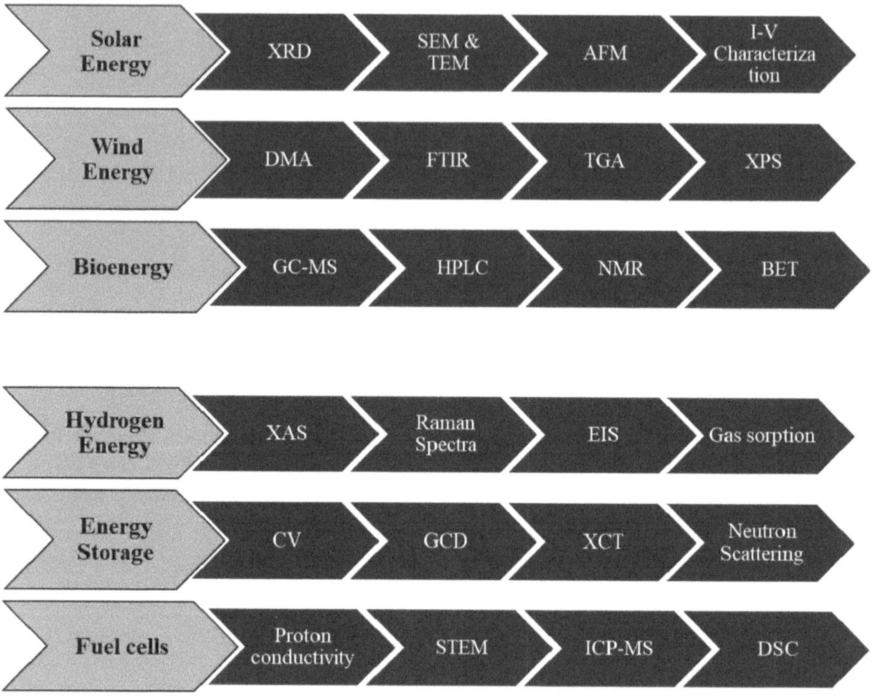

FIGURE 2.7 Types of renewable energy and corresponding characterizations.

phase purity, and the presence of any unwanted secondary phases. Interpreting the crystallinity of the polymer matrix helps optimize the mechanical properties of the composites. [57–60].

2.5.2 Scanning Electron Microscopy (SEM) and Transmission Electron Microscopy (TEM)

SEM and TEM are powerful tools for the characterization of materials used in renewable energy technologies. SEM and TEM provide comprehensive insights into the micro- and nanoscale morphology, microstructure, and elemental composition of materials, which are essential for maximizing their performance and understanding their behavior in various energy applications. SEM offers detailed surface morphology and microstructural analysis of materials like silicon wafers in PV cells, where surface roughness and texture affect light absorption efficiency, which could be further exploited for optimization. Additionally, it examines the composite materials used in wind turbine blades, evaluating matrix bonding and fiber distribution. TEM provides atomic-scale resolution, revealing the internal structure of materials, which is critical for understanding defects, grain boundaries, and phase distributions in thin-film and perovskite solar cells. TEM's ability to provide high-resolution images and electron diffraction patterns makes it invaluable for analyzing nanostructured catalysts in biofuel production, hydrogen generation, and storage materials. The dispersion and crystallographic orientation of the nanoparticles can be visualized by TEM, which directly impacts their stability and performance. Together, SEM and TEM fully characterize renewable energy materials, thus helping to develop robust and efficient energy conversion and storage systems [61–63].

2.5.3 Atomic Force Microscopy (AFM)

AFM provides high-resolution, three-dimensional surface profiles of materials at the nanometer scale. It is employed to observe the surface topography and roughness of PV materials, such as silicon wafers, thin films, and perovskite layers used in solar energy. AFM assesses the detailed surface morphology to optimize light absorption and minimize reflection losses, which is crucial for enhancing solar cell efficiency. AFM also studies the mechanical characteristics and structural integrity, surface area, and porosity of biomass and biofuel catalysts. It evaluates the surface characteristics of electrode materials in energy storage devices such as lithium-ion batteries and supercapacitors, revealing information about particle size distribution, aggregation, and surface defects. This data is vital for improving charge/discharge rates and overall energy density. AFM is also used to study nanostructured materials for hydrogen production and storage, such as photocatalysts and metal–organic frameworks, where surface characteristics significantly impact performance. AFM can also be used to measure mechanical properties, stiffness, and adhesion forces, which predict the behavior of the material under operational conditions. [61,64,65]. AFM contributes to efficient and durable energy solutions due to its critical role in the development and optimization of advanced materials for renewable energy technologies by providing high-resolution imaging and mechanical property measurements.

2.5.4 Current–Voltage (I-V) Characterization

I-V characterization is the most basic and sought-after technique to assess the performance of renewable energy devices such as fuel cells, solar cells, PV cells, and batteries. It helps to calculate short-circuit current (Isc), open-circuit voltage (Voc), fill factor (FF), and efficiency of energy conversion. These parameters are used to estimate the cells' ability to convert sunlight into electrical energy and identify energy loss due to series and shunt resistance. The cell is exposed to simulated sunlight, and the output, i.e., current, is measured as a function of the applied voltage, resulting in an I-V curve. The shape of the curve can indicate the efficiency of the electrochemical reactions, the effectiveness of the catalysts, overall power output, as well as defects in the cell. These parameters are essential for optimization regarding design, material, and fabrication to increase the efficiency of fuel cells. I-V characterization provides data on capacity, resistance, and charging–discharging cycle stability for energy storage systems [61,66]. Analyzing these measurements helps to predict how batteries will perform in real-world applications.

2.5.5 Dynamic Mechanical Analysis (DMA)

DMA is a basic technique used for characterizing the mechanical properties of composite materials or polymers by applying periodic stress and strain to assess deformation. It provides insights into the thermal and mechanical properties of the materials, such as stiffness, damping, viscoelasticity, storage modulus (elastic behavior), and loss modulus (viscous behavior), which helps evaluate the performance and durability of the material when exposed to different temperatures and environmental conditions. This information can be beneficial for predicting the longevity of solar cells or wind turbines under environmental conditions [61,67].

DMA is used to analyze the mechanical and viscoelastic properties of biomass and bio-composites and their conversion into biofuels and bioenergy. It also helps characterize the materials used in hydrogen storage tanks and fuel cell components [68]. DMA of the electrode materials and separators of batteries and supercapacitors gives insights into their charge–discharge cycles and how these materials perform under mechanical stress. It helps in the development of robust and durable energy storage devices.

2.5.6 Fourier Transform Infrared (FTIR) Spectroscopy

FTIR can also provide detailed insights into the chemical composition and molecular structure of the materials used in renewable energy production. It is used to analyze the polymer blends in solar cells or perovskite cells and the organic and inorganic materials of PV cells. It identifies functional groups, chemical bonds, impurities, and degradation, which helps to predict the efficiency and stability of solar cells [69,70].

FTIR also provides information on the composition of biomass feedstocks, including the content of cellulose, hemicellulose, and lignin, and monitors their transformation during the pre-treatment and conversion processes. FTIR detects ester functional groups and traces or residual reactants in biodiesel and biofuels to ensure their purity. It can also help identify reaction intermediates, absorbed species, and active sites of the catalysts used in hydrogen production and storage. FTIR is used to analyze the chemical composition, stability, and degradation mechanisms of electrode materials as well as electrolytes used in energy storage devices such as batteries and supercapacitors [65,71]. Characterization techniques provide insights into the properties of materials used and help optimize performance, functionality, stability, and reliability. They play an important role in addressing the challenges of renewable energy production, conversion, and storage.

2.6 CHALLENGES AND FACTORS AFFECTING NANO-RENEWABLE ENERGY RESOURCES

The efficiency and advancement of nanotechnology in the field of renewable energy resources are influenced by many factors, including technological, financial, environmental, and regulatory aspects.

2.6.1 Material and Fabrication

Optimizing the performance of nanomaterials in energy applications requires an intricate understanding and control of their atomic and molecular characteristics. High surface area-to-volume ratios of nanomaterials can improve efficiency and reactivity in energy-related applications. As a result, surface area can be maximized for optimal performance. The effectiveness of nanomaterials in PV cells and other solar energy applications is influenced by their optical properties, including their capacity to absorb and manipulate light. Nanomaterials used in energy storage and conversion devices, such as batteries and fuel cells, should have high electrical conductivity [72]. Developing reliable and reproducible fabrication techniques for nanomaterials is technically challenging. Variations in the size, shape, and distribution of nanomaterials can significantly impact their performance.

2.6.2 Scalability and Precision

It is challenging to reduce the manufacturing cost of nanomaterials while maintaining their quality at the same time. It is also difficult to maintain the quality and performance of nanomaterials while shifting production from the laboratory to the industrial level. To achieve this, we require a method that can be replicated and scaled up. The size, shape, and distribution of nanomaterials must be accurately maintained and uniform in order to ensure consistent functionality in renewable energy applications [73].

2.6.3 Synergy

The overall performance of energy devices depends on the interaction of nanomaterials with the other components of the device, so they need to be carefully managed. Nanomaterials should be compatible with the conventional infrastructure of renewable energy. This involves cost-effective integration of nanomaterials with existing technology, which can be technically challenging. To achieve optimal performance after integration, precise control over material quality and device design is needed. The overall performance of the system, including battery capacity, the lifespan of wind turbine blades, and the efficiency of solar cells, should improve with the use of nanomaterials [74].

2.6.4 Economic Viability

The economic benefits of using nanomaterials – including initial production, widespread adoption, and long-term maintenance – should outweigh those of conventional materials. The production of high-quality nanomaterial s is complex and, thus, expensive. The synthesis and processing procedures for nanomaterials require advanced techniques, which contribute to their higher production costs compared to conventional materials. This could act as a barrier to their widespread adoption in the field of sustainable and renewable energy. It is crucial to find ways to reduce the production and maintenance costs of nanomaterials without compromising quality [74].

2.6.5 Environmental and Health Impacts

It is necessary to thoroughly assess the potential toxicity of nanomaterials to both humans and the environment to ensure that their use does not introduce new environmental problems. The environmental impact and potential health risks of nanomaterials are not fully understood. Some nanomaterials may be toxic to humans and ecosystems. Procedures for the safe handling, application, and disposal of such nanomaterials must be established to minimize the possibility of toxicity. One should also aim to reduce the environmental footprint of producing and using nanomaterials so that the positives outweigh the negatives [75].

2.6.6 Regulatory Issues

The field of nanomaterials is still developing, and so are the regulations regarding it. To maintain safety, clear and uniform laws governing the use of nanomaterials in renewable energy sources are required. The regulations should cover production, use, and disposal guidelines, as well as standardization and qualitative techniques to ensure safety; only then can it earn acceptance in the market [76].

2.6.7 Technological Advancements

Ongoing research and development are crucial for improving existing nanomaterials and discovering new ones. There is a need to optimize energy efficiency, storage capacity, and material durability to advance in the field of nanotechnology-based renewable energy. Multidisciplinary collaborative research is essential to achieve this [76].

2.6.8 Demand and Competition

The adoption of nanotechnology-based solutions may be affected by competing technologies and the competitive dynamics of the renewable energy market. The demand for sustainable energy solutions can drive investment and innovation in nanotechnology. The presence of alternative technologies and competition in the market can negatively impact the adoption of nanotechnology-based solutions [74].

2.6.9 Public Awareness and Acceptance

Public awareness and understanding of the benefits and associated risks can influence market acceptance of nanotechnology in renewable energy. Gaining public confidence through open and honest communication regarding the effectiveness and safety of nanomaterials is crucial for wider adoption. It is necessary to demonstrate that new nanotechnology-based renewable energy solutions outperform current ones in terms of performance, reliability, and affordability in order to win over the market [77].

2.6.10 Performance Assessment

One important point of discussion is how the use of nanomaterials can enhance the efficiency of energy conversion and storage, and to what extent. Their performance needs to be quantified to justify their usage. The nanomaterials should show a significant advantage over conventional materials in terms of reliability, durability, and working conditions [78].

Nanotechnology is significantly advancing in the field of renewable energy. A strategy is needed to address several challenges, such as technological innovation, commercial viability, environmental sustainability, and regulatory compliance, for it to flourish at its full potential. To overcome these obstacles and utilize nanotechnology in renewable energy to its fullest, stakeholders must invest in and collaborate on multidisciplinary research [79].

2.7 CONCLUSION

Nanostructured-based technology (nanotechnology) and renewable energy are intrinsically interconnected, with nanotechnology offering significant potential to enhance the efficiency and effectiveness of renewable energy sources. This chapter explores how nanotechnology addresses various renewable energy types, provides unique solutions for ongoing challenges, and paves the way for a more sustainable energy sector. Nanotechnology plays a significant role in improving solar panels, enabling them to collect and convert sunlight into electricity with higher efficiency. It enhances hydrogen energy production, storage, and utilization, resulting in more stable systems. In wind energy, nanomaterials are used to improve turbine manufacturing in terms of weight, strength, cost-effectiveness, durability, and performance.

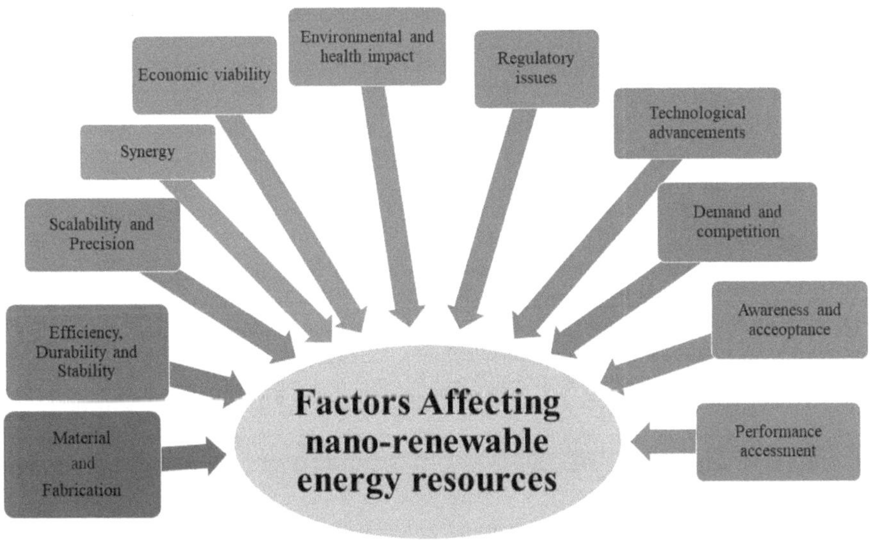

FIGURE 2.8 Factors affecting nano-renewable energy.

In geothermal energy, advanced nanomaterials are effective in extreme conditions and enhance extraction efficiency and durability. In biomass energy, nanotechnology optimizes biofuel production and minimizes waste, while in hydropower, it enhances turbine component efficiency and lifespan. The synergy between nanotechnology and renewable energy sources promises innovative methods for efficient energy generation and storage, reduced costs, and minimized environmental impacts, driving progress toward a cleaner, more sustainable energy future.

REFERENCES

1. P. Moriarty, D. Honnery, What is the global potential for renewable energy? *Renew. Sustain. Energy Rev.*, 16 (2012) 244–252. https://doi.org/10.1016/j.rser.2011.07.151
2. I. Dincer, Renewable energy and sustainable development: a crucial review, *Renew. Sustain. Energy Rev.*, 4 (2000) 157–175. https://doi.org/10.1016/S1364-0321(99)00011-8
3. S.R. Bull, Renewable energy today and tomorrow, *Proc. IEEE*, 89 (2001) 1216–1226. https://doi.org/10.1109/5.940290
4. A. Garg, R.K. Ratnesh, Solar cell trends, and the future: a review, *J. Pharm. Negat. Results*, 13, (2022) 2051–2060. https://doi.org/10.47750/pnr.2022.13.S06.268
5. G. Walker, Renewable energy and the public, *Land Use Policy*, 12 (1995) 49–59. https://doi.org/10.1016/0264-8377(95)90074-C
6. A. Harjanne, J.M. Korhonen, Abandoning the concept of renewable energy, *Energy Policy*, 127 (2019) 330–340. https://doi.org/10.1016/j.enpol.2018.12.029
7. M. Nasrollahzadeh, S.M. Sajadi, M. Sajjadi, Z. Issaabadi, An Introduction to Nanotechnology, 1st ed., Elsevier Ltd., 2019. https://doi.org/10.1016/B978-0-12-813586-0.00001-8
8. E. Serrano, G. Rus, J. García-Martínez, Nanotechnology for sustainable energy, *Renew. Sustain. Energy Rev.*, 13 (2009) 2373–2384. https://doi.org/10.1016/j.rser.2009.06.003
9. R.K. Ratnesh, M.K. Singh, J. Singh, Enhancing ZnO/Si heterojunction photodetector performance for ultra high responsivity across wide spectral range, *J. Mater. Sci.: Mater. Electron.*, 35 (2024) 756. https://doi.org/10.1007/s10854-024-12516-5
10. J.E. Contreras, E.A. Rodriguez, J. Taha-Tijerina, Nanotechnology applications for electrical transformers – a review, *Electr. Power Syst. Res.*, 143 (2017) 573–584. https://doi.org/10.1016/j.epsr.2016.10.058
11. R.K. Ratnesh, M.K. Singh, V. Kumar, S. Singh, R. Chandra, M. Singh, J. Singh, Mango leaves (Mangifera Indica) derived highly florescent green graphene quantum dot nanoprobes for enhanced on-off dual detection of cholesterol and Fe^{2+} ions based on molecular logic operation, *Biomater. Sci. Eng.*, 7 (2024) 4417–4426. https://doi.org/10.1021/acsabm.4c00292
12. A.M. Stoneham, J.L. Gavartin, Dynamics at the nanoscale, *Mater. Sci. Eng. C*, 27 (2007) 972–980. https://doi.org/10.1016/j.msec.2006.09.020
13. N. Baig, I. Kammakakam, W. Falath, I. Kammakakam, Nanomaterials: a review of synthesis methods, properties, recent progress, and challenges, *Mater. Adv.*, 2 (2021) 1821–1871. https://doi.org/10.1039/d0ma00807a
14. M.K. Yadav, R. Kumar, R.K. Ratnesh, J. Singh, R. Chandra, A. Kumar, V. Vishnoi, G. Singh, A.K. Singh, Revolutionizing technology with spintronics: devices and their transformative applications, *Mat. Sci. Eng. B*, 303 (2024) 117293. https://doi.org/10.1016/j.mseb.2024.117293
15. N. Abid, A.M. Khan, S. Shujait, K. Chaudhary, M. Ikram, M. Imran, J. Haider, M. Khan, Q. Khan, M. Maqbool, Synthesis of nanomaterials using various top-down and bottom-up approaches, influencing factors, advantages, and disadvantages: a review, *Adv. Colloid Interface Sci.*, 300 (2022) 102597. https://doi.org/10.1016/j.cis.2021.102597

16. R. Wüstenhagen, M. Wolsink, M.J. Bürer, Social acceptance of renewable energy innovation: an introduction to the concept, *Energy Policy*, 35 (2007) 2683–2691. https://doi.org/10.1016/j.enpol.2006.12.001
17. N.K. Sharma, P.K. Tiwari, Y.R. Sood, Solar energy in India: strategies, policies, perspectives and future potential, Renew. *Sustain. Energy Rev.*, 16 (2012) 933–941. https://doi.org/10.1016/j.rser.2011.09.014
18. K. Blok, Renewable energy policies in the European Union, *Energy Policy*, 34 (2006) 251–255. https://doi.org/10.1016/j.enpol.2004.08.032
19. A. Jäger-Waldau, Photovoltaics and renewable energies in Europe, *Renew. Sustain. Energy Rev.*, 11 (2007) 1414–1437. https://doi.org/10.1016/j.rser.2005.11.001
20. N.K.M.A. Alrikabi, Renewable energy types, *J. Clean Energy Technol.*, 2 (2014) 61–64. https://doi.org/10.7763/jocet.2014.v2.92
21. Z. Abdin, M.A. Alim, R. Saidur, M.R. Islam, W. Rashmi, S. Mekhilef, A. Wadi, Solar energy harvesting with the application of nanotechnology, *Renew. Sustain. Energy Rev.*, 26 (2013) 837–852. https://doi.org/10.1016/j.rser.2013.06.023
22. D. Ghernaout, Nanotechnology phenomena in the light of the solar energy, *J. Energy, Environ. Chem. Eng.*, 3 (2018) 1. https://doi.org/10.11648/j.jeece.20180301.11
23. A. Garg, R.K. Ratnesh, Design and simulation of GaAs/InP and Si/SiC hetrojunction solar cells, *ICCCES-2022, Lecture notes in Electrical Engineering (LNEE)*, Springer, (2022), 867–875. https://doi.org/10.1007/978-981-19-7753-4_66
24. C. Pelosi, Light concentration increases solar-cell efficiency and reduces cost, *SPIE Newsroom.*, 2 (2007) 77–80. https://doi.org/10.1117/2.1200704.0689
25. H. Ramsurn, R.B. Gupta, Nanotechnology in solar and biofuels, *ACS Sustain. Chem. Eng.*, 1 (2013) 779–797. https://doi.org/10.1021/sc400046y
26. S. V. Vassilev, C.G. Vassileva, V.S. Vassilev, Advantages and disadvantages of composition and properties of biomass in comparison with coal: an overview, *Fuel*, 158 (2015) 330–350. https://doi.org/10.1016/j.fuel.2015.05.050
27. S. Firoz, A review: advantages and disadvantages of biodiesel, *Int. Res. J. Eng. Technol.*, 04 (2017) 530–535. www.irjet.net
28. M.H. Ahmadi, M. Ghazvini, M.A. Nazari, M.A. Ahmadi, F. Pourfayaz, G. Lorenzini, T. Ming, Renewable energy harvesting with the application of nanotechnology: a review, *Int. J. Energy Res.*, 43 (2019) 1387–1410. https://doi.org/10.1002/er.4282
29. P.K. Pandey, R. Kumar, V. Gupta, et al. An innovative hybrid controller-based combined grid-connected hybrid renewable energy system. *Electr. Eng.*, (2024). https://doi.org/10.1007/s00202-024-02363-2
30. Y. Ding, M. Greiner, R. Schlögl, S. Heumann, A metal-free electrode: from biomass-derived carbon to hydrogen, *ChemSusChem.*, 13 (2020) 4064–4068. https://doi.org/10.1002/cssc.202000714
31. S. Raza, Y. Orooji, E. Ghasali, A. Hayat, H. Karimi-Maleh, H. Lin, Engineering approaches for CO_2 converting to biomass coupled with nanobiomaterials as biomediated towards circular bioeconomy, *J. CO_2 Util.*, 67 (2023) 102295. https://doi.org/10.1016/j.jcou.2022.102295
32. E. Barbier, Geothermal energy technology and current status: an overview, *Renew. Sustain. Energy Rev.*, 6 (2002) 3–65. https://doi.org/10.1016/S1364-0321(02)00002-3
33. V. Stefansson, the renewability of geothermal energy, *Proc. World Geotherm. Congr.*, 2000 (2000) 883–888
34. I.B. Fridleifsson, Geothermal energy for the benefit of the people, *Renew. Sustain. Energy Rev.*, 5 (2001) 299–312. https://doi.org/10.1016/S1364-0321(01)00002-8
35. M. Soltani, F. Moradi Kashkooli, M. Alian Fini, D. Gharapetian, J. Nathwani, M.B. Dusseault, A review of nanotechnology fluid applications in geothermal energy systems, *Renew. Sustain. Energy Rev.*, 167 (2022) 112729. https://doi.org/10.1016/j.rser.2022.112729

36. A.M. Abdalla, B.E. Elnaghi, S. Hossain, M. Dawood, O. Abdelrehim, A.K. Azad, Nanotechnology utilization in energy conversion, storage and efficiency: a perspective review, *Adv. Energy Convers. Mater.*, 1 (2020) 30–54. https://doi.org/10.37256/aecm.11202075
37. Y. Cui, S. Tian, S. Zoras, J. Zhu, Recent advances in various nanomaterials utilized in geothermal heat exchangers, *Nano Energy*,122 (2024) 109309. https://doi.org/10.1016/j.nanoen.2024.109309
38. M. Faizal, A. Bouazza, R.M. Singh, Heat transfer enhancement of geothermal energy piles, *Renew. Sustain. Energy Rev.*, 57 (2016) 16–33. https://doi.org/10.1016/j.rser.2015.12.065
39. G.M. Joselin Herbert, S. Iniyan, E. Sreevalsan, S. Rajapandian, A review of wind energy technologies, *Renew. Sustain. Energy Rev.*, 11 (2007) 1117–1145. https://doi.org/10.1016/j.rser.2005.08.004
40. A.D. Sahin, Progress and recent trends in wind energy, *Prog. Energy Combust. Sci.*, 30 (2004) 501–543. https://doi.org/10.1016/j.pecs.2004.04.001
41. F. Blaabjerg, K. Ma, Wind energy systems, reviews application of power electronics in wind energy systems, *IEEE Power Energy Soc.*, 105 (2017) 2116–2131. https://linkinghub.elsevier.com/retrieve/pii/B9780123972705000133
42. S.S. John, P. Malviya, N. Sharma, V.K. Manjhi, K. Sudhakar, Nanotechnology for solar and wind energy applications recent trends and future development, *Proceedings of 2015 International Conference on Green Computing Internet Things, ICGCIoT 2015*, (2016) 1223–1227. https://doi.org/10.1109/ICGCIoT.2015.7380650
43. N. Rathore, F. Yettou, A. Gama, Improvement in wind energy sector using nanotechnology, *SIENR 2021 – 6th International Symposium on New and Renewable Energy*, (2021) 1–5. https://doi.org/10.1109/SIENR50924.2021.9631927
44. P.P. Edwards, V.L. Kuznetsov, W.I.F. David, Hydrogen energy, *Philos. Trans. R. Soc. A Math. Phys. Eng. Sci.*, 365 (2007) 1043–1056. https://doi.org/10.1098/rsta.2006.1965
45. M.A. Rosen, S. Koohi-Fayegh, The prospects for hydrogen as an energy carrier: an overview of hydrogen energy and hydrogen energy systems, *Energy, Ecol. Environ.*, 1 (2016) 10–29. https://doi.org/10.1007/s40974-016-0005-z
46. A. Pugazhendhi, S. Shobana, D.D. Nguyen, J.R. Banu, P. Sivagurunathan, S.W. Chang, V.K. Ponnusamy, G. Kumar, Application of nanotechnology (nanoparticles) in dark fermentative hydrogen production, *Int. J. Hydrogen Energy*, 44 (2019) 1431–1440. https://doi.org/10.1016/j.ijhydene.2018.11.114
47. K. Byrappa, T. Adschiri, Hydrothermal technology for nanotechnology, *Prog. Cryst. Growth Charact. Mater.*, 53 (2007) 117–166. https://doi.org/10.1016/j.pcrysgrow.2007.04.001
48. Y. Lang, R.R. Arnepalli, A. Tiwari, A review on hydrogen production: methods, materials and nanotechnology, *J. Nanosci. Nanotechnol.*, 11 (2011) 3719–3739. https://doi.org/10.1166/jnn.2011.4157
49. J. Tang, T. Liu, S. Miao, Y. Cho, Emerging energy harvesting technology for electro/photo-catalytic water splitting application, *Catalysts*, 11 (2021) 1–24. https://doi.org/10.3390/catal11010142
50. T. Mukhopadhyay, A. Ghosh, Hydrogen production from wastewater by photo- electrolysis using ligand-free semiconductor nano-composite of TiO_2 modified by Au: a brief review, *International Journal of Chemical Studies*, 11 (2023) 46–75.
51. H. Shao, G. Xin, J. Zheng, X. Li, E. Akiba, Nanotechnology in Mg-based materials for hydrogen storage, *Nano Energy*, 1 (2012) 590–601. https://doi.org/10.1016/j.nanoen.2012.05.005
52. R. Darmawi, S.M. Sipahutar, M.S. Bernas, Imanuddin, Renewable energy and hydropower utilization tendency worldwide, *Renew. Sustain. Energy Rev.*, 17 (2013) 213–215. https://doi.org/10.1016/j.rser.2012.09.010

53. N.F. Yah, A.N. Oumer, M.S. Idris, Small scale hydro-power as a source of renewable energy in Malaysia: a review, *Renew. Sustain. Energy Rev.*, 72 (2017) 228–239. https://doi.org/10.1016/j.rser.2017.01.068
54. V. Prakash Sharma, U. Sharma, M. Chattopadhyay, V.N. Shukla, Advance applications of nanomaterials: a review, *Mater. Today Proc.*, 5 (2018) 6376–6380. https://doi.org/10.1016/j.matpr.2017.12.248
55. R.H. Fernando, Nanocomposite and nanostructured coatings: recent advancements, *ACS Symp. Ser.*, 1008 (2009) 2–21. https://doi.org/10.1021/bk-2009-1008.ch001
56. X. Xiao, D. Wang, Y. Li, E. Jackson, Y. Fang, Y. Zhang, N. Xie, X. Shi, Investigation into the synergistic effect of nano-sized materials on the anti-corrosion properties of a waterborne epoxy coating, *Int. J. Electrochem. Sci.*, 11 (2016) 6023–6042. https://doi.org/10.20964/2016.07.66
57. S.R. Kothiyal, R.K. Ratnesh, A.K. Shukla, Enhancing sensing with surface plasmons resonance: theoretical insights and design strategies in photonic crystal fiber, *International Conference Artificial Intelligence and Information Technologies*, Taylor & Francis, (2023) 279–285. https://doi.org/10.1201/9781003510833-46
58. S. Sohni, N.A.N. Norulaini, R. Hashim, S.B. Khan, W. Fadhullah, A.K. Mohd Omar, Physicochemical characterization of Malaysian crop and agro-industrial biomass residues as renewable energy resources, *Ind. Crops Prod.*, 111 (2018) 642–650. https://doi.org/10.1016/j.indcrop.2017.11.031
59. S.K. Satyarthi, V.P. Singh, H. Verma, A.K. Singh, Electrical and optical properties of environmental friendly $Li_{(1-x)}Smx/_3NbO_3$ ceramics for high-temperature energy storage applications, *Energy Storage*, 6 (2024) 1–21. https://doi.org/10.1002/est2.642
60. S.K. Satyarthi, The study of MgO nano-composite: synthesis, characterization and probable application submitted in partial fulfillment of the requirement. Thesis, Master of Technology, Department of Applied Physics, Delhi Technological University (2016).
61. P.H. Salame, V.B. Pawade, B.A. Bhanvase, *Characterization Tools and Techniques for Nanomaterials*, Elsevier Inc., 2018. https://doi.org/10.1016/B978-0-12-813731-4.00003-5
62. C.J. Liu, U. Burghaus, F. Besenbacher, Z.L. Wang, Preparation and characterization of nanomaterials for sustainable energy production, *ACS Nano.*, 4 (2010) 5517–5526. https://doi.org/10.1021/nn102420c
63. S.K. Satyarthi, V.P. Singh, C.B. Singh, A.K. Singh, Enhanced dielectrics, ferroelectric and optical properties of lithium niobate for high temperature applications using potassium oxide (K_2O) additive, *Ceram. Int.*, 50 (2024) 20376–20390. https://doi.org/10.1016/j.ceramint.2024.03.161
64. Z. Lou, Y. Zhang, Y. Li, L. Xu, Study on microscopic physical and chemical properties of biomass materials by AFM, *J. Mater. Res. Technol.*, 24 (2023) 10005–10026. https://doi.org/10.1016/j.jmrt.2023.05.176
65. V.P. Singh, S.K. Satyarthi, A. Dwivedi, A. Dwivedi, A.K. Singh, Boosting energy storage of poly (vinylidene difluoride) nanocomposite based flexible self-standing film with low amount of hydroxylated V_2O_5, (2022). https://doi.org/10.1021/acsaem.2c02425
66. B. Paviet-Salomon, J. Levrat, V. Fakhfouri, Y. Pelet, N. Rebeaud, M. Despeisse, C. Ballif, Accurate determination of photovoltaic cell and module peak power from their current-voltage characteristics, *IEEE J. Photovoltaics*, 6 (2016) 1564–1575. https://doi.org/10.1109/JPHOTOV.2016.2590940
67. M. Cristea, D. Ionita, M.M. Iftime, Dynamic mechanical analysis investigations of pla-based renewable materials: how are they useful? *Materials (Basel)*, 13 (2020) 1–24. https://doi.org/10.3390/ma13225302
68. Z. Huang, Y. Wang, J. Long, J. Hu, Mechanical and dynamic mechanical analysis of PBO paper-based composites, *Text. Res. J.*, 92 (2022) 1454–1465. https://doi.org/10.1177/00405175211056382

69. R.B. Poruba A., Vanecek M., Rosa J., Feitknecht L., Wyrsch N., Meier J., Shah A., Repmann T., Fourier transform photocurrent spectroscopy in thin film silicon solar cells, In *Proceedings of the 17th European Photovoltaic Solar Energy Conference*, 2981 (2001) 3–6.
70. V.P. Singh, C.B. Singh, S.K. Satyarthi, D. Kumar, A.K. Singh, Highly enhanced energy storage properties of H_2O_2-hydroxylated rare earth ferrites ($LaFeO_3$ and $GdFeO_3$) nanofillers in poly(vinylidene fluoride)-based nanocomposite films, *J. Mater. Sci. Mater. Electron.*, 33 (2022) 20170–20184. https://doi.org/10.1007/s10854-022-08836-z
71. R.K. Ratnesh, Hot injection blended tunable CdS quantum dots for production of blue LED and a selective detection of Cu^{2+} ions in aqueous medium, *J. OpticsLaser Technol.*, 116, (2019) 103–111. https://doi.org/10.1016/j.optlastec.2019.03.001
72. K.S. Ahmad, M. Nawaz, S.B. Jaffri, Role of renewable energy and nanotechnology in sustainable desalination of water: mini review, *Int. J. Environ. Anal. Chem.*, 102 (2022) 7700–7719. https://doi.org/10.1080/03067319.2020.1837121
73. R. Vakulchuk, I. Overland, D. Scholten, Renewable energy and geopolitics: a review, *Renew. Sustain. Energy Rev.*, 122 (2020) 109547. https://doi.org/10.1016/j.rser.2019.109547.
74. A. Menegaki, Valuation for renewable energy: a comparative review, *Renew. Sustain. Energy Rev.*, 12 (2008) 2422–2437. https://doi.org/10.1016/j.rser.2007.06.003
75. K. Shivarama Krishna, K. Sathish Kumar, A review on hybrid renewable energy systems, *Renew. Sustain. Energy Rev.*, 52 (2015) 907–916. https://doi.org/10.1016/j.rser.2015.07.187
76. O. Ellabban, H. Abu-Rub, F. Blaabjerg, Renewable energy resources: current status, future prospects and their enabling technology, *Renew. Sustain. Energy Rev.*, 39 (2014) 748–764. https://doi.org/10.1016/j.rser.2014.07.113
77. Ş.E. Can Şener, J.L. Sharp, A. Anctil, Factors impacting diverging paths of renewable energy: a review, *Renew. Sustain. Energy Rev.*, 81 (2018) 2335–2342. https://doi.org/10.1016/j.rser.2017.06.042
78. A.G. Olabi, M.A. Abdelkareem, Renewable energy and climate change, *Renew. Sustain. Energy Rev.*, 158 (2022) 112111. https://doi.org/10.1016/j.rser.2022.112111
79. E. Taibi, D. Gielen, M. Bazilian, The potential for renewable energy in industrial applications, *Renew. Sustain. Energy Rev.*, 16 (2012) 735–744. https://doi.org/10.1016/j.rser.2011.08.039

3 Advancement Challenges and Applications of Nanomaterials

Nisha, Himanshu, Yashaswi Sharma, Ratneshwar Kumar Ratnesh, Manisha Tiwari, Ramesh Chandra, Sanjay Kumar Singh, and Snigdha Singh

3.1 INTRODUCTION

Materials that have at least one dimension ranging from 1 to 100 nm are considered nanomaterials. Because of their extremely small size and high surface-to-volume ratio, nanomaterials exhibit unique physical, chemical, and biological traits. The ultimate aim of nanotechnology is to integrate the fields of physics, chemistry, biology, and medicine to create new materials with exceptionally unique properties that can drive innovation and change.[1] Nanotechnology has influenced the advancement of various fields, ranging from improved drug delivery in biomedical applications to the development of sustainable measures for a cleaner and better future by providing water purification techniques and helping to curb environmental pollution. Nanotechnology has also played an important role in the advancement of clean energy by employing nanomaterials in solar cells, batteries, etc.[2] Aquaculture and fishery industries have also employed nanotechnology by using nanosensors that help control food quality by sensing any contamination that may be present in fish and water.[3]

The history of nanomaterials extends over several decades, marked by significant advancements in scientific understanding and technological applications. Natural sources of nanosized materials include volcanic ash, sea spray, forest fires, the breakdown of radioactive elements, and the erosion of rocks that consist of metals or anions1. It is believed that nanoscale particles originated from meteorites formed during the Big Bang event, which led to the creation of the universe and our planet. Nanosized diamonds found in meteorites are believed to be the most common type of grains predating the solar system.[4] Although the idea of nanomaterials was introduced in the late 19th century, people have been creating and utilizing nanomaterials for centuries across various ancient civilizations. In the old days of Egypt, the dark color of soot from burning oil was turned into dark, highly dense, and durable paints for inscribing on papyrus. Back then, the Egyptians were oblivious to the fact that soot is made up of carbon nanomaterials. Clay minerals, which are microscopic

DOI: 10.1201/9781003634553-3

particles with thicknesses of a nanometer or less,[5] were used in Cyprus for bleaching clothes and refining wool.[6] Since the thirteenth to the eighteenth century, blades made from "Damascus" steel were renowned for their hardness, durability, and sharpness, accredited to the existence of cementite nanowires (NWs) and carbon nanotubes (CNTs).[7] The recent era of progress in nanomaterials growth spanned from the 1950s to 1959, and the modern era of nanotechnology began in the 1960s. The father of nanotechnology, Richard Feynman's lecture "There's Plenty of Room at the Bottom," is often cited as the beginning of the concept of nanotechnology. The opportunity to manipulate individual atoms and molecules was discussed by him.[8]

Nanotechnology is based on the handling of atoms following quantum theories. Due to the increase in surface area of material, their properties change by diminishing the dimensions of nanomaterials. Nanomaterial synthesis may be designed *via* two approaches: bottom–up and top–down methods.[9] The production of nanomaterials has led to substantial changes in their properties, including their magnetic, electrical, optical, mechanical, and catalytic properties, which are totally dissimilar from their bulk counterparts. Nanomaterial properties can be altered by just monitoring the size, shape, reaction conditions, and surface functionalization. Recently, nanoparticles have been employed in various applications such as reinforcing building materials, safety sensing components, and biomedical uses like biosensors and drug delivery systems.[10,11] Nanotechnology has helped in the development of much more sensitive sensors that are used in software and data management. Semiconductor nanomaterials have revolutionized the field of quantum computing by enhancing the accuracy and data transmission speeds of computers.[12] The world's population is increasing rapidly each year, raising significant concerns about food safety. Nanomaterials are emerging as a crucial solution, aiding in the advancement of food industries.[13] Nanotechnology has enabled the development of better food packaging by using smart sensors to monitor food quality and check for the presence of any contamination.[14]

Advancing nanomaterials presents several significant challenges despite their immense potential in various fields. There are challenges of nanomaterials such as synthesis, characterization techniques, environmental impact, and economic and market challenges. Challenges include understanding the possible health and environmental impacts of nanosized materials, ensuring the scalability of production methods, and developing regulations to govern their use.[15] One of these is the design of nanovaccines, which inadequately considers the interface between nanomaterials and immune cells. Secondly, there is the challenge of transforming these treatments from minimal scale synthesis to large-scale commercial production in the laboratory. Additionally, nanovaccines may be costly and complex to synthesize, posing challenges for medical applications. Although there are further advancements in enhancing the ease of synthesis and fabrication of nanomaterials, they will offer new modes for their improvements in medical therapies.[16] Nanotechnology, by providing wide applications and helping in the advancement of various fields, has become a boon to the world as well as to the future by its application in the field of sustainable development. However, their tiny size also raises concerns regarding pure synthesis, environmental safety, regulatory conditions, etc.[17], which we have briefly discussed in this chapter (Figure 3.1).

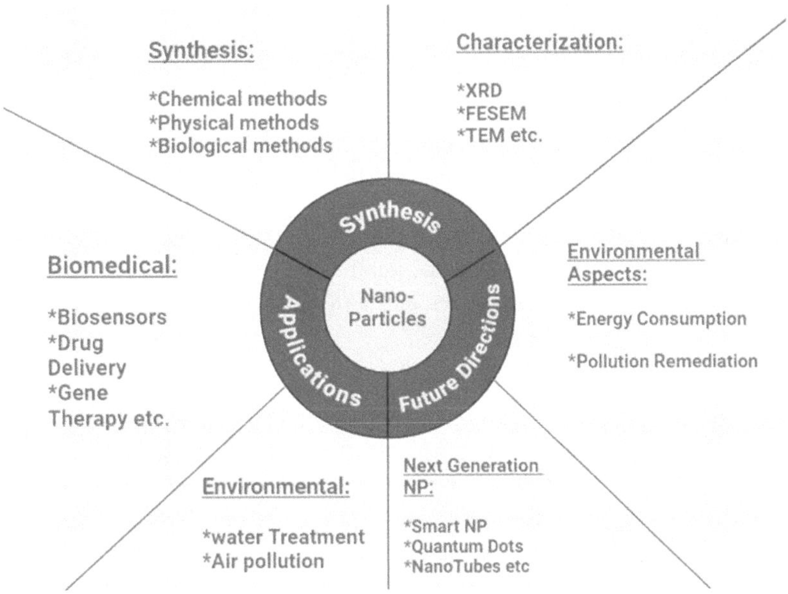

FIGURE 3.1 Applications, synthesis, and future directions of nanoparticles.

3.2 ADVANCEMENT CHALLENGES OF NANOMATERIALS

3.2.1 Synthesis and Fabrication

For the preparation of nanomaterials, different approaches are used, including chemical methods, physical methods, and biological methods. Physical methods consist of techniques such as ball milling, thermal evaporation, ultrasonication, and photoirradiation. Chemical methods include the sol–gel method, solvothermal methods, reverse micelles, co-precipitation, etc., and biological methods involve the use of bacteria, fungi, algae, viruses, etc.[18] Recently, new methods for the preparation of nanoparticles have also been developed, such as using microfluids, which help to synthesize nano oxides, quantum dots, biopolymer nanoparticles, etc.[19] There are several challenges faced when using chemical and physical methods. The reducing and capping agents used in them are expensive, hazardous, and highly toxic.[20] Moreover, many nanomaterials are synthesized in small batches, limiting their practical applications. Developing scalable and cost-effective synthesis methods without compromising quality is crucial for widespread adoption. Imperfections in nanomaterials can significantly impact their performance and compromise their intrinsic features. For example, while CNTs are renowned for their strength, impurities, varying lengths, and arbitrary orientations may greatly reduce their tensile strength.[21] Achieving cost-effective synthesis of nanomaterials presents a significant challenge. High-quality nanomaterials typically require sophisticated instrumentation and stringent conditions, which hinder their bulk production. This challenge is particularly pronounced in the synthesis of two-dimensional nanomaterials. Many existing methods for bulk production are cost-efficient but often yield materials of inferior quality due to inherent defects. The controlled synthesis of nanomaterials remains a complex task. For example, achieving optimal conductivity, chiral selectivity, and precise control over lengths and widths are critical hurdles in the synthesis of CNTs.[22,23]

Beyond synthesis, many nanomaterials are prone to aggregation or degradation over time, affecting their stability and performance. Upon contact, nanomaterials often agglomerate due to electrostatic interactions, large surface energy, or physical entanglement.[24] Electrostatic stabilization can mitigate this aggregation of the nanoparticles.[25] Functionalizing nanomaterials to enhance stability, biocompatibility, and functionality without altering their core properties remains a significant challenge.[26]

3.2.2 Characterization Techniques

Traditional analytical techniques often face limitations when dealing with nanoparticles due to their unique properties. Nanoparticles typically have extremely small sizes (ranging from 1 to 100 nm) and a large surface area; these characteristics can lead to challenges. It has been observed that acquiring specific information about individual nanoparticles presents significant challenges.[27] There is an urgent requirement for approaches to effectively gather data on the purity and properties of large quantities of nanomaterials.[28] Initial findings regarding nanoparticle properties often assumed that their physicochemical properties remained constant once they were formed. However, it has been observed that the properties of ceria particles in aqueous solutions can differ depending on the composition of the solution and the duration of exposure[29] (Figure 3.2).

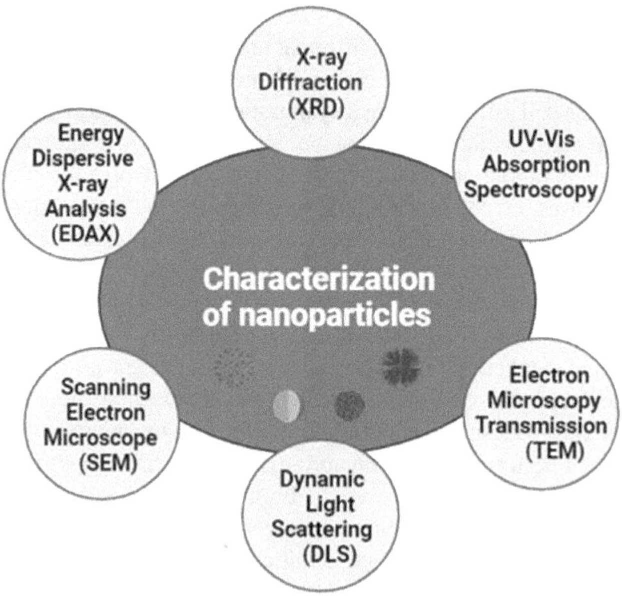

FIGURE 3.2 Various techniques for characterization of nanoparticles.

Various characterization methods have been utilized for the analysis of synthesized nanoparticles including microscopy-based techniques such as transmission electron microscopy (TEM), X-ray diffraction (XRD), photoluminescence spectroscopy, scanning TEM, and ultraviolet–visible spectroscopy. These techniques provide valuable information about the nanoparticles; for example, XRD elucidates information about crystal structure and composition, while TEM reveals nanoparticle size, shape, and monodispersity.[30] Additional techniques such as asymmetric-flow field-flow fractionation coupled to multiple detectors have recently been employed for the characterization of nanoparticles in drug delivery applications.[31] Nanoparticles can also be examined using Raman spectroscopy, optical spectroscopy, X-ray fluorescence (XRF), and solid-state nuclear magnetic resonance.[32] However, these characterization techniques pose several challenges. For instance, XRD is not suitable for amorphous substances and it tends to produce overly broad patterns for particles smaller than 3 nm. When the particle size exceeds 50 nm, XRD may not clearly distinguish the multiple surface boundaries. Similarly, other techniques like field emission scanning electron microscopy require an extreme vacuum environment and necessitate decontamination steps. TEM has limitations in depth of resolution, XRF exhibits low excitation efficiency, and both techniques are time-consuming.[33]

Some other characterization challenges are listed below:

- When the surface characteristics are not known, particles lack distinct definition or characterization.[34]
- Several widely used analysis techniques for characterizing nanomaterials do not detect surface coatings.[35]
- Various synthesis methods often produce nanoparticles of distinct types.[36]
- Nanomaterials display dynamic behavior, evolving over time and responding to their surroundings.[37,38]
- Nanoparticles are sensitive to damage and can undergo changes due to fluctuations in temperature and exposure to light.[39,40]

According to Nurmi et al., to enhance characterization refinement, several factors need attention:

- Removing residual solutes to reduce salt or other component deposition.
- Eliminating solvents and co-solvents to minimize interference with characterization procedures.
- Removing solvents without altering particle phases.
- Minimizing surface abrasion or dissolution.
- Preventing reactions with the medium or impurities that could alter samples immediately or over time.

3.2.3 Safety and Environmental Impact

The toxicity of nanomaterials is a significant concern that cannot be overlooked and remains poorly understood, posing substantial implications for their environmental, household, and industrial applications. The impact of nanoparticle-based materials on cellular toxicity is not well-defined,[41] making it essential to understand their potential health and environmental effects. Certain nanoparticles can pose risks owing to their small size and capability to penetrate biological barriers.[42] For instance, silica nanoparticles used in drug delivery have been associated with numerous side effects such as blood toxicity, immunotoxicity, and genotoxicity.[43] Nanotoxicity can manifest through mechanisms like cellular stress induced by nanoparticle interactions, potentially affecting organs such as the liver, brain, and kidneys.[44] The aggregation of boron nitride (BN) nanoparticles on the surface of algae cells leads to toxicity in aquatic ecosystems, and toxicity is concentration-dependent.[45] Orally ingested Ag nanoparticles can damage epithelial integrity, reduce the width of the mucosal layer, and alter the duodenal microbiota.[46]

Antibacterial nanomaterials released in significant quantities may disrupt beneficial bacteria present in wastewater and sewage, potentially polluting recycled water.[47] Additionally, certain metal or noble metal nanoparticles can be environmentally toxic if exposed.[48] Hence, establishing standards for safe handling and disposal is crucial for their responsible use. Studies indicate that lipid-based nanomaterials and polymer nanostructures typically present fewer evident threats because of their biocompatibility and ability to degrade naturally.[49,50] However, research by Borowska and Brzoska underscores potential risk factors associated with metal-based nanoparticles in cosmetics.[51]

The unique properties of nanomaterials raise regulatory and ethical concerns, necessitating the development of appropriate guidelines to ensure their safe use. Addressing ethical considerations regarding privacy, security, and societal impacts is also crucial. Moreover, gaining comprehensive knowledge of how nanomaterials interact with tissues, cells, and proteins is essential for fostering their safe, widespread adoption of nanotechnology development.[52]

3.2.4 Economic and Market Challenges

Many advanced nanomaterials are currently expensive due to costly raw materials or complex synthesis processes, which hinders their widespread adoption. Finding cost-effective methods for large-scale production and ensuring affordability while maintaining quality is crucial yet challenging. Bridging the gap between nanomaterial research and practical applications remains a significant challenge. Designing nanomaterials that can integrate seamlessly into existing technologies and manufacturing processes is essential for their commercialization.

The involvement of technical organizations, government, industries, and the workforce is essential for shaping green nanotechnology through processes such as opinion formation, education, and decision-making. Green economy policies must strike a balance between harnessing the environmental, societal, occupational, and health benefits of nanotechnology while also considering its costs and risks.[53] Protecting intellectual property (IP) in nanotechnology presents unique challenge, as the materials and design of nanoproducts can be reverse engineered using various methods. Unlike processes or specific commercial methods, which can typically be safeguarded as trade secrets, there is no singular method or material for producing certain nano-based products, posing significant challenges for IP protection.

If researchers and product developers choose to utilize a patented technology or method without obtaining a license, they must pursue a declaratory judgment to challenge the patent's validity, or wait to be sued for patent infringement.[54] Moreover, determining infringement often requires costly tools such as electron microscopes and particle size analyzers, making it impractical to deploy for every suspected case of IP violation. Regulatory offices overseeing IP in nanotechnology are not universally present across all countries. Specifically, these offices are established in the US, Japan, and Europe, but they are not readily accessible to all stakeholders, particularly innovators from developing nations. These innovators often face challenges in preparing and submitting applications that comply with the stringent requirements of these regulatory bodies, which may differ from those in their home countries.[55]

Monitoring the environmental fate of nanomaterials, which originate from various sources and pathways, presents significant challenges. Reliable techniques for accurately quantifying exposure to nanomaterials are currently lacking, necessitating proactive approaches like quantitative structure–activity relationship (QSAR)-adverse outcome pathway for risk assessment.[56] New technology-based firms encounter significant obstacles to commercialization, such as limited access to the corresponding resources held by large businesses and organizations, as well as hurdles in prioritizing development objectives and forming alliances accordingly.[57]

3.2.5 Conclusion and Concern

The synthesis of nanoparticles may be achieved using two approaches namely top–down and bottom–up approaches. Achieving cost-effective synthesis of nanomaterials presents a significant challenge. Defective nanomaterials can impact their functional properties and stability at high temperatures. There are also challenges regarding the characterization techniques of nanomaterials to acquire specific information about nanoparticles. Although some advanced techniques have been discovered to improve the barriers faced in the synthesis, fabrication, and characterization of nanomaterials for large-scale production.

3.3 APPLICATIONS OF NANOMATERIALS

3.3.1 Applications of Nanomaterials in Biomedical Field

Over time and through ongoing research, nanomaterials have demonstrated vast potential in the field of medical sciences. They are utilized extensively, from drug delivery using nanocarriers such as liposomes, polymeric micelles, and nano-emulsions, to the development of biosensors. These biosensors play a crucial role in biomedicine by effectively detecting diseases such as swine flu and liver cancer, facilitating early treatment and improving healthcare outcomes for people worldwide (Figure 3.3).

3.3.1.1 Biosensors

The introduction of nanotechnology in biosensors comes along with a lot of advantages such as electrochemical properties, higher surface to volume ratio etc. To monitor biological reactions, biosensors work in a way that they first analyze the

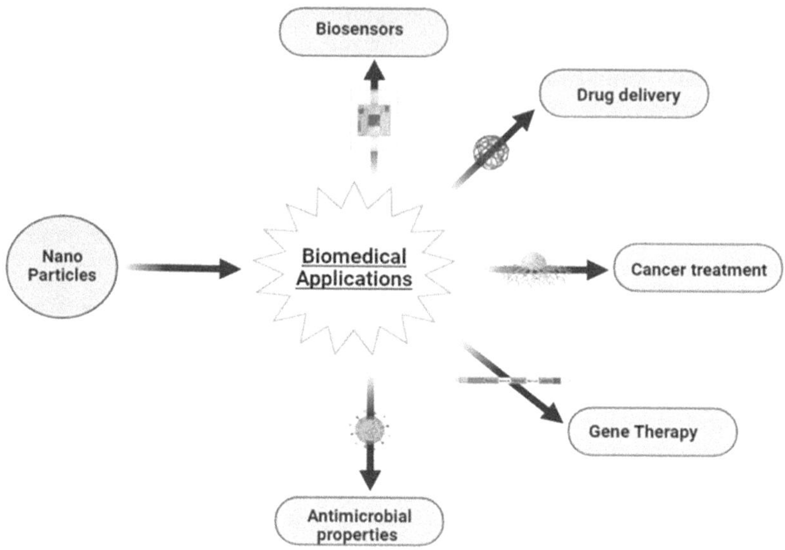

FIGURE 3.3 Biomedical applications of nanoparticles.

interactions in the biological reaction and then convert them into electrical impulses.[58] Nanobiosensors have become the main center of attention and are extensively studied because of the properties of nanoparticles associated with them, such as optical properties and the ability of nanoparticles to couple with fluorescent markers, making them suitable candidates for biosensors.[59] Fluorescence properties depend on size and form, so quantum dots provide many applications in biosensors and imaging, offering advantages such as a higher yield of molar extinction, a wider range of absorption, and more. Currently, biosensors are employed in the detection of various disorders, such as diabetes and cancer, utilizing serum analysis as a basis. The introduction of nanomaterials in biosensors has also allowed the enzyme detection system to be immobilized, which helps in the recycling and reuse of expensive enzymes.[60] Biosensors use biological molecules to identify the target and subsequently use output elements that convert the biorecognition event into optical, electrical, or mass-sensitive signals. Focusing on specific applications of biosensors, deoxyribonucleic acid (DNA)-based biosensors can be viewed as a subset within the field of biosensors. DNA-based biosensors use DNA strands containing short oligonucleotides as probes for target recognition.[61] The utilization of nanomaterials has accelerated the development of DNA biosensors for easier and less expensive target detection. By exploiting the double helix structure of DNA and base pairings, a large number of DNA nanostructures with known shapes, sizes, and geometries have been developed with high nanoscale precision. Recent research has highlighted the application of nanostructured conducting polymer electrospun materials in electrochemical biosensors.[62] One of the most important factors in these kinds of nano-based biosensors is the use of inexpensive methods for creating nanofibrous structures, i.e., the electrospinning technique, which is an affordable method for creating nanofibrous structures from different polymers. The combination of conducting polymers and these nanofibers leads to the formation of bio-composites, which have enhanced conductivity and stability that are essential for the accuracy of devices used for biosensing. Borophene exhibits properties such as thermal conductivity, enhanced tensile strength, and electrical conductivity, which are the reasons behind its use in biosensors.[63] Recently, a cotton thread with high conductive properties has been developed as a substrate for the development of a biosensor for swine flu detection.[64] Zirconium trisulfide nanodots have been synthesized for use in a biosensor for the detection of lung cancer serum biomarkers.[65]

3.3.1.2 Drug Delivery

Nanoparticles such as polymeric, inorganic, and lipid-based types have the ability to increase the therapeutic efficiency of drugs by providing a better and easier drug delivery system at target sites.[66] Polymeric nanoparticles are derived from polymer building blocks and offer benefits such as controllable size, shape, and surface charge, which are crucial factors in drug delivery. Some recent works involving the use of different nanomaterials in drug delivery are listed in Table 3.1. These nanoparticles have primarily been investigated for delivering four main types of payloads: (1) small molecules, (2) peptides, (3) aptamers, and (4) antibodies. Lipid-based nanoparticles are usually spherical, containing at least one bilayer lipid surrounding an internal aqueous cavity. Lipid-based nanoparticles working as delivery systems offer various advantages, such as high bioavailability, formulation simplicity, self-assembly, biocompatibility, and the ability to act as carriers for larger molecules or drugs, along with numerous chemical properties that can be manipulated to monitor their biological activity.67 These factors explain why lipid-based nanoparticles are the most widely adopted class of nanomedicines approved by the Food and Drug Administration (FDA). The major problem these nanoparticles face in drug delivery is biological barriers.68 Biological barriers reduce the portion of administered nanoparticles that are supposed to reach the therapeutic target site. Overcoming these biological barriers is a challenging task, as they vary from disease to disease and also from patient to patient.

When discussing drug delivery to the central nervous system, it is important to note that the human brain is protected by a membrane known as the blood–brain barrier, which lacks intracellular gaps, making it difficult for substances from the environment to penetrate. However, nanoparticles coated with polymers can overcome this

TABLE 3.1
Common Nanocomposites Used in Drug Delivery

S. No.	Nanocomposite	Biological Application	Reference
1.	SiO_2 nanoparticles	Fracture healing	Li et al.[67]
2.	Magnetic chitosan	Anticancer drug	Salmanian et al.[68]
3.	Polylactic–glycolic acid nanoparticles	Pulmonary drug delivery	Chintapula et al.[69]
4.	CeO_2 nanoparticles	Drug delivery and antimicrobial	Sanmugam et al.[70]
5.	Hydrogel nanocomposite	Brain tumor treatment	Kang et al.[71]

challenge by crossing the blood–brain barrier, facilitating drug delivery to the central nervous system. This advancement in nanotechnology contributes significantly to improving treatments for various nervous system diseases, including brain cancer.[72]

3.3.1.3 Treatment of Cancer

Cancer is one of the fastest-growing diseases in the world and has the highest mortality rate. The most commonly used treatments for cancer include surgical methods, chemotherapy, and radiation therapy.[73] In recent times, immunotherapy and photothermal therapy (PTT) have also emerged as alternatives. These traditional methods for the treatment of cancer come with several disadvantages or side effects; for example, chemotherapeutic drugs are cytotoxic as they involve modifications to the chemical structure of DNA, inhibit the synthesis of nucleic acids, and interfere with mitotic tubulin synthesis. The specificity of chemotherapeutic targeted drugs is critical because non-selective or incorrect targeting can cause permanent damage to normal body cells, resulting in significant side effects such as hair loss and gastrointestinal toxicity. Addressing these issues remains a constant pursuit in the search for safer drug alternatives. In recent years, natural products and nanotechnology have therefore become the primary focus in anticancer or antitumor drug research because natural drugs are much safer to use and have fewer side effects compared to traditional chemotherapeutic drugs. Natural drugs also enhance immunity and sensitivity to chemotherapy. Drug delivery systems involving nanoparticles help to some extent in overcoming the target-specific drug delivery issue associated with traditional chemotherapy. Recently, co-delivery systems have also been developed for the treatment of cancers, such as polyethylene imine polymer-based co-delivery systems for breast cancer.[74] Many nanomaterial properties like high surface-to-charge ratio can increase the accuracy of chemical drug complexes, which is helpful in drug therapy and can eventually increase the effectiveness of nanomaterial-based treatments while minimizing the risk of side effects such damaging normal body cells. Photodynamic therapy (PDT) and PTT are two new optical methods for the treatment of cancer. In PDT, a photosensitizer is applied at cancerous sites, leading to the production of singlet oxygen and eventually causing cell death. PTT makes use of materials with a high ability for photoconversion which helps raise the temperature at cancerous sites and eventually causes cell death.[75] PDT and PTT are recent methods for cancer treatment in which nanoparticles can be used because of their unique fluorescence properties. Selenium nanoparticles also exhibit anticancer property and are used in prostate, breast, and lung cancer treatments.[76] Arsenic trioxide nanomaterials are useful in the treatment of leukemia, and realgar nanoparticles consisting of As_4S_4 are also employed.[77] The superparamagnetic behavior of nanomaterials has certain advantages that are helpful in cancer diagnosis and treatment. Recently, calcium ion (Ca^{2+})-based nanomaterials have been developed for tumor therapy. Liu et al. have reviewed the importance of these anticancer nanomaterials on a Ca^{2+} homeostasis.[78] A very common example is iron oxide superparamagnetic nanoparticles (SPIONs), which help in the treatment of cancer hyperthermia due to properties such as high specificity, smaller size, controllable speed, and the capability to influence immunity.[79] To enhance the efficiency of SPIONs, targeting species such as aptamers, transferrin, and peptides are being used to coat the surface of SPIONs.[80]

3.3.1.4 Gene Therapy

Gene therapy is primarily utilized for treating genetic disorders and hair-related issues. This includes the development of gene-suppressing oligonucleotides like antisense DNA oligonucleotides, as well as ribonucleic acid (RNA) interference utilizing small interfering RNAs and micro RNAs, which are designed to complementarily target specific genetic sequences.[81] Nanoparticle carriers have two special abilities; (1) stem cell-targeted delivery of DNA and siRNA to protect them from degradation. Au nanoparticles provide an attractive and important scaffold that is helpful for nucleic acids to reach targeted sites. Ding et al.[82] reviewed the utilization of silver nanoparticles, which can be either covalent or non-covalent, for use in gene delivery and RNA technologies. Recently, many reports have emerged regarding the use of nanoparticles to distribute viral vectors, including nucleic acids and retroviruses, which ultimately led to the development of the modern technique known as magnetofection. The superparamagnetic property of nanoparticles is used in magnetofection to improve gene delivery under the influence of a magnetic field.[83] Nanoparticles have been of great use in breast cancer gene therapy as they are effective gene carriers, enhancing circulation time and improving bioavailability.[84] The drug and gene co-delivery by nanoparticles has a synergistic effect on breast tumors and increases cellular uptake.[85] Ya Gong et al. conducted a review of iron-based nanomaterials for their potential applications in gene therapy for tumors and gene delivery. Iron nanoparticles are highly regarded for their properties such as magnetic resonance imaging, PTT, and magnetic-assisted drug delivery.[86] Recently, Trojan horse lipid nanoparticles have been developed, introducing a novel approach to brain gene therapy.[87]

3.3.1.5 Antimicrobials and Wound Healing

Selenium nanoparticles have generated significant interest recently in the treatment of bacteria, fungi, and parasites with almost no side effects.[88] Silver compounds like metallic silver and silver nitrate are in use for the treatment of wounds, skin infections, and burns.[89] However, there is a major disadvantage associated with them: silver ions in higher microdoses are toxic. Antimicrobial resistance has become one of the major problems in healthcare.[90] Nanoparticles have emerged as a solution to this problem. Recently, the antimicrobial activity of polymeric nanoparticles like chitosan has also been found.[91] Silver nanoparticles show better antimicrobial activity compared to silver compounds. The confirmation of the antibacterial and antifungal properties of silver nanoparticles has been established, along with the poor skin penetration of silver nanoparticles, which may be due to the large surface area they possess, providing better contact with microorganisms. Wound dressings that contain nanocrystalline silver coating are widely available in the market. The modifications made by antibiotics on the surface of silver nanoparticles enhance their antimicrobial activities. TiO_2 nanoparticles also show antimicrobial properties when photoinduced by UV light.[92] Conventional parasitic treatments lead to a rise in drug resistance, so an extensive range of nano-based delivery systems has been effective in the treatment of various parasitic infections.[93] Nanomaterial-based bio-films also exhibit antibacterial actions, which have significant potential for development in the future.[94] Metal-based nanomaterials have also been currently employed in plant pathogens, offering advantages such as low toxicity, selectivity, and biodegradability.[95]

3.3.2 Application of Nanoparticles to Cope with Environmental Pollution

Nanotechnology is emerging as a boon to the world with its diverse applications, notably in addressing environmental degradation caused by water and air pollution. This area of application requires extensive study, especially in the current era where pollution's detrimental effects are significantly impacting quality of life.

3.3.2.1 Waste Water Treatment

Due to the disposal of various wastes in water, such as sewage, industrial chemicals, and agricultural waste, wastewater contains a large amount of harmful and toxic materials. There are various nanomaterials that are useful for the treatment of wastewater, including polymeric nanoparticles, metallic nanoparticles, and carbon-based nanoparticles. Several pathways are used in nanotechnology to treat wastewater, such as adsorption, photocatalysis, and nanofiltration.[96] One of the widely used methods for removing harmful contaminants from water is adsorption. This process involves allowing contaminants to adhere to nanoparticles, which can then be extracted.[97] It uses electrostatic interactions, π–π stacking, and van der Waals interactions.[98] It is one of most extensively used technique due to its low cost and high efficiency in treating a variety of water pollutants. The adsorption technique can be very useful in the removal of metallic pollutants like lead, mercury, and cadmium from industrial waste. Nanoparticles can also be utilized to extract heavy metals from water. The most commonly used nanoparticles in the adsorption technique are nanotubes and metal-based nanoparticles. The introduction of metal-based nanoparticles allows the easy removal of adsorbent materials from wastewater, making the treatment cost-effective.[99] Magnetic nanoparticles are highly effective in removing fluoride concentrations, which are responsible for various toxic diseases.[100]

3.3.2.2 Metallic Nano-Adsorbents

Metallic nano-adsorbents are affordable nanoparticles with excellent absorbent properties, making them crucial for extracting heavy metals from wastewater. X. Chen et al. conducted a review on metal–organic framework (MOF)-based magnetic nanomaterials and their use in the degradation of dyes, toxic chemicals, and heavy metal ions.[116] The best examples of these metal-based nanoparticles are ferric oxide, magnesium oxide, manganese oxide, and titanium oxides, as they all possess a great surface-to-volume ratio that enhances the chemical interactions between the adsorbent and adsorbate.[101] Luca Burrat et al., reviewed metallic nanoparticles embedded in a polymeric matrix, such as hydrogels, as a viable treatment for wastewater.[102] TiO_2 nanoparticles are being used for the removal of Cd from wastewater.[103] Ag-based nanoparticles are also used to disinfect contaminated water due to their antimicrobial properties.

3.3.2.3 CNTs Acting as Adsorbents

CNTs have a cylindrical shape and are graphene sheets that can have a single or multiwalled structure. They exhibit very strong adsorbing power for metal ions and pollutant molecules because they have a hollow structure, are highly porous, have a huge surface area, and low mass density.[104] CNTs are used to adsorb pollutants from wastewater, such as dyes, heavy metals, and aromatic hydrocarbons.[105] CNT-based membranes have been developed that enhance the adsorbing capacity compared to normal CNTs, making them helpful in wastewater treatment.[106] NF270-Ppy-CNTs are used to develop membranes for ion rejection in water treatment, and the modified membranes show better ion rejection.[107] CNTs are employed to regulate microalgal cells present in wastewater, thereby enhancing the efficiency of water treatment. CNTs significantly enhance the growth and photosynthesis of the consortium.[108] MXene–CNT composites also find environmental applications owing to their large surface area, unique architecture, and ease of functionalization.[109]

3.3.2.4 Photocatalytic Nanoparticles

Photocatalytic degradation is a very reliable method for treating many toxic and harmful compounds. TiO_2–biochar-based nano-photocatalysts are used for removing organic contaminants, which is useful for sewage remediation.[110] Recently, nanocellulose has been developed; however, nanocellulose itself shows very little photocatalytic activity under UV light. ZnO and TiO2 are added to increase its photocatalytic activity.[111] TiO_2 nanoparticle-based photocatalytic processes are highly useful due to their low cost and environmental compatibility, as recently Gd-doped TiO2 nanopowders have been used for the photocatalytic degradation of methylene blue dye.[112] ZnO nanoparticles show strong photocatalytic activity against methyl orange and are therefore useful in water purification or treatment.[113] The photocatalytic and adsorptive properties of boron-nitride-based nanomaterials have proven to be useful in the removal of toxic metals and dyes from wastewater.[114]

3.3.2.5 Air Pollution Control

Industries release a significant number of toxic pollutants into the air, such as NOx, SOx, particulate matter, and more. Nanomaterials are used to target specific pollutants depending on their type and characteristics. TiO2 nanoparticles are used to capture and absorb pollutants in the air and convert them into harmless inorganic ions in the presence of UV light.[115] $ZnSn(OH)_6$ nanoparticles are used as photocatalysts for the treatment of contaminated air.[116] Recently, electrospun nanofibrous polymer membranes have been used for the filtration of tiny particles and toxic gases from the air, as they possess high separation efficiency and flexibility.[117] Ag-TiO_2 nanorods loaded onto nanofibers show good particulate matter filtration.[118]

3.3.2.6 Conclusion

Nanomaterials possess unique properties that make them suitable for diverse applications. One significant field benefiting from nanomaterials is biomedicine, where they contribute to the development of healthcare devices such as biosensors. These devices aid in the detection and diagnosis of various diseases. Nanomaterials have also been used to enhance the efficiency of drugs by providing better drug delivery. Similarly, nanoparticles have presented a helpful role in gene therapies and the antimicrobial action of drugs. The applications of nanoparticles are not limited to the biomedical field ; they extend to addressing major environmental issues such as wastewater treatment through the removal of toxic and heavy metals, and curbing air pollution.

3.4 FUTURE DIRECTIONS AND EMERGING TRENDS

3.4.1 Next Generation Nanomaterials

3.4.1.1 Smart Nanomaterials

It is an intriguing prospect to develop nanomaterials with better efficient characteristics designed for effective therapeutics, especially in biomedical applications. Recently, smart materials have attracted substantial attention to reduce the drawbacks shown by ordinary nanomaterials.[119] Smart materials can be defined as those that respond to stimuli from external agents, such as light, electric fields, temperature, magnetic fields, and pH, leading to an innovative class of functional properties that are useful in many treatments, such as controlled drug delivery, curing various diseases, and biosensing. The availability of different smart materials depends on the properties exhibited by different types of stimulating agents, as shown in Figure 3.4. Thermo-responsive smart materials are applicable to chronic diseases, and some pH-sensitive nanocarriers are advantageous for tumor imaging and analysis in future biological therapies.[120] Photo-responsive smart nanomaterials are used for fluorescence imaging and tumor therapeutics.[121] Electro-responsive nanomaterials have unique redox properties that are useful for electro-controlled drug release to treat numerous illnesses.[122] Multi-stimuli responsive smart nanomaterials are found to be more effective than single-stimuli responsive smart nanomaterials.[123] In the future, certain requirements such as light absorption, biocompatibility, low toxicity, and easy functionalization will be necessary to design a better photo-therapeutic platform for nanomaterial-based systems.

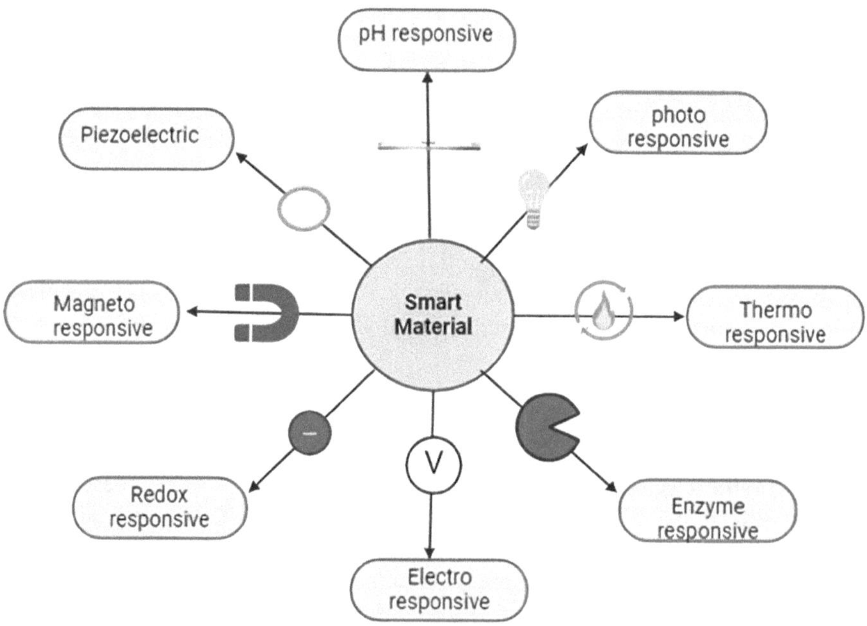

FIGURE 3.4 Classification of smart nanomaterials.

3.4.1.2 Quantum Dots

Quantum dots are defined as small crystals that have variable electrons occupying discrete quantum states. Sulfur has proven useful in a variety of biomedical applications for several decades. Sulfur-based quantum dots (SQDs) are reported as multifunctional and sustainable materials with great potential for large-scale biological applications. SQDs are safe, biocompatible, photoluminescent, easy to use, stable, economically synthesized, and easily surface functionalized.[124] Colloidal QDs show prodigious potential for display technologies because of their unique optical properties. Therefore, QD-based display technologies and QD enhancement films and quantum dot-organic LEDs have been successfully marketed, indicating the dominance of quantum dots in display technologies.[125]

3.4.1.3 Nanotubes and NWs

Graphene nanoribbons and CNTs have gained significant attention in next-generation integrated circuits because of their mechanical, thermal, and electrical properties. Recently, single, double, and multi-walled CNTs have shown improved results compared to copper. Additionally, the interconnection of copper with CNTs has shown better results than graphene nanoribbons s.[126] Flexible transparent electrodes are of great importance in various fields such as solar cells, smart windows, heaters, and LEDs. Metallic NWs are considered a good substitute for flexible transparent electrodes due to their high conductivity. Currently, efforts are underway to design next-generation nanomaterials to replace common indium tin oxide.[127]

3.4.1.4 2D Metal Oxide Nanomaterials

The incorporation of two-dimensional metal oxide nanomaterials has improved the performance of zero-generation rechargeable batteries, which have reduced ion diffusion lengths, improved kinetics, and increased chemically active boundaries. These innovations have also contributed to the advancement of nanotechnology and the use of rechargeable batteries. The use of two-dimensional nanomaterials for energy storage devices aims to achieve large energy density, stable cyclability, and high power density, which is detailed and summarized along with the potential and challenges that still lie ahead. The use of two-dimensional metal oxide nanoparticles in these energy devices is expected to present significant prospects for addressing issues arising from the world's growing energy demand. Recently, advancements in the use of two-dimensional metal oxide nanomaterials in rechargeable lithium-ion batteries and nickel-ion batteries have proven useful.[128]

3.4.2 Environmental Aspects

3.4.2.1 Energy Consumption and Storage

The long lifespan, safety, affordability, and high theoretical energy density of zinc-air batteries make them a viable option for large-scale energy storage. On the other hand, their low power density has prevented them from being widely used. A capable approach to this problem is using Ni-doped cobalt oxide nanosheets along with structural engineering; these have been shown to enhance endurance and boost energy density in Zn-air batteries.[129] The combination of these characteristics has produced outstanding performance in Zn-air batteries using a developed nickel-doped cobalt oxide nanosheet electrode. On the nanoscale, the strong 2D structure and many nanopores inside the nanosheets provide a large electrocatalytic area as well as enable efficient oxygen transport. Smart windows in modern buildings have the capacity to significantly lower energy utilization by effectively controlling the amount of light and heat that enters the structure. Because of their special qualities, inorganic nanomaterials are essential to the development of smart windows. The amount of sunlight and visible light that passes through smart windows can be changed by using electrochromic materials with specific nanostructures.[130] One of the most widely used electrochromic materials, tungsten trioxide, has attracted a lot of interest lately and is thought to be a great choice for smart window applications.[131] Compared to organic electrochromic materials, it possesses more advanced properties such as exceptional electrochromic activity, stability, and a reversible color shift from deep blue to optical transparency. Furthermore, in order to further improve performance, phosphorus-doped tungsten oxide films with tunneling capabilities are presently being developed.[132]

3.4.2.2 Water and Air Pollution Remediation

A major worry for public and environmental health is surface and groundwater contamination, and as the population grows and the climate changes, pressures on water resources and usage will also rise. Modified nanomaterials have potential for environmental remediation, but there are current knowledge gaps and upcoming challenges in this field, including the facilitation of water remediation and the use of nanomaterials in the remediation of air pollution.[133]MOFs consist of metal ions organized in a porous, crystalline structure with organic ligands. These structures often have dimensions that can be adjusted at the nanoscale level. MOFs can be functionalized using a wide range of ligands to improve selectivity for particular contaminants. The removal of arsenic using nano adsorbents generated by doping two metals is more effective, with a removal rate of around 96.5%.[134] Prepared MXene-based nanosheets are applied in the reduction of microorganisms in polluted water and in the removal of heavy metals like chromium. We can achieve greater sustainability and effectiveness in treating wastewater while simultaneously safeguarding the environment and public health by fully using nanotechnologies. The use of nanotechnology to recycle resources from wastewater and capture renewable energy benefits the environment. The full potential of nanotechnology for wastewater cleanup and public health requires responsible development, efficient regulation, and public awareness.

3.4.3 Integration with Other Technologies

The integration of nanotechnology with machine learning provides various benefits, including superior cancer treatment, advanced drug delivery approaches, and early diagnosis methods. This intersection holds potential for a great future in cancer

TABLE 3.2
Future and Emerging Trends in Recent Nanocomposites

S.No.	Nanocomposite	Nature/Morphology of Material	Applications	References
1.	TiO_2/Al_2O_3	Nanowires	Energy storage	Peng et al.[136]
2.	MoS_2-Mo_2N	Nanowires	Flexible supercapacitors	Ranjan and Kaur[137]
3.	TiO_2	Carbon dots	Photodegradation of methylene blue	Sendão et al.[138]
4.	Transition metal-based chalcogenides/oxides	Two-dimensional metal oxides	Supercapacitor	Devi et al.[139]
5.	Metal–oxide semiconductor/gC_3N_4	Nanosheets	Detection of different gases	Khoramshahi et al.[140]
6.	M/SnO_2 (M = Al, Mn, Ni)	Nanosheets/nano spherical	Optoelectronics, supercapacitors, Li batteries, and solar cell devices	Sedky et al.[141]
7.	Magnetic halloysite nanotubes@ polyethylene imine	Nanotubes	Determination and extraction of gallic acid from green tea	Tang et al.[142]

therapy with a focus on integrating multi-omics information via machine learning and extreme learning techniques for the discovery of novel biomarkers, diagnosis of patient responses to ongoing therapies, and identification of medicinal targets. Furthermore, machine learning algorithms are being utilized for the analysis of advanced imaging techniques like optical and magnetic resonance imaging, which are integrated with nanotechnology-based targeted agents for the early diagnosis of cancer and which will aid in enhancing therapeutic outcomes.[135] Some recent developments in this field are summarized in Table 3.2.

3.4.3.1 Machine Learning and Artificial Intelligence

Nowadays, artificial intelligence and machine learning are used to enhance and advance the procedures for nanotoxicology and nanotherapeutics. A three-dimensional nano-QSAR model was used to develop novel nanotherapeutics by correlating small energy conformations. For nano-cytotoxicity, two factors, such as enthalpy of formation (correlated to energy band gap) and electronegativity (related to stability), were supported by the nano-QSAR model. Artificial intelligence and machine learning-aided nanorobots are also utilized to target drug delivery, minimizing the drawbacks of typical chemotherapy.[143] Artificial intelligence was found to be effective in identifying the characteristics, nanomanufacturing processes, and behavior of nanomaterials by improving efficiency and reducing defects. Additionally, two tools, such as support vector machines and artificial neural networks, were found effective for connecting raw materials with nano products.[144]

3.4.3.2 Biotechnology

Roco et al. defined nanobiotechnology as the field that utilizes nanoscale principles and biological components and materials synthesized at the nanoscale.[145] Nanobiotechnology plays a vital role in the production of biofuel, employing biological systems designed at the nanoscale to transform renewable resources into clean energy sources.[145]

3.4.3.3 Conclusion

Next-generation nanomaterials consist of advanced materials engineered at the nanoscale (1–100 nm). These materials have unique biological characteristics different from their classical counterparts due to their quantum properties and large surface area. Smart nanomaterials, which respond to external stimuli such as pH, temperature, and electromagnetic fields, are effective for controlling drug delivery and the treatment of various diseases. Furthermore, SQDs are biocompatible, easy to use, photoluminescent, and economical to synthesize. Advancements in 2D metal oxide nanomaterials have been found useful in energy storage devices. Nanomaterials are integrated with other technologies, namely, AI, ML, and biotechnology, to enhance their potential in various fields.

3.5 SUMMARY

Nanomaterials have unique biological characteristics different from their classical counterparts due to their quantum properties and large surface area. The characteristics of nanoparticles may be altered by monitoring their shape, size, and surface functionalization. Nanomaterials are synthesized using various methods, including chemical, physical, and biological methods. Once synthesized, nanoparticles undergo characterization using techniques such as TEM, scanning electron microscopy, XRD, UV–visible spectroscopy, and others to gather information on size, shape, crystal structure, and composition. However, these characterization techniques are not without challenges; for instance, XRD is unsuitable for amorphous substances and shows broad patterns for nanoparticles smaller than 3nm. TEM has a limited depth of resolution, while XRF is time-consuming. Nanoparticles find extensive application in biomedical fields, including targeted drug delivery, early disease diagnosis, biomarkers, gene therapy, antimicrobials, biosensing, and cancer treatment. Moreover, nanoparticles are utilized in other fields such as energy storage devices, wastewater treatment, and air pollution control. Although nanomaterials have numerous applications, nanotoxicity includes distressing cells due to interactions with nanoparticles and has adverse effects on the liver, brain, and kidneys. Next-generation nanomaterials, such as smart nanomaterials, belong to advanced materials engineered on the nanoscale (1–100 nm). Smart nanomaterials are dependent on external agents like pH, light, temperature, and electromagnetic fields. These nanomaterials are employed in treating numerous illnesses and in biosensing. The integration of nanotechnology with machine learning, artificial intelligence, and biotechnology has led to enhancements in the methodology for nanotoxicology and nanotherapeutics.

ACKNOWLEDGEMENTS

The authors are thankful to the Department of Chemistry, University of Delhi, for providing support and facilities. S.S. and R.C are thankful to DST Indo-Russia: INT/RUS/RFBR/389 for providing financial assistance. Nisha is obliged to USIC, University of Delhi, for instrumental facilities and CSIR for awarding her a Junior Research Fellowship (09/045(1792)2020-EMR-I).

NOTES

1 Z. R. Dai, J. P. Bradley, D. J. Joswiak, D. E. Brownlee, H. G. M. Hill, and M. J. Genge, *Nature* **418**/6894, 157–159 (2002). https://doi.org/10.1038/nature00897
2 B. Elzein, *Heliyon* **10**/10, e31393 (2024). https://doi.org/10.1016/j.heliyon.2024.e31393
3 A. K. Kolupula, S. P. Gora, C. Bhanu Prakash, S. Nagaraju, J. Pagala, and S. Battapothula, *Proc. Indian Natl. Sci. Acad.* 1–22 (2024). https://doi.org/10.1007/s43538-024-00238-5
4 S. Simakov, V. Scribano, N. Melnik, V. Pechnikov, I. Drozdova, V. Vyalov, and M. Novikov, *Nano and Micro Diamond Formation in Nature: Ultrafine Carbon Particles on Earth and Space*, Springer Nature (2023). https://doi.org/10.1007/978-3-031-43278-1
5 F. J. Heiligtag, and M. Niederberger, *Mater. Today* **16**/7–8, 262–271 (2013). https://doi.org/10.1016/j.mattod.2013.07.004
6 B. Mubeen, A. N. Ansar, R. Rasool, I. Ullah, S. S. Imam, S. Alshehri, M. M. Ghoneim, S. I. Alzarea, M. S. Nadeem, and I. Kazmi, *Antibiotics (Basel)* **10**/12, 1473 (2021). https://doi.org/10.3390/antibiotics10121473
7 M. Reibold, P. Paufler, A. A. Levin, W. Kochmann, N. Pätzke, and D. C. Meyer, *Nature* **444**/7117, 286–286 (2006). https://doi.org/10.1038/444286a
8 J. Jeevanandam, A. Barhoum, Y. S. Chan, A. Dufresne, and M. K. Danquah, *Beilstein J. Nanotechnol.* **9**/1, 1050–1074 (2018). https://doi.org/10.3762/bjnano.9.98
9 G. Paramasivam, A. Sanmugam, V. V. Palem, M. Sevanan, A. B. Sairam, N. Nachiappan, B. Youn, J. S. Lee, M. Nallal, and K. H. Park, *Int. J. Biol. Macromol.* **254**/2, 127904 (2024). https://doi.org/10.1016/j.ijbiomac.2023.127904
10 F. M. Abdel-Haleem, M. Saad, A. Barhoum, M. Bechelany, and M. S. Rizk, *Mater. Sci. Eng. C Mater. Biol. Appl.* **89**, 140–148 (2018). https://doi.org/10.1016/j.msec.2018.04.001
11 R. Rasouli, and A. Barhoum, Advances in nanofibers for antimicrobial drug delivery, in *Handbook of Nanofibers*, edited by A. Barhoum, M. Bechelany, and A. Makhlouf, Springer, Cham (2019). https://doi.org/10.1007/978-3-319-53655-2_33
12 S. Malik, K. Muhammad, and Y. Waheed, *Molecules* **28**/2, 661 (2023). https://doi.org/10.3390/molecules28020661
13 A. Soozanipour, F. Ejeian, Y. Boroumand, A. Rezayat, and S. Moradi, *Chemosphere* **312**/1, 137185 (2023). https://doi.org/10.1016/j.chemosphere.2022.137185
14 M. S. de Sousa, A. E. Schlogl, F. R. Estanislau, V. G. L. Souza, J. S. dos Reis Coimbra, and I. J. B. Santos, *Coatings* **13**/8, 1411 (2023). https://doi.org/10.3390/coatings13081411
15 B. H. Abbasi, H. Fazal, N. Ahmad, M. Ali, N. Giglioli-Guivarch, and C. Hano, *Nanocosmetics*, 79–108 (2020). https://doi.org/10.1016/b978-0-12-822286-7.00005-x
16 S. Shetty, P. C. Alvarado, D. Pettie, and J. H. Collier, *Annu. Rev. Biomed. Eng.* **26**/1, 273–306 (2024). https://doi.org/10.1146/annurev-bioeng-110122-124359
17 N. Baig, I. Kammakakam, and W. Falath, *Mater. Adv.* **2**/6, 1821–1871 (2021). https://doi.org/10.1039/D0MA00807A
18 P. Szczyglewska, A. Feliczak-Guzik, and I. Nowak, *Molecules* **28**/13, 4932 (2023). https://doi.org/10.3390/molecules28134932
19 F. Yao, P. Zhu, J. Chen, S. Li, B. Sun, Y. Li, M. Zou, X. Qi, P. Liang, and Q. Chen, *Mikrochim. Acta* **190**/7, 256 (2023). https://doi.org/10.1007/s00604-023-05838-4

20 I. Ijaz, E. Gilani, A. Nazir, and A. Bukhari, *Green Chem. Lett. Rev.* **13**/3, 223–245 (2020). https://doi.org/10.1080/17518253.2020.1802517

21 Y. Bai, R. Zhang, X. Ye, Z. Zhu, H. Xie, B. Shen, D. Cai, B. Liu, C. Zhang, Z. Jia, S. Zhang, X. Li, and F. Wei, *Nat. Nanotechnol.* **13**/7, 589–595 (2018). https://doi.org/10.1038/s41565-018-0141-z

22 F. Yang, X. Wang, D. Zhang, J. Yang, D. Luo, Z. Xu, J. Wei, J. Q. Wang, Z. Xu, F. Peng, X. Li, R. Li, Y. Li, M. Li, X. Bai, F. Ding, and Y. Li, *Nature* **510**/7506, 522–524 (2014). https://doi.org/10.1038/nature13434

23 Z. Li, M. Xiao, C. Jin, and Z. Zhang, *Biosensors* **13**/3, 326 (2023). https://doi.org/10.3390/bios13030326

24 C. R. Vandenabeele, and S. Lucas, *Mater. Sci. Eng. R Rep.* **139**, 100521 (2020). https://doi.org/10.1016/j.mser.2019.100521

25 V. Selvamani, Stability studies on nanomaterials used in drugs, in *Characterization and Biology of Nanomaterials for Drug Delivery*, Elsevier (2019, pp. 425–444). https://doi.org/10.1016/b978-0-12-814031-4.00015-5

26 D. R. Baer, P. Munusamy, and B. D. Thrall, *Biointerphases* **11**/4, 04B401 (2016). https://doi.org/10.1116/1.4964867

27 D. R. Baer, M. H. Engelhard, G. E. Johnson, J. Laskin, J. Lai, K. Mueller, P. Munusamy, S. Thevuthasan, H. Wang, N. Washton, A. Elder, B. L. Baisch, A. Karakoti, S. V. N. T. Kuchibhatla, and D. Moon, *J. Vac. Sci. Technol. A* **31**/5, 50820 (2013). https://doi.org/10.1116/1.4818423

28 NNI Interagency Workshop, Instrumentation and Metrology for Nanotechnology Grand Challenge Workshop. https://metrology.nano.gov

29 S. V. Kuchibhatla, A. S. Karakoti, D. R. Baer, S. Samudrala, M. H. Engelhard, J. E. Amonette, S. Thevuthasan, and S. Seal, *J. Phys. Chem. C Nanomater. Interfaces* **116**/26, 14108–14114 (2012). https://doi.org/10.1021/jp300725s

30 S. Mourdikoudis, R. M. Pallares, and N. T. Thanh, *Nanoscale* **10**/27, 12871–12934 (2018). https://doi.org/10.1039/c8nr02278j

31 J. Bian, N. Gobalasingham, A. Purchel, and J. Lin, *Molecules* **28**/10, 4169 (2023). https://doi.org/10.3390/molecules28104169

32 D. Mitra, P. Adhikari, R. Djebaili, P. Thathola, K. Joshi, M. Pellegrini, N. O. Adeyemi, B. Khoshru, K. Kaur, A. Priyadarshini, A. Senapati, M. Del Gallo, P. K. Das Mohapatra, A. K. Nayak, V. Shanmugam, and P. Panneerselvam, *Plant Physiol. Biochem.* **196**, 103–120 (2023). https://doi.org/10.1016/j.plaphy.2023.01.017

33 S. Gangadoo, A. Taylor-Robinson, and J. Chapman, *Mater. Technol.* **30**/sup5, B44-B56 (2015). https://doi.org/10.1179/1753555714Y.0000000201

34 J. Jie, W. Zhang, K. Peng, G. Yuan, C. S. Lee, and S. T. Lee, *Adv. Funct. Mater.* **18**/20, 3251–3257 (2008). https://doi.org/10.1002/adfm.200800399

35 E. C. Scher, L. Manna, and A. P. Alivisatos, *Philos. Trans. R. Soc. Lond. A* **361**/1803, 241–257 (2003). https://doi.org/10.1098/rsta.2002.1126

36 D. R. Baer, *J. Surf. Anal.* **17**/3, 163–169 (2011). https://doi.org/10.1384/jsa.17.163

37 P. M. Ajayan, *Nature* **427**/6973, 402–403 (2004). https://doi.org/10.1038/427402a

38 S. Helveg, C. López-Cartes, J. Sehested, P. L. Hansen, B. S. Clausen, J. R. Rostrup-Nielsen, F. Abild-Pedersen, and J. K. Nørskov, *Nature* **427**/6973, 426–429 (2004). https://doi.org/10.1038/nature02278

39 J. P. Zhao, Z. Y. Chen, X. J. Cai, and J. W. Rabalais, *J. Vac. Sci. Technol. B Microelectron. Nanometer Struct. Process. Meas. Phenom.* **24**/3, 1104–1108 (2006). https://doi.org/10.1116/1.2188410

40 D. R. Baer, M. H. Engelhard, D. J. Gaspar, D. W. Matson, K. H. Pecher, J. R. Williams, and C. M. Wang, *J. Surf. Anal.* **12**/2, 101 (2005).

41 P. Tucci, G. Porta, M. Agostini, D. Dinsdale, I. Iavicoli, K. Cain, A. Finazzi-Agró, G. Melino, and A. Willis, *Cell Death Dis.* **4**/3, e549–e549 (2013). https://doi.org/10.1038/cddis.2013.76

42 M. Simkó, and M. O. Mattsson, *Part. Fibre Toxicol.* **7**, 42 (2010). https://doi.org/10.1186/1743-8977-7-42
43 Y. Tan, D. Yu, J. Feng, H. You, Y. Bai, J. He, H. Cao, Q. Che, J. Guo, and Z. Su, *Drug Deliv. Transl. Res.* **13**/9, 2213–2238 (2023). https://doi.org/10.1007/s13346-023-01312-z
44 N. Sharma, B. D. Kurmi, D. Singh, S. Mehan, K. Khanna, R. Karwasra, S. Kumar, A. Chaudhary, V. Jakhmola, A. Sharma, S. K. Singh, K. Dua, and D. Kakkar, *J. Drug Target.* **32**/5, 457–469 (2024). https://doi.org/10.1080/1061186X.2024.2316785
45 P. Saxena, A. K. Gupta, V. Saharan, and Harish, *Environ. Sci. Pollut. Res. Int.* **30**/9, 23646–23654 (2023). https://doi.org/10.1007/s11356-022-23912-x
46 Q. Ren, J. Ma, X. Li, Q. Meng, S. Wu, Y. Xie, Y. Qi, S. Liu, and R. Chen, *ACS Appl. Mater. Interfaces* **15**/23, 27774–27788 (2023). https://doi.org/10.1021/acsami.3c05692
47 J. Chen, X. Dong, J. Zhao, and G. Tang, *J. Appl. Toxicol.* **29**/4, 330–337 (2009). https://doi.org/10.1002/jat.1414
48 M. N. Moore, *Environ. Int.* **32**/8, 967–976 (2006). https://doi.org/10.1016/j.envint.2006.06.014.
49 Shariatinia, Z. (2018). *Int J Bio. Macromol.*, **120**, 1406–1419.
50 A. Jimtaisong, and N. Saewan, *Int. J. Cosmet. Sci.* **36**/1, 12–21 (2014). https://doi.org/10.1111/ics.12102
51 S. Borowska, and M. M. Brzóska, *J. Appl. Toxicol.* **35**/6, 551–572 (2015). https://doi.org/10.1002/jat.3129
52 S. Sharifi, S. Behzadi, S. Laurent, M. L. Forrest, P. Stroeve, and M. Mahmoudi, *Chem. Soc. Rev.* **41**/6, 2323–2343 (2012). https://doi.org/10.1039/c1cs15188f
53 I. Iavicoli, V. Leso, W. Ricciardi, L. L. Hodson, and M. D. Hoover, *Environ. Health* **13**, 78 (2014). https://doi.org/10.1186/1476-069X-13-78
54 T. K. Tullis, *Nanotechnol. Rev.* **1**/2, 189–205 (2012). https://doi.org/10.1515/ntrev-2012-0501
55 C. Keswani, editor, *Intellectual Property Issues in Nanotechnology*, CRC Press (2020). https://doi.org/10.1201/9781003052104
56 R. W. Lai, K. W. Yeung, M. M. Yung, A. B. Djurišić, J. P. Giesy, and K. M. Leung, *Environ. Sci. Pollut. Res. Int.* **25**/4, 3060–3077 (2018). https://doi.org/10.1007/s11356-017-9489-0
57 E. M. A. Maine, and E. W. Garnsey, CTM working paper, *SSRN J.* **2004**/02 (2004). https://doi.org/10.2139/ssrn.1923131
58 M. Ramesh, R. Janani, C. Deepa, and L. Rajeshkumar, *Biosensors* **13**/1, 40 (2022). https://doi.org/10.3390/bios13010040
59 P. Ratre, N. Nazeer, R. Kumari, S. Thareja, B. Jain, R. Tiwari, A. Kamthan, R. K. Srivastava, and P. K. Mishra, *Biosensors* **13**/2, 226 (2023). https://doi.org/10.3390/bios13020226.
60 S. Pandit, D. Dasgupta, N. Dewan, and A. Prince, *J. Pharm. Innov.* **5**/6A, 18 (2016).
61 J. Chao, D. Zhu, Y. Zhang, L. Wang, and C. Fan, *Biosens. Bioelectron.* **76**, 68–79 (2016). https://doi.org/10.1016/j.bios.2015.07.007
62 C. R. Gawali, E. Daweshar, A. Kolhe, and S. Kumar, *Microchem. J.* 110326 (2024). https://doi.org/10.1016/j.microc.2024.110326
63 G. H. Gupta, S. Kadakia, D. Agiwal, T. Keshari, and S. Kumar, *Mater. Adv.* **5**/5, 1803–1816 (2024). https://doi.org/10.1039/D3MA00829K
64 S. Shankar, Y. Kumar, D. Chauhan, N. Sharma, R. Chandra, and S. Kumar, *Biosens. Bioelectron.* **242**, 115722 (2023). https://doi.org/10.1016/j.bios.2023.115722
65 S. Shankar, Y. Kumar, D. Chauhan, P. Tiwari, N. Sharma, R. Chandra, and S. Kumar, *Microchem. J.* **189**, 108555 (2023). https://doi.org/10.1016/j.microc.2023.108555
66 M. A. Beach, U. Nayanathara, Y. Gao, C. Zhang, Y. Xiong, Y. Wang, and G. K. Such, *Chem. Rev.* **124**/9, 5505–5616 (2024). https://doi.org/10.1021/acs.chemrev.3c00705
67 C. Li, Q. Wang, X. Gu, Y. Kang, Y. Zhang, Y. Hu, T. Li, H. Jin, G. Deng, and Q. Wang, *Int. J. Nanomed.* **14**, 3845–3860 (2019). https://doi.org/10.2147/IJN.S202741

68 G. Salmanian, S. A. Hassanzadeh-Tabrizi, and N. Koupaei, *Int. J. Biol. Macromol.* **184**, 618–635 (2021). https://doi.org/10.1016/j.ijbiomac.2021.06.108
69 U. Chintapula, S. Yang, T. Nguyen, Y. Li, J. Jaworski, H. Dong, and K. T. Nguyen, *ACS Appl. Mater. Interfaces* **14**/51, 56498–56509 (2022). https://doi.org/10.1021/acsami.2c15204
70 A. Sanmugam, S. Abbishek, S. L. Kumar, A. B. Sairam, V. V. Palem, R. S. Kumar, A. I. Almansour, N. Arumugam, and D. Vikraman, *J. Mech. Behav. Biomed. Mater.* **145**, 106033 (2023). https://doi.org/10.1016/j.jmbbm.2023.106033
71 T. Kang, G. D. Cha, O. K. Park, H. R. Cho, M. Kim, J. Lee, D. Kim, B. Lee, J. Chu, S. Koo, T. Hyeon, D. H. Kim, and S. H. Choi, *ACS. Nano* **17**/6, 5435–5447 (2023). https://doi.org/10.1021/acsnano.2c10094
72 M. Saeedi, M. Eslamifar, K. Khezri, and S. M. Dizaj, *Biomed. Pharmacother.* **111**, 666–675 (2019). https://doi.org/10.1016/j.biopha.2018.12.133
73 B. Li, H. Shao, L. Gao, H. Li, H. Sheng, and L. Zhu, *Drug Deliv.* **29**/1, 2130–2161 (2022). https://doi.org/10.1080/10717544.2022.2094498
74 A. I. Fahira, R. Amalia, M. I. Barliana, V. A. Gatera, and R. Abdulah, *Breast Cancer* **14**, 71–83 (2022). https://doi.org/10.2147/BCTT.S350403
75 Z. Cheng, M. Li, R. Dey, and Y. Chen, *J. Hematol. Oncol.* **14**, 1 (2021). https://doi.org/10.1186/s13045-021-01096-0
76 F. Martínez-Esquivias, M. Gutiérrez-Angulo, A. Pérez-Larios, J. A. Sánchez-Burgos, J. S. Becerra-Ruiz, and J. M. Guzmán-Flores, *Anti-Cancer Agents Med. Chem. (Formerly Current Medicinal Chemistry-Anti-Cancer Agents)* **22**/9, 1658–1673 (2022). https://doi.org/10.2174/1871520621666210910084216
77 S. E. Hollow, and T. C. Johnstone, *Curr. Opin. Chem. Biol.* **72**, 102229 (2023). https://doi.org/10.1016/j.cbpa.2022.102229
78 J. Liu, Y. Xie, J. Ma, and H. Chu, *Chem. Biol. Interact.* 393, 110948 (2024). https://doi.org/10.1016/j.cbi.2024.110948
79 J. Jahanbani, M. Ghotbi, F. Shahsavari, E. Seydi, S. Rahimi, and J. Pourahmad, *J. Biochem. Mol. Toxicol.* **34**/10, e22557 (2020). https://doi.org/10.1002/jbt.22557
80 D. Zhi, T. Yang, J. Yang, S. Fu, and S. Zhang, *Acta Biomater.* **102**, 13–34 (2020). https://doi.org/10.1016/j.actbio.2019.11.027
81 M. Raszewska-Famielec and J. Flieger, *Int. J. Mol. Sci.* **23**/24, 15980 (2022). https://doi.org/10.3390/ijms232415980
82 Y. Ding, Z. Jiang, K. Saha, C. S. Kim, S. T. Kim, R. F. Landis, and V. M. Rotello, *Mol. Ther.* **22**/6, 1075–1083 (2014). https://doi.org/10.1038/mt.2014.30
83 F. Herranz, E. Almarza, I. Rodríguez, B. Salinas, Y. Rosell, M. Desco, J. W. Bulte, and J. Ruiz-Cabello, *Microsc. Res. Tech.* **74**/7, 577–591 (2011). https://doi.org/10.1002/jemt.20992
84 Z. Mirza, and S. Karim, *Semin. Cancer Biol.* **69** (2021). https://doi.org/10.1016/j.semcancer.2019.10.020
85 M. Ashrafizadeh, A. Zarrabi, A. Bigham, A. Taheriazam, Y. Saghari, S. Mirzaei, M. Hashemi, K. Hushmandi, H. Karimi-Maleh, E. Nazarzadeh Zare, E. Sharifi, Y. N. Ertas, N. Rabiee, G. Sethi, and M. Shen, *Med. Res. Rev.* **43**/6, 2115–2176 (2023). https://doi.org/10.1002/med.21971
86 Y. Gong, X. Hu, M. Chen, and J. Wang, *J. Nanobiotechnol.* **22**/1, 309 (2024). https://doi.org/10.1186/s12951-024-02550-0
87 W. M. Pardridge, *Trends Mol. Med.* **29**/5, 343–353 (2023). https://doi.org/10.1016/j.molmed.2023.02.004
88 F. Martínez-Esquivias, J. M. Guzmán-Flores, A. Pérez-Larios, N. González Silva, and J. S. Becerra-Ruiz, *J. Nanosci. Nanotechnol.* **21**/11, 5383–5398 (2021). https://doi.org/10.1166/jnn.2021.19471

89 I. X. Yin, J. Zhang, I. S. Zhao, M. L. Mei, Q. Li, and C. H. Chu, *Int. J. Nanomed.* **15**, 2555–2562 (2020). https://doi.org/10.2147/IJN.S246764
90 K. B. Liew, A. K. Janakiraman, R. Sundarapandian, S. H. Khalid, F. A. Razzaq, L. C. Ming, A. Khan, A. Kalusalingam, and P. W. Ng, *J. Med. Life* **15**/3, 328–335 (2022). https://doi.org/10.25122/jml-2021-0097
91 N. A. S. Rozman, W. Y. Tong, C. R. Leong, W. N. Tan, M. A. Hasanolbasori, and S. Z. Abdullah, *J. Microbiol. Biotechnol.* **29**/7, 1009–1013 (2019). https://doi.org/10.4014/jmb.1904.04065
92 A. B. Younis, Y. Haddad, L. Kosaristanova, and K. Smerkova, *Wiley Interdiscip. Rev. Nanomed. Nanobiotechnol.* **15**/3, e1860 (2023). https://doi.org/10.1002/wnan.1860
93 R. H. Alsharedeh, M. Rezigue, R. M. Bashatwah, H. Amawi, A. A. Aljabali, M. A. Obeid, and M. M. Tambuwala, *Curr. Drug Deliv.* **21**/6, 828–851 (2024). https://doi.org/10.2174/1567201820666230223085403
94 B. Li, J. Mao, J. Wu, K. Mao, Y. Jia, F. Chen, and J. Liu, *Small* **20**/7, e2306135 (2024). https://doi.org/10.1002/smll.202306135
95 A. S. Kashyap, N. Manzar, S. K. Vishwakarma, C. Mahajan, and U. Dey, *World J. Microbiol. Biotechnol.* **40**/3, 104 (2024). https://doi.org/10.1007/s11274-024-03911-5
96 K. Jain, A. S. Patel, V. P. Pardhi, and S. J. S. Flora, *Molecules* **26**/6, 1797 (2021). https://doi.org/10.3390/molecules26061797
97 S. Cheriyamundath and S. L. Vavilala, *Water Environ. J.* **35**/1, 123–132 (2021). https://doi.org/10.1111/wej.12610
98 F. Lu, and D. Astruc, *Coord. Chem. Rev.* **408**, 213180 (2020). https://doi.org/10.1016/j.ccr.2020.213180
99 S. Ramu, I. Kainthla, L. Chandrappa, J. M. Shivanna, B. Kumaran, and R. G. Balakrishna, *Environ. Sci. Pollut. Res. Int.* **31**/1, 167–190 (2024). https://doi.org/10.1007/s11356-023-31162-8
100 S. Panhwar, H. A. Keerio, N. H. Khokhar, M. Muqeet, Z. Ali, M. Bilal, and A. Ul Rehman, *J. Water Health* **22**/1, 123–137 (2024). https://doi.org/10.2166/wh.2023.116
101 M. Hua, S. Zhang, B. Pan, W. Zhang, L. Lv, and Q. Zhang, *J. Hazard. Mater.* **211–212**, 317–331 (2012). https://doi.org/10.1016/j.jhazmat.2011.10.016
102 L. Burratti, E. Sgreccia, F. Bertelà, and F. Galiano, *Chemosphere* **362**, 142636 (2024). https://doi.org/10.1016/j.chemosphere.2024.142636
103 T. M. Joseph, H. E. Al-Hazmi, B. Śniatała, A. Esmaeili, and S. Habibzadeh, *Environ. Res.* **238**/1, 117114 (2023). https://doi.org/10.1016/j.envres.2023.117114
104 G. Yu, Y. Lu, J. Guo, M. Patel, A. Bafana, X. Wang, B. Qiu, C. Jeffryes, S. Wei, Z. Guo, and E. K. Wujcik, *Adv. Compos. Hybrid Mater.* **1**/1, 56–78 (2018). https://doi.org/10.1007/s42114-017-0004-3
105 Y. Mishra, V. Mishra, A. Chattaraj, A. A. Aljabali, M. El-Tanani, M. R. Farani, Y. S. Huh, Ã. Serrano-Aroca, and M. M. Tambuwala, *Environ. Res.* **238**/1, 117088 (2023). https://doi.org/10.1016/j.envres.2023.117088
106 M. Baratta, A. V. Nezhdanov, A. I. Mashin, F. P. Nicoletta, and G. De Filpo, *Sci. Total Environ.* **924**, 171578 (2024). https://doi.org/10.1016/j.scitotenv.2024.171578
107 C. Vargas-Figueroa, L. Pino-Soto, A. Beratto-Ramos, Y. Tapiero, B. L. Rivas, M. E. Berrio, M. F. Melendrez, and R. M. Bórquez, *Membranes* **13**/7, 616 (2023). https://doi.org/10.3390/membranes13070616
108 J. Liu, L. Liu, S. Wang, S. Sun, C. Hu, and Y. Zhao, *J. Environ. Manag.* **347**, 119120 (2023). https://doi.org/10.1016/j.jenvman.2023.119120
109 F. Mohajer, G. M. Ziarani, A. Badiei, S. Iravani, and R. S. Varma, *Nanomaterials (Basel)* **13**/2, 345 (2023). https://doi.org/10.3390/nano13020345
110 Y. Liu, X. Dai, J. Li, S. Cheng, J. Zhang, and Y. Ma, *RSC Adv.* **14**/1, 478–491 (2024). https://doi.org/10.1039/d3ra06910a

111 A. Salama, R. Abouzeid, W. S. Leong, J. Jeevanandam, P. Samyn, A. Dufresne, M. Bechelany, and A. Barhoum, *Nanomaterials (Basel)* **11**/11, 3008 (2021). https://doi.org/10.3390/nano11113008

112 M. Ben Chobba, M. Messaoud, M. L. Weththimuni, J. Bouaziz, M. Licchelli, F. De Leo, and C. Urzì, *Environ. Sci. Pollut. Res. Int.* **26**/32, 32734–32745 (2019). https://doi.org/10.1007/s11356-019-04680-7

113 A. Spoială, C. I. Ilie, R. D. Trușcă, O. C. Oprea, V. A. Surdu, B. Ș. Vasile, A. Ficai, D. Ficai, E. Andronescu, and L. M. Dițu, *Materials (Basel)* **14**/16, 4747 (2021). https://doi.org/10.3390/ma14164747

114 V. Gadore, S. R. Mishra, A. K. Singh, and M. Ahmaruzzaman, *RSC Adv.* **14**/5, 3447–3472 (2024). https://doi.org/10.1039/d3ra08323c

115 K. F. Salama, and M. Zafar, *Int. J. Prev. Med.* **13**/1, 67 (2022). https://doi.org/10.4103/ijpvm.IJPVM_586_20

116 X. Peng, M. Jiang, X. Wang, D. Li, S. Sun, Y. Shao, and Y. Zheng, *Environ. Sci. Pollut. Res. Int.* **28**/24, 31770–31777 (2021). https://doi.org/10.1007/s11356-021-12970-2

117 M. A. Abdulhamid, and K. Muzamil, *Chemosphere* **310**, 136886 (2023). https://doi.org/10.1016/j.chemosphere.2022.136886

118 S. J. Wang, X. Y. Zhang, D. Su, Y. F. Wang, C. M. Qian, X. R. Zhou, Y. Z. Li, and T. Zhang, *Molecules* **25**/15, 3369 (2020). https://doi.org/10.3390/molecules25153369

119 S. Thangudu, *Appl. Nanomater. Hum. Health* 287–312 (2020). https://doi.org/10.1007/978-981-15-4802-4_15

120 M. Fathi, E. D. Abdolahinia, J. Barar, and Y. Omidi, *Nanomedicine (Lond)* **15**/22, 2171–2200 (2020). https://doi.org/10.2217/nnm-2020-0146

121 X. Chen, S. Hussain, A. Abbas, Y. Hao, A. H. Malik, X. Tian, H. Song, and R. Gao, *Mikrochim. Acta* **189**/3, 83 (2022). https://doi.org/10.1007/s00604-021-05153-w

122 P. P. Upare, H. S. Shin, J. H. Lee, and B. G. Park, Development of efficient strategies for physical stimuli-responsive programmable nanotherapeutics, in *Nanorobotics and Nanodiagnostics in Integrative Biology and Biomedicine*, Springer International Publishing, Cham (2022, 201–228). https://doi.org/10.1007/978-3-031-16084-4_9

123 J. Majumder, and T. Minko, *Expert Opin. Drug Deliv.* **18**/2, 205–227 (2021). https://doi.org/10.1080/17425247.2021.1828339

124 R. Priyadarshi, S. Pourmoslemi, A. Khan, Z. Riahi, and J. W. Rhim, *Colloids Surf. B Biointerfaces* **237**, 113863 (2024). https://doi.org/10.1016/j.colsurfb.2024.113863

125 J. Kim, J. Roh, M. Park, and C. Lee, *Adv. Mater.* **36**/20, e2212220 (2024). https://doi.org/10.1002/adma.202212220

126 H. Li, C. Xu, N. Srivastava, and K. Banerjee, *IEEE Trans. Electron Devices* **56**/9, 1799–1821 (2009). https://doi.org/10.1109/TED.2009.2026524

127 S. Ding, J. Chen, K. Li, and H. Zhang, *J. Mater. Chem. A* (2024). https://doi.org/10.1039/D4TA03027C

128 J. Mei, T. Liao, L. Kou, and Z. Sun, *Adv. Mater.* **29**/48, 1700176 (2017). https://doi.org/10.1002/adma.201700176

129 O. Gohar, M. Z. Zubair Khan, I. Bibi, N. Bashir, U. Tariq, M. Bakhtiar, M. Ramzan Abdul Karim, F. Ali, M. Bilal Hanif, and M. Motola, *Mater. Des.* **241**, 112930 (2024). https://doi.org/10.1016/j.matdes.2024.112930

130 A. V. Shchegolkov, S. H. Jang, A. V. Shchegolkov, Y. V. Rodionov, A. O. Sukhova, and M. S. Lipkin, *Nanomaterials (Basel)* **11**/9, 2376 (2021). https://doi.org/10.3390/nano11092376

131 Z. Hai, Z. Wei, C. Xue, H. Xu, and F. Verpoort, *J. Mater. Chem. C* **7**/42, 12968–12990 (2019). https://doi.org/10.1039/C9TC04489B

132 K. Bon-Ryul, K. H. Kim, and H. J. Ahn, *Nanoscale* **11**/7, 3318–3325 (2019). https://doi.org/10.1039/c8nr08793h

133 M. A. Ahmad, M. Adeel, N. Shakoor, R. Javed, M. Ishfaq, Y. Peng, M. Zain, I. Azeem, I. Ali, M. Usman, Z. Wu, G. Gohari, M. Xu, Y. Rui, Z. Zhang, J. C. White, and X. Deng, *Sci. Total Environ.* **894**, 164861 (2023). https://doi.org/10.1016/j.scitotenv.2023.164861
134 P. Thamarai, R. Kamalesh, A. Saravanan, P. Swaminaathan, and V. C. Deivayanai, *Environ. Nanotechnol. Monit. Manag.* **21**, 100913 (2024). https://doi.org/10.1016/j.enmm.2024.100913
135 A. V. Singh, M. H. D. Ansari, D. Rosenkranz, R. S. Maharjan, F. L. Kriegel, K. Gandhi, A. Kanase, R. Singh, P. Laux, and A. Luch, *Adv. Healthc. Mater.* **9**/17, e1901862 (2020). https://doi.org/10.1002/adhm.201901862
136 B. Peng, H. Luo, H. Xie, D. Zhai, R. Guo, Y. Liu, M. Li, H. Chen, J. Cao, and D. Zhang, *ACS Appl. Polym. Mater.* **6**/11, 6747–6757 (2024). https://doi.org/10.1021/acsapm.4c01006
137 B. Ranjan, and D. Kaur, *ACS Appl. Mater. Interfaces* **16**/12, 14890–14901 (2024). https://doi.org/10.1021/acsami.4c00067
138 R. M. Sendão, M. Algarra, E. Ribeiro, M. Pereira, A. Gil, N. Vale, J. C. G. Esteves da Silva, and L. Pinto da Silva, *Adv. Sustain. Syst.* **8**/1, 2300317 (2024). https://doi.org/10.1002/adsu.202300317
139 S. B. Devi, S. Sekar, S. Sathish, S. Dhanasekaran, R. Nirmala, D. Y. Kim, Y. Lee, S. Lee, and R. Navamathavan, *J. Alloys Compd.* 173481 (2024). https://doi.org/10.1016/j.jallcom.2024.173481
140 V. Khoramshahi, M. Azarang, M. Nouri, A. Shirmardi, and R. Yousefi, *Talanta Open* **9**, 100290 (2024). https://doi.org/10.1016/j.talo.2024.100290
141 A. Sedky, N. Afify, A. Almohammedi, M. A. Sayed, A. M. Ali, and A. M. Abd-Elnaiem, *Ceram. Int.* **50**/2, 3409–3421 (2024). https://doi.org/10.1016/j.ceramint.2023.11.088
142 X. Tang, G. I. Urujeni, X. Ni, Z. Lu, D. Wang, J. Gao, F. Meriem, H. He, D. Xiao, and P. Dramou, *Int. J. Biol. Macromol.* **265**/1, 130914 (2024). https://doi.org/10.1016/j.ijbiomac.2024.130914
143 M. Agboklu, F. A. Adrah, and P. M. Agbenyo, *J. Nanotechnol. Res.* **6**, 16–26 (2024). https://www.doi.org/10.26502/jnr.2688-85210040
144 M. Nandipati, O. Fatoki, and S. Desai, *Materials (Basel)* **17**/7, 1621 (2024). https://doi.org/10.3390/ma17071621
145 M. C. Roco, *Curr. Opin. Biotechnol.* **14**/3, 337–346 (2003). https://doi.org/10.1016/s0958-1669(03)00068-5

4 Nanotechnology Inventions Geared toward Manufacturing

Rajesh Kumar Thakur, Ishteyaque Ahmad, Bhawna Mallick, Ratneshwar Kumar Ratnesh, and Sangita Ranjan

4.1 INTRODUCTION

Industry 4.0 was originally introduced at the Hannover Fair in 2011 and was officially announced in 2013 as a German strategic initiative to play a leading role in industries that are now revolutionizing the manufacturing sector. Industry 4.0, often known as Industry 4.0, marks the beginning of the Fourth Industrial Revolution [1]. Industry 4.0 has a broad impact on the entire industry. CPS in Industry 4.0 manufacturing combines digital and physical systems, allowing computers to dynamically and autonomously create and update facilities to fulfill production schedules while eliminating the need for human interaction at all stages. Cloud computing, one of today's most widely utilized technologies, has reduced the need for most enterprises to maintain their own physical IT infrastructure for computation and storage. Rather, they use cloud computing services from companies like Google, Microsoft, and Amazon via the Internet to reduce their operational expenses and adjust their computing resources in response to their evolving business needs. The concept of cloud factories will be widely adopted in Industry 4.0. Sensors and actuators exchange data with automation and control units, which are often industrial programmable logic controllers, in order to collect process information and adjust their behavior. Furthermore, to track the process's development, supervisory and monitoring supervisory control and data acquisition systems display the most important magnitudes in real time. Information that is graphical and numerical, together with the creation of alarms [2]. Technology advancements are changing not only corporate practices but also traditional items. Technology is being converted to digital format via the digital revolution. Industry 4.0 is a union between tangible assets and cutting-edge technology like cloud computing, Internet of Things (IoT), robots, 3D printing, artificial intelligence (AI), etc. Organizations that have embraced Industry 4.0 are adaptable and ready to make decisions based on data [3]. Currently, the Fourth Industrial Revolution is ongoing. This, often known as "Industry 4.0," refers to the use of information and communication technology in industry. Its foundation is the progress of the Third Industrial Revolution. Computer-based production

DOI: 10.1201/9781003634553-4

systems are improved by a network connection and, to some extent, possess an online counterpart. They make it possible for systems to communicate with one another and to generate data about themselves. This represents the latter stage of manufacturing automation [4]. An Industry 4.0 system's performance is closely linked to lean manufacturing's success and the expansion of the business. Both elements affect performance [5–7]. The digitalization of industrial processes and factories is being facilitated by new methods that automate data calculations more accurately and efficiently, therefore improving tasks. Industry 4.0 is often referred to as the Fourth Industrial Revolution, integrating cyber-physical systems (CPS) into factories within logistics, warehousing, and production processes, making it easier to accept new digital assets and integrate them with corporate values [8]. Industry 4.0 will make the disparities that now exist between individuals, organizations, and nations more apparent since each nation will adopt its technologies in a unique manner [9–10]. The poll also showed that the choice to deploy Industry 4.0 is negatively impacted by technological challenges. The goal of Industry 4.0 is to combine and take advantage of the convergence of developing technologies [11]. The introduction of "Industry 4.0" brought with it great expectations for addressing the most recent issues in production systems. With the use of its technology, Industry 4.0 is aiding and reinforcing this trend by changing lifestyles, introducing new business models and manufacturing techniques, and reviving the sector in preparation for the so-called digital transformation [12].

4.2 TOOLS AND TECHNIQUES FOR INDUSTRY 4.0

The CPS is a fundamental component of Industry 4.0 Helen Gill of the US NSF created the phrase "cyber-physical system" to promote research on on-demand manufacturing to attain resource efficiency and personalization improved by data analytics in intelligent and self-governing systems. The entire industry is significantly impacted by Industry 4.0 [10]. Systems in Industry 4.0 connect digital and physical systems, allowing computers to dynamically and independently build and change facilities to meet production schedules while eliminating the need for human engagement throughout the manufacturing process. Cloud computing, one of today's most popular technologies, reduces the requirement for most enterprises to have physical IT infrastructure for computation and data storage. Instead, they use cloud computing services over the Internet from businesses like Amazon, Microsoft, and Google to cut operational expenses and adapt computer resources to changing company needs [13–15]. A cloud factory concept similar to this will be common in Industry 4.0. Specialized companies, such as Google, Microsoft, and Amazon in "cloud computing," will maintain a pool of production facilities for which regular manufacturers would pay on a metered basis. This allows a company to utilize more facilities during peak seasons while freeing up more cloud space for use by other manufacturers during the off-season. Meanwhile, specialist companies that offer cloud computing with integrated physical and computational components are linked [13]. While the terms Industry 4.0 and CPS are frequently used interchangeably, Industry 4.0 refers to the full business cycle of using CPS to dynamically alter facilities and operational

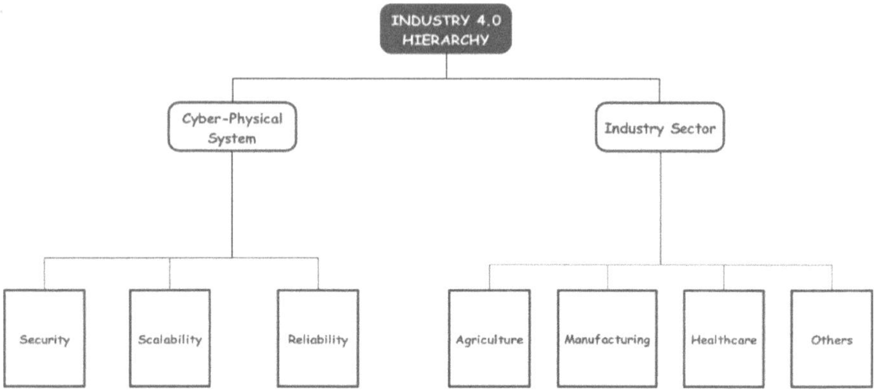

FIGURE 4.1 Industry 4.0 hierarchy.

techniques to meet client needs while reducing human interaction as shown in Figure 4.1. The CPS is an essential component of Industry 4.0, boosting the adaptability, resilience, scalability, and security of physical facilities via feedback loops that incorporate embedded sensors, processors, and actuators that are controlled and monitored by computers.

There is significant potential in the manufacturing sector to use AI techniques to improve decision-making, as well as various applications of AI and machine learning (ML) in the manufacturing business, as shown in Figure 4.2.

There is now a paradigm shift in the manufacturing industry. The traditional manufacturing era is giving way to Industry 4.0's smart manufacturing era, thanks to improvements in big data and ML. This paradigm shift leads to new opportunities. IoT stands for intelligent connected objects and flexible automation. Businesses may expand and succeed by utilizing AI and technology wisely. The largest businesses in the world are using AI and ML in manufacturing and are spending enormous sums of money on its advancement. According to McKinsey's, AI and ML technologies account for 40% of the value that analytics can deliver today; ML alone is estimated to have an annual value of between $3.5 trillion and $5.8 trillion.

4.3 VARIOUS TECHNIQUES USED IN COMPUTING 4.0

Industry 4.0's underlying "hyper-connectivity" of various technologies in the manufacturing process doubles the amount of data collected in real time from different sources and formats, which must be saved and evaluated using cutting-edge technology and complicated algorithms. The strategy is mainly based on big data and cloud computing. Both are seen as critical needs for Industry 4.0, offering significant and vital components for the strategy's implementation. Figure 4.3 shows the connection between Industry 4.0 and these two enablers.

Computing, ML, and more sophisticated AI systems all use data to generate control decisions, including forecasts, simulations, and a variety of other Industry 4.0-related outputs. Most of these algorithms, along with their project portfolio

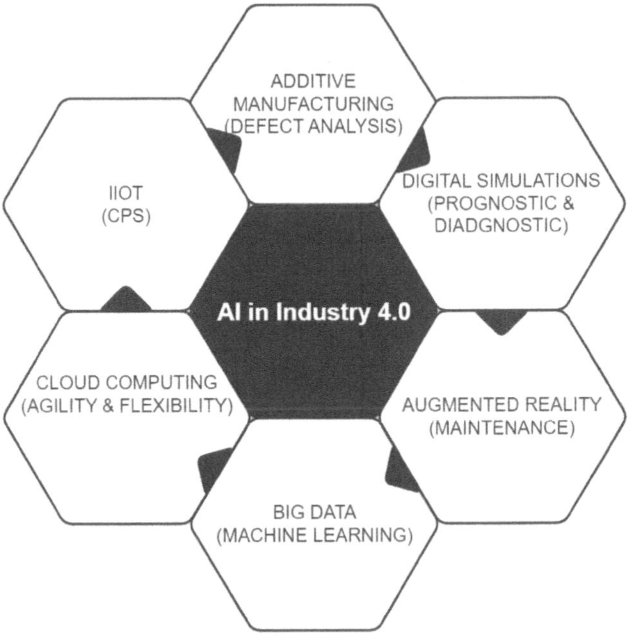

FIGURE 4.2 Artificial intelligence and machine learning.

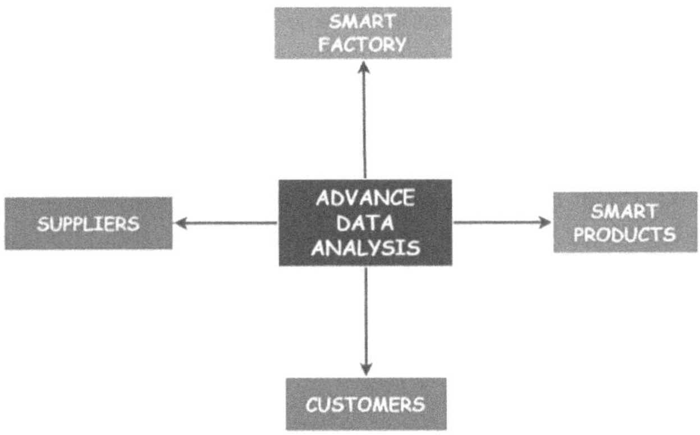

FIGURE 4.3 Big data, cloud computing, Industry 4.0.

management, encompass technologies such as Industry 4.0: IoT-enabled digital and physical systems, Big Data, ML, and AI. These technologies also include drones, robotics, and cloud-based software for 5G and 6G networks. 3D Printing Process: Although theoretical frameworks for blockchains have always existed, the most significant distinctions today are the increased availability of data and advancements in modern computing processing power. AI and ML programs process data on current CPUs and/or specialized general-purpose graphics processing units [16]. IoT is a new technology paradigm that envisions a worldwide network of compatible gadgets and products. IoT is generating significant attention from various businesses and is often regarded as one of the most crucial areas of future technology. IoT is the cornerstone of Industry 4.0 operations. By retrofitting IoT sensors, equipment can be modernized and industry capabilities can be enhanced. IoT sensors are installed in boards and utilized in accordance with market requirements:

- **Visual computing:** Several common Industry 4.0 technologies, such as big data, sophisticated HMI, 3D models and simulations, and cloud computing. While machine-to-machine communication, IoT and services, computing, and CPS technologies can all be applied individually to certain industrial problems, visual computing frequently serves as the "glue factor" that allows for a comprehensive solution that places the user at the center (in the roles of interpretation, analysis, and interaction) [17].
- **Using Industry 4.0 technologies and intelligent manufacturing:** Smart manufacturing consists of real-time, integrated information technology operations. Digital technology uses automatically generated electric fields to detect and record tags. The chips have been put to use, and new design advancements have increased their dependability to the point where they are now suitable for mass production. Chips are currently used by many businesses to increase equipment utilization, decrease unplanned downtime, and check inventories. Industry 4.0 enables a seamless combination of people, processes, and information. As a result, they must be combined at a pace that would not confuse employees [18].
- **Classification based on technology:** The goal of smart manufacturing systems is to increase productivity, quality, and efficiency in the manufacturing process by incorporating cutting-edge digital technologies. The following are some significant categories in smart manufacturing systems depending on technology.
- **AI systems:** AI is utilized in smart manufacturing systems to foresee equipment breakdowns, increase product quality, and streamline production processes.
- **IoT:** With software, sensors, and connections enabling data exchange with other devices and systems, IoT is a network that connects physical objects, transportation, and other technology.
- **Industrial IoT (IIoT):** The IIoT is crucial for monitoring and optimizing manufacturing processes in smart industrial systems. It does this by utilizing sensors, devices, and other cutting-edge technologies for data gathering and exchange.
- **Cloud-based software:** With cloud computing, manufacturers may use massive amounts of data produced by intelligent manufacturing systems to make data-driven decisions that increase production quality and efficiency.
- **Cybersecurity:** For smart manufacturing systems to be secure against security threats and data breaches, cybersecurity is essential. This includes encrypting data, implementing secure network protocols, and restricting access to private information.

- **Augmented Reality (AR):** Smart manufacturing systems use AR to provide employees with real-time information and guidance on tools and production processes. This can increase worker productivity, reduce errors, and improve the quality of the final output.
- **Digital Twin:** A digital twin is a computerized replica of a real object or system that is used in smart manufacturing to optimize production processes, predict equipment failure, and improve product design.
- **Automation technology:** The design, construction, and operation of robots to carry out diverse tasks is the field of robotics. Applications for robotics include manufacturing, healthcare, and exploration.
- **Three-dimensional printing and additive manufacturing:** When additive manufacturing, such as 3D printing, is used in smart manufacturing systems, complex components can be produced more cheaply and efficiently than with traditional manufacturing methods, increasing production flexibility and reducing waste.

4.4 TYPES OF COMPUTER 4.0 AND ADVANCED COMPUTING

The program's purpose is to provide a new perspective on the integrated logistics industry by using technologies such as computer vision, virtual reality, AR, and ML to tackle issues. Warehouse Logistics Processes:

- **Module for AR:** The platform's AR solution provides every employee with digital information in their line of sight, such as their task list. Furthermore, workers can see the most efficient route through the storage facility to complete given tasks, saving overall operational time. The AR viewer mechanisms built into the Mad Gaze smart glasses also scan common visual data seen on a container, such as barcodes and quick response codes, as well as specialized logistical markings.
- **Component for virtual reality:** Virtual reality can help in both designing layouts and managing operations to maximize the efficiency of the logistics process. It allows for the seamless integration of digitally produced three-dimensional models into the real world. Using virtual reality, it became feasible to install visualizations of interacting digital content relevant to the project concept, as well as reorganize and monitor the warehouse's physical structure on a 1:1 basis [19].
- **Quantum world:** Quantum computing could revolutionize a number of industries, including finance, chemistry, and encryption. Quantum computers, for instance, can simulate the behavior of molecules and chemical interactions, which may help in the creation of novel materials and medications. Quantum computing could be utilized in finance to maximize investment portfolios and enhance risk management. Quantum computers also have the potential to break many encryption techniques currently used to protect data. Quantum computing has much to offer Industry 4.0, particularly in areas that need complex simulation and optimization [20].

- **Parallel computing:** A computer architecture known as parallel computing uses several processors or threads to accomplish multiple tasks simultaneously. In this paradigm, multiple processes operate simultaneously, and their combined outputs. Work can be done in parallel rather than sequentially, which could shorten the time it takes to complete.
- **Edge AI:** The next wave of AI computing is being driven by edge AI applications, which will enhance people's lives in a variety of contexts, including the workplace, home, school, and car. The use of AI on physical devices is referred to as the "edge." As opposed to keeping all of an organization's data in one location, such a private data warehouse or the data center of a cloud provider, "Edge AI" enables AI computations to be performed at the internet's edge, near to users. Since the Internet is available everywhere, any location might be considered its periphery. Ubiquitous traffic signals, mobile phones, and self-sufficient devices are a couple of examples. It could also include any kind of establishment, such as a factory, store, or medical center. All sizes of businesses aim to increase the number of operations they automate since it boosts productivity [21].
- **Edge computing to accelerate private 5G:** The trend toward cloud computing in manufacturing is consistent with the current movement toward cloud computing in many systems. When the cloud is the center, edge computing refers to being situated outside or at the edge of the cloud. Stated differently, it defines computing systems that operate in close proximity to the consumer, both physically and temporally. Delays can be minimized since different kinds of data are processed close to the user. Telecommunication companies are creating specialized data centers and servers with edge computing features in response to this era of edge computing [22].
- **Neuromorphic computing:** In the past few years, a number of massive neuromorphic systems have emerged, taking advantage of the large supply of transistors now available on a single microchip and, in one case, a whole silicon plate. With the use of scalable designs and technological capabilities, neuromorphic capabilities can be enhanced to support neural networks with millions of neurons and billions of connections. These new capabilities enable modelers to consider representing entire animal brains, ranging from insects to small mammals, as well as important segments of the human nervous system. Additionally, the same technologies provide platforms that can accommodate new cognitive architecture scales. These platforms are currently in use, sometimes on a large scale, yet their full potential is still untapped. However, they already show a wide variety of methods driven by a shared belief in the potential benefits of emerging brain-inspired computational approaches. The massive structures shown below have a mix of competing aims and complementing techniques: The distributed digital neural models that underpin the IBM TrueNorth processor are designed for real-time cognitive applications. Sub-threshold analog neural circuits are used in real-time on the Stanford Neurogrid. To understand biological processes, particularly long-term learning, the Heidelberg Brain Scales

technology operates ten thousand times faster than physiological actual time, leveraging wafer-scale supra limit analogous neural pathways. Once more, with the primary goal of modeling biological nervous systems, the previously mentioned SpiNNaker deployment machine is a real-time digitally multi-core computer that incorporates neurological or synapse simulation software running on smaller embedded processing devices. All these strategies need sacrifices or trade-offs among multiple desired objectives. This includes comparisons between software and devices, analogies verses computational algorithms related to scalability and reconfigurability, energy efficiency, and system density [23,24].

- **Cloud computing:** By 2024, cloud computing will have become a fundamental component of the modern information technology landscape, supporting a wide range of industry-specific operational tasks. According to Gartner forecasts, by 2025, 95% of future workloads are expected to reside in the cloud. In response, companies are adjusting their tactics to capitalize on the cost-effectiveness and flexibility built into cloud infrastructure, which are critical for maintaining competitiveness. Forecasts indicate that global spending on online computing facilities will exceed one trillion dollars by the year 2024, driven by several current trends, including rising demand for innovative platforms and as-a-service approaches, particularly those incorporating machine intelligence. Businesses increasingly view the cloud as a strategic facilitator for fostering innovation, improving agility, and achieving success across various industries, rather than merely as a way to save costs. The upward trend in cloud costs and the increasing uptake of cutting-edge AI technologies like ChatGPT are shaping the future of cloud computing. Companies are proactively migrating to cloud-based solutions to maximize productivity and improve consumer interactions. Additionally, businesses will continue to use cloud services to gain access to cutting-edge technologies, which will improve the overall effectiveness of their processes. By eliminating the need to build or purchase expensive infrastructure, which is essential for supporting these resource-intensive operations, cloud computing accelerates the adoption of new technologies [25].

4.5 APPLICATION OF ADVANCED COMPUTING TECHNIQUE

Advanced computing techniques have significantly impacted the computing field, providing previously unheard-of capabilities and creative potential. Advanced computing approaches are driving substantial breakthroughs across sectors, from quantum computing's potential to revolutionize cryptography to AI-powered decision-making and beyond. Recognizing different forms of AI, the major goals of using AI are to understand and perform intelligent tasks such as reasoning, acquiring new skills, and responding to unforeseen conditions or challenges. As a result, AI is regarded as a branch of study and technology that focuses on mimicking a wide range of tasks and activities associated with human cognitive capacities [26]. Nonetheless, due to the diversity and dynamic nature of the real world, creating an efficient AI model necessitates a significant amount of effort and data. To better understand the concept

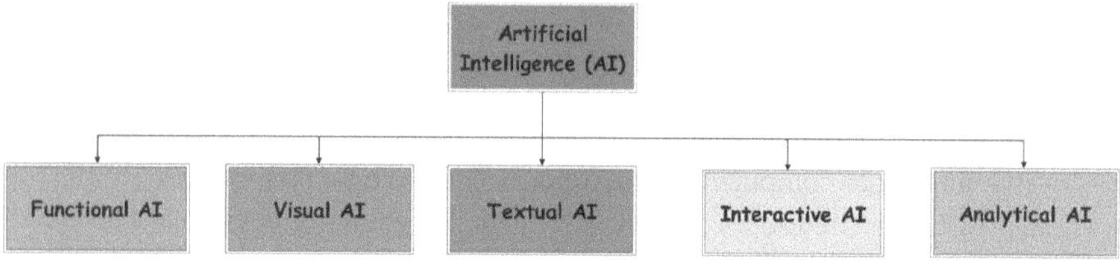

FIGURE 4.4 AI technology and practical operations.

of AI's power, we examine various types of AI, such as analytical, functional, interactive, textual, and visual, and how they can be used to solve problems in the Fourth Industrial Revolution, as shown in Figure 4.4.

- **Visual AI:** Generally speaking, visual AI can identify, categorize, and arrange objects in addition to providing insights from pictures and videos.
- **Textual AI:** Text analytics, often called natural language processing (NLP), allows businesses to leverage text recognition and speech-to-text conversion.
- **Interactive AI:** Interactive AI enables effective communication automation, which is widely used in many aspects of daily life, especially at work.
- **Functioning AI:** Functional AI, like analytical AI, examines enormous amounts of data for patterns and connections.
- **Analytical AI:** Generally speaking, analytics refers to the process of finding, analyzing, and sharing significant data patterns.
- **NLP:** NLP is the application of ML to analyze and comprehend human discourse. It enables speech recognition, machine translation, sentiment analysis, and AI assistants such as Siri and Alexa.
- **Image and video analysis:** Computer vision and other AI technologies enable the analysis and understanding of images and videos. This has applications in facial recognition, object recognition and monitoring, content filtering, medical imaging, and self-driving vehicles.
- **Robotics and automation:** Automation and robotics rely heavily on AI. Robots with AI algorithms can perform complex tasks in manufacturing, healthcare, logistics, and exploration. They can adapt to changing environments, learn from experience, and collaborate with others.
- **Recommendation systems:** AI-powered recommendation algorithms are used to personalize user experiences in e-commerce, streaming video services, and social networking websites. Companies leverage consumer preferences, behavior, and past interactions to suggest appropriate products, films, music, or content.
- **Financial services:** AI is commonly used in the financial services industry for detecting fraudulent transactions, trading algorithms, credit scoring, and risk assessment. ML models can analyze vast amounts of financial data to identify trends and predict outcomes.
- **Healthcare:** AI applications in healthcare include disease evaluation, diagnostic imaging research, pharmaceutical development, personalized

therapy, and patient monitoring. AI can help identify trends in healthcare data and generate predictions to improve treatment and diagnosis.

- **Virtual assistants and chatbots:** AI is driving the interaction. Artificially intelligent assistants and chatbots engage with humans, evaluate their inquiries, and deliver relevant information or instructions. They are employed to provide customer service, collect data, and offer personalized assistance.
- **Gaming:** ML techniques are utilized in gaming to create realistic avatars, hostile behavior, and intelligent decisions. AI is also used to improve game visuals, physics modeling, and gameplay testing.
- **Smart homes and IoT:** AI enables the development of smart home systems that can automate tasks, manage devices, and learn from user preferences. AI can enhance the efficiency and effectiveness of IoT networks and components.
- **Facial recognition:** Aside from individual use, face recognition is a common intelligence technique in many countries, particularly in high-security contexts.
- **Recommendation system:** Many of the technological spaces we use daily, such as online shopping, entertainment websites, social media platforms, and streaming video services like YouTube, utilize recommendation engines to collect user data and provide personalized suggestions to enhance engagement. This AI program is employed in virtually every business.
- **Navigation:** The implementation of GPS provides users with current, accurate, and complete data, hence boosting safety. To enhance the user experience, the system employs convolutional and graph neural network models to automatically detect the number of lanes and road types beneath road barriers. Uber and other transportation businesses rely extensively on AI to increase operational efficiency, monitor traffic, and optimize routes.
- **Robotics:** AI-enabled robots detect obstacles in real time and plan their route accordingly. They can be used to move items around medical, manufacturing, and warehouse facilities, clean workspaces and large machinery, and manage inventories.
- **Human resource:** Robots that use neural networks detect obstacles in real time and arrange their routes accordingly. They are useful for transferring goods in medical, manufacturing, and warehousing contexts, cleaning workplaces and huge machinery, and keeping track of stock levels.
- **Agriculture:** AI-powered applications such as image recognition, automation, and ML are utilized to detect soil problems and nutrient deficits. AI can detect where weeds are growing. AI bots can pick crops more efficiently and quickly than humans.

4.5.1 ML Integration's Advantages for Banks

There are many advantages to integrating ML into banking that have the potential to revolutionize the sector. Among the principal advantages are: Improved Customer Experience: Banks can obtain a comprehensive understanding of their clients by analyzing large amounts of data. This enables the offering of customized experiences and services, such as targeted marketing campaigns, chatbots that provide prompt and accurate customer care, and personalized product suggestions. Banks may better anticipate client needs, increase customer satisfaction, and forge closer bonds with their clientele thanks to ML integration. Better Risk Management: ML algorithms can analyze vast volumes of historical and current data to identify trends, anomalies, and potential hazards. The use of ML integration improves risk management procedures in areas like fraud and credit risk assessment, regulatory compliance, and anti-money laundering (AML). Banks can detect fraudulent activity in real time, make more accurate risk assessments, and more effectively adhere to constantly changing regulations by utilizing ML. Enhanced Operational Efficiency: By automating repetitive and manual operations, ML integration reduces the risk of human error and boosts operational efficiency. Processes such as data entry, fraud detection, document verification, and customer onboarding can all be automated with ML learning algorithms. In addition to saving time and costs, this frees up bank staff members to focus on higher-value tasks like addressing complex problems and offering personalized customer care. Improved Fraud Detection and Prevention: ML models can detect trends and fraudulent actions by analyzing large volumes of data from various sources. Anomalies and patterns that are challenging to identify in conventional rule-based systems can be found using ML methods. Banks can greatly increase their capacity to identify and stop fraud, reducing financial losses and safeguarding clients' funds, by incorporating ML into their current fraud detection systems [27].

- **Grid computing:** Grid computing enables the integration of information assets such as computational power, connection speeds, and storage capacity from several computer systems into a unified system. This integration allows users and applications to quickly access all available resources. A grid user sees only one large virtualized machine, whereas an Internet user receives an integrated version of material on the internet.
- **High-performance computing (HPC):** However, HPC applications differ greatly from standard web and service-based applications. First, HPC typically has rather high computer resource usage. This runs counter to the advantageous characteristics of low average usage and high unpredictability that support the viability of the cloud business model. Additionally, an HPC cloud user typically prefers a dedicated instance, but for a cloud provider, this means fewer opportunities for multi-tenancy and higher prices in order to rent a dedicated computing resource to a single tenant profitably. Our investigation clearly shows that multi-tenancy and virtualization noise can have a major impact on the scalability and predictability of HPC systems [28].

4.5.2 Machines and Deep Learning Techniques for Industrial Applications

Digital technologies and the growth of Industry 4.0 are revolutionizing industries, with AI playing an important role. AI has evolved and is currently employed in a wide range of applications in industrial production and automation. Integrating AI and smart technologies into the industrial ecosystem can provide firms with a competitive advantage. The concept of 4.0 manufacturing automation incorporates distributed control systems or SCADA technology. Smart manufacturing, on the other hand, is the integration of AI, IIoT, and analytics technologies into traditional manufacturing systems to improve automation, quality, process optimization, and cost savings.

4.5.3 Advance Computing in Nanotechnology

Nanotechnology is based on microengineering ideas from physics and materials science. Nanoscaling is not a new notion in the computer business; technologists and technicians have long strived to create customized forms of computer-based technologies that require the least amount of space while performing optimally. As a result, nanotubes are increasingly being studied as an alternative to silicon chips in computer devices. Feynman and Drexler's work influenced computer scientists' efforts to develop breakthrough nano computers capable of providing large-scale advantages. A few years ago, it seemed impossible to imagine laptops, mobile phones, and other essential devices being as compact as they are now [29].

- **Computational nanotechnology:** Computational nanotechnology refers to the application and development of algorithms and computer systems to enhance nanoscience and nanotechnology. To help describe and understand computational nanotechnology, we divided it into five distinct sections, as seen in Figure 4.5. Some proposed subfields focus on computational support for nanoscience and nanotechnology, while other disciplines investigate how nanotechnology findings influence the evolution of computational approaches and tools. Each area will be briefly discussed below.

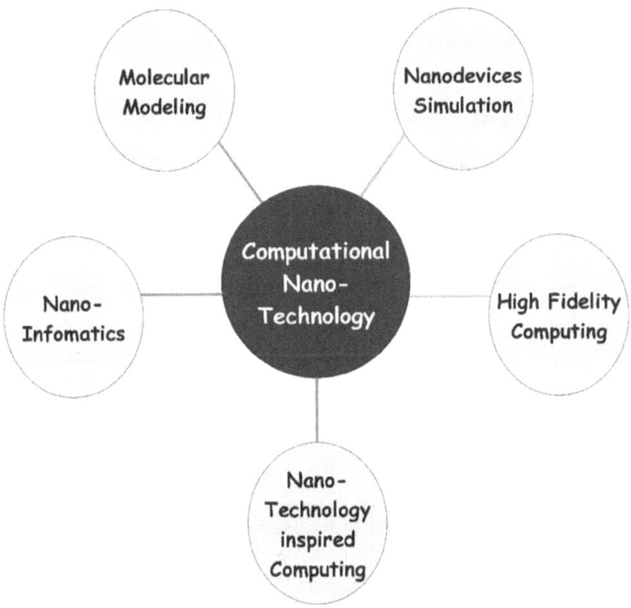

FIGURE 4.5 Computational nano technology.

- **Molecular modeling:** Molecular modelling refers to approaches and tactics for simulating the behavior of molecules or molecular systems. Computational approaches have transformed molecular modelling, enabling previously unachievable calculations.
- **Nanodevice simulation:** These technologies and approaches enable community members to evaluate data and conduct nanotechnology research and development in complete secrecy. These findings can be utilized to create new products, integrate nanotechnology into industrial processes, and assess the environmental, health, and safety consequences of nanomaterial production.
- **Nano informatics:** Scientists can define and analyze potential nanotechnology, nanotechnology devices, and applications using computational models of physical, chemical, and biological processes, as well as increased computer processing capacity. Nano informatics is a new area of study in computational nanotechnology.
- **Computer inspired by nanotechnology:** Researchers should expect nanoscience and nanotechnology to accelerate the development of new computational intelligence capabilities. Nanotechnology has the potential to produce faster CPU architectures than existing ones. This development will facilitate the creation of nanostructure simulators, resulting in important technological advances.
- **High-fidelity computing:** Recent advances in computer science have made computational methods suitable for nanoscience and nanotechnology research. Many simulations, especially those involving molecular modelling, need a large amount of computer power. Field programmable gate arrays provide high-speed calculations by storing them in dedicated hardware.

4.5.4 Approaches to Selecting Vendors in the Age of Industry 4.0

The success of enterprises in the age of Industry 4.0 is based on choosing the proper suppliers. Established approaches are being transformed to incorporate new technology such as data analytics, AI, and blockchain, allowing for smarter data-driven supplier selection procedures [30].

- **Implementing multi-agent theory for intelligent supplier selection:** Industry 4.0 ideals are supported by an innovative and intelligent approach to locating suppliers that employs multi-agent theory. In this approach, each potential provider is viewed as a separate agent acting within a broader network of agents. These agents can collect, process, and exchange information independently, resulting in a cooperative environment for supplier selection. The cornerstone of this method is its capacity to use the influence of intelligent computers that collaborate to investigate and assess supplier features against established standards. The ability of multi-agent theory to gather and disseminate information is one of its main advantages in supplier selection. These agents have access to a wealth of crucial information, such as details regarding price, quality, lead time, digital readiness, and other factors. They can also converse and exchange this knowledge with one another, making decision-making more thorough and thoughtful. The collaborative nature of multi-agent systems is critical in this context.

- **Selecting providers using adaptive neuro-fuzzy techniques:** An innovative method of supplier selection called adaptive neuro-fuzzy methods is made possible by the combination of fuzzy logic and neural networks. Industry 4.0's data-rich and dynamic environment is a perfect fit for this strategy. To create an accurate vendor selection system, fuzzy logic—a mathematical foundation for handling ambiguity and inaccuracy—is combined with neural networks, which are capable of learning and adapting from data. This method uses fuzzy logic to define the characteristics and requirements of each supplier as linguistic variables. This enables the system to manage the inherent uncertainty in the supplier assessment procedure, particularly when working with parameters that are difficult to measure. This approach results in a highly intelligent and adaptable framework that can make decisions about supplier selection in accordance with the changing demands of modern supply chain management. It reduces an organization's reliance on manual and static decision-making procedures, helping them stay competitive in the quickly evolving Industry 4.0 environment. To sum up, the adaptive neuro-fuzzy approach combines the benefits of neural networks with fuzzy logic to effectively address the challenges and intricacies of the digital age, offering a robust solution for selecting suppliers.
- **Using multifaceted mixed decision-making techniques to choose a supplier:** Cost is taken into consideration along with effectiveness, adaptability, technological readiness, environmental responsibility, and a host of other aspects that make it incredibly flexible in response to the complex demands of modern supply chain management. The hybrid MCDM technique differs from other methods in that it can appropriately analyze and weigh these factors, frequently using several decision-making algorithms such as TOPSIS, AHP, and others. By combining different approaches, organizations can lessen the disadvantages of using a single strategy and make decisions on supplier selection that are more reliable and sound. Moreover, the hybrid MCDM technique embraces the information-rich Industry 4.0 environment. It can process and analyze massive volumes of data, fusing historical and modern insights to assist users in making informed vendor selections. The digital transformation of Industry 4.0 and the incorporation of several decision-making approaches facilitate the alignment of vendor selection with the requirements of a state-of-the-art, dynamic supply chain. This method uses both quantitative and qualitative data to improve an organization's capacity to choose significant vendors in a rapidly changing market.
- **Using hybrid fuzzy MCDM methods to choose suppliers:** This data-driven approach ensures that supplier selection will be flexible and responsive to shifts in supplier performance and market conditions. Organizations can make informed supplier decisions that align with the intricacies of a constantly changing, technologically advanced supply

chain environment by combining the advantages of MCDM approaches with fuzzy logic. This methodology provides a more thorough understanding of supplier attributes, enabling the business to make more strategic supplier decisions, and improves an organization's capacity to manage incomplete and ambiguous data.

- **Hybrid fuzzy multiple-criteria decision-making approaches for sustainable source selection:** The objective of the fuzzy technique, commonly referred to as the "green picking method," is to find and work with suppliers who are not only economical and efficient, but also ecologically concerned. Furthermore, this strategy fully utilizes Industry 4.0's data-rich environment. Massive amounts of data, such as details on supplier policies and environmental outcomes, can be handled and evaluated. This data-driven feature ensures that the supplier selection process is flexible enough to adapt to shifting market conditions, legal requirements, and seller sustainability programs.

4.5.5 Acquiring and Data Processing

Predictive repair leverages multiple data sources. Sensor data, historical maintenance logs, environmental data, and even information from external sources like the weather are included. A thorough understanding of the health of the equipment is produced by integrating data from multiple sources [31].

- **Data difficulties:** In a production setting, gathering and organizing data is not without challenges. Unpredictability, incompleteness, and noise can all affect data quality. Building reliable maintenance prediction models requires ensuring the quality and integrity of the data.
- **Methods of pre-processing data:** The initial examination of the data involved using attribute engineering, identification, and standardization techniques. Estimation processes fill in missing data, while feature engineering adds significant features to raw data to improve the predictive power of models, and standardization reduces data to a common range.
- **Real-world information sources:** Vibrating sensors on rotating machinery, temperature sensors in HVAC systems, and pressure sensors in hydraulic systems are examples of useful data sources. These sources offer valuable information regarding the condition of the equipment.

4.5.6 Manufacturing Driven by Data

Information-driven industrialization is a novel approach to manufacturing that leverages insights from massive amounts of real-time data analysis to improve productivity and decision-making. In conclusion, data-driven manufacturing uses AI, ML, and advanced analytics to produce pertinent knowledge, encourage continuous improvement, and increase overall efficiency. Proactive maintenance is an essential part of data-driven regular utilization because it identifies and corrects irregularities and inefficiencies in machinery performance before they result in expensive downtime. By using sensor data and AI systems, manufacturers can accurately track the state and production of critical resources, spot problems, and plan maintenance operations more effectively. By using this budgeting strategy, businesses can reduce maintenance costs, lengthen asset lifespans, improve uptime, and enhance production output and overall equipment effectiveness. Utilizing state-of-the-art analytical and AI capabilities, data-driven production also helps businesses streamline their manufacturing processes and enhance the quality of their products. To identify patterns, behaviors, and opportunities for growth, businesses can examine data from a variety of sources within the manufacturing ecosystem, such as sensors, equipment, and production control systems. These were likely previously unnoticed sources of insight. Businesses can cut waste, manage resources as effectively as possible, and modify production parameters to meet customer demands and quality requirements by employing this data-driven strategy. Early detection and process optimization are made possible by data-driven manufacturing. Moreover, it makes predictive quality control possible, which is the process of detecting and resolving errors and quality variations in real-time throughout the manufacturing cycle [32]. By analyzing sensor data and historical quality data, AI systems can identify patterns and anomalies that might indicate potential quality issues. This enables manufacturers to make necessary adjustments before creating defective products [33].

This proactive approach reduces the likelihood of costly scrap and rework while increasing overall item quality and customer satisfaction. Additionally, data-driven manufacturing helps businesses optimize resource allocation and scheduling of production processes by leveraging state-of-the-art data analysis and ML insights. Manufacturers can better meet customer needs while reducing costs and increasing productivity by optimizing production schedules, minimizing transition times, and strategically allocating resources. Analyzing production data, demand projections, and supply chain dynamics can help achieve this. Businesses may become more responsive to shifting consumer tastes and market conditions by employing this data-driven strategy, which will eventually boost their profitability and competitiveness. Using digital twins—virtual copies of real assets or processes—to replicate and improve production processes in a virtual setting is another facet of data-driven manufacturing. Manufacturers can conduct virtual experiments, simulate different scenarios, and optimize operations without compromising actual output by establishing digital twins of their production lines, industrial machinery, and entire buildings. Manufacturers may reduce risk and accelerate innovation by using this virtual simulation to identify potential bottlenecks, adjust production parameters, and test new ideas before implementing them in the real world [34]. To fully realize the promise of information-driven production, businesses need a team of highly skilled professionals, strong data facilities, and a data-driven culture. Manufacturers need to invest in data-gathering technology such as sensors and IoT devices to obtain real-time data from industrial machines and processes [35]. To manage and analyze enormous volumes of data, they also need to invest in processing and storage infrastructure, such as cloud-based computing and analytics for big data platforms. In addition, producers need to develop and employ state-of-the-art statistical methods and AI to extract meaningful insights from large and diverse data sets. Finally, producers must collaborate with other teams. Businesses must also create an environment where data is used to inform decisions made at all levels of the organization, from the shop floor to the executive office. In summary, real-time statistical insights result in a paradigm shift in the manufacturing process by influencing data-driven production decision-making and optimization. Manufacturers may improve production scheduling and resource allocation, manage equipment maintenance proactively, optimize production processes, and increase product quality by utilizing advanced analytics, AI, and ML. Businesses require advanced analytical tools, robust data infrastructures, and an organizational culture that highly values data in order to reap the benefits of information-driven production [33].

4.6 APPLICATION OF NANOTECHNOLOGY IN DIFFERENT MANUFACTURING SECTORS

Nanotechnology is making remarkable and extensive contributions across a wide range of industries.

4.6.1 Nanotechnology and the Computer Industry

Nanotechnology, stemming from microengineering in physics and material sciences, has revolutionized the computer industry. By replacing silicon chips with nanotubes, inspired by Feynman and Drexler, ultra-thin devices like laptops and mobiles have become possible. Since the early 21st century, nanotechnology has significantly advanced information and communication technology, promoting interconnected communities, economic competitiveness, and environmental sustainability. Innovations like smart sensors, nanochips, memory storage devices, optoelectronics, quantum computing, and lab-on-a-chip technologies are revolutionizing traditional, costly methods into efficient and sustainable solutions. Both private and public spending in nanocomputing is increasing, driving rapid growth in the biotechnology and computing industries. Integrating nanotechnology with informatics and computational industries can address societal problems and achieve sustainable goals. The fourth industrial revolution, based on AI, IoT, robotics, cloud computing, and blockchain, relies on advancements in nanotechnology. Advancements in computing materials through nanotechnology are driving innovation across multiple sectors, including automotive, aerospace, renewable energy, IT, bioinformatics, and environmental management. Sensors using nanomaterials benefit from increased sensitivity and durability, being chemically inert and corrosion-resistant. The integration of semiconductor nanomaterials in quantum computing has dramatically improved processing speeds, accuracy, and data transmission capabilities. These advancements facilitate the development of nano-level components and communication protocols, often referred to as the Internet of Nano Things. This evolving field has potential applications in telecommunications, industry, and medicine. Derived from the concept of IoT, it encompasses sensors, software, and associated technologies enabling widespread communication. The Internet of Nano Things enhances data manipulation, collection, and sharing globally.

Nanotechnology also enhances drones and robotics for maintenance, inspections, transportation, and data analysis. Drones, which are more mobile but human-reliant, and self-operational robots boost industrial automation and precision. This integration, called the Internet of Robotics and Nano Things, improves manufacturing flexibility. Drones equipped with nano sensors enhance detection for safety profiling, security surveys, and mapping, despite speed and legal concerns. Additionally, nanotechnology-based 3D printing streamlines production with minimal infrastructure and waste, enabling customization and sustainability using nanoscale materials.

Nanotechnology in computers is integral to modern industries, connecting electronics, manufacturing, processing, and packaging. Silicon tubes, pivotal in the 20th-century industrial revolution, are being replaced by more efficient nanotubes. Innovations like lab-on-a-chip and nanoscale memory chips enhance storage capacity while minimizing space, benefiting various industries with compact and flexible computing solutions.

4.6.2 Nanotechnology and the Bioprocessing Industry

Nanotechnology is pivotal in advancing the bioprocessing industry, particularly in enhancing the manufacturing of high-quality food and biomaterials. Researchers are exploring how nanomaterials can revolutionize food production by enabling pathogen detection, monitoring food quality, developing biosensor devices, and creating smart, reusable, and biodegradable packaging materials. Nanoencapsulation plays a crucial role in delivering active food compounds efficiently and improving their bioavailability. With global food demands rising alongside concerns for safe and sustainable food production, nanotechnology offers solutions to enhance sensitivity in food quality monitoring and combat food fraud through robust brand protection measures. Moreover, integrating nanotechnology into food production processes can lead to significant reductions in energy consumption, contributing positively to environmental sustainability. These advancements ensure safer, healthier food products and enhance the bioactivity and solubility of nutrients, ultimately boosting human health benefits. However, stringent safety regulations are essential to ensure the safe consumption of nano-enhanced foods, given their direct interaction with the human body.

4.6.3 Nanotechnology and the Agriculture Industry

Agriculture is crucial for global economies and sustains populations by providing essential nutrition. With climate change threatening agricultural stability, proactive measures are essential. Nanotechnology plays a crucial role in modern agriculture by improving the production, processing, storage, packaging, and transportation of agricultural products. It enables precision farming techniques to improve nutrient absorption and detect agricultural diseases more effectively. Nanoclays and zeolites boost fertilizers and restore soil fertility, while smart seeds coated with nano-polymeric mixtures ensure optimal germination. Additionally, nanotechnology is transforming herbicides, pesticides, fungicides, and insecticides, while advancements in poultry and animal husbandry treatments are underway.

4.6.4 Nanotechnology and the Food Industry

Nanotechnology in the food industry encompasses diverse applications such as manufacturing, packaging, safety measures, and targeted drug delivery, as shown in Figure 4.6. Polymer/clay nanocomposites are used in packaging due to their strong barrier properties, while nanoparticle mixtures serve as antimicrobial agents, protecting food from spoilage during storage, especially in canned goods. Nanosensors and nanoassemblies aid in detecting microbes and residues, ensuring food safety. Despite these advancements, concerns about nanoparticle toxicity hinder direct use in food products, necessitating ongoing research. Nanotechnology is rapidly advancing in food packaging, sensing, storage, and antimicrobial applications, including innovations in the dairy industry for processing, shelf life, safety, and nutrient delivery. However, addressing public concerns about the safety of nano-based food products remains crucial for broader market acceptance.

4.6.5 Nanotechnology in the Packaging Industry

The packaging industry is undergoing significant transformation, driven by heightened environmental awareness globally. Key concerns include ensuring food safety and reducing pollution from non-biodegradable plastics. To address these challenges, researchers are investigating eco-friendly biopolymer-based materials processed at the nanoscale, leveraging nanotechnology's capabilities. This method not only improves food safety and production efficiency but also addresses environmental concerns by reducing waste and pollution. Nanotechnology enables the creation of advanced packaging solutions such as anticaking agents, nutraceutical delivery systems, and antimicrobial nanomaterials blended with polymers to preserve food

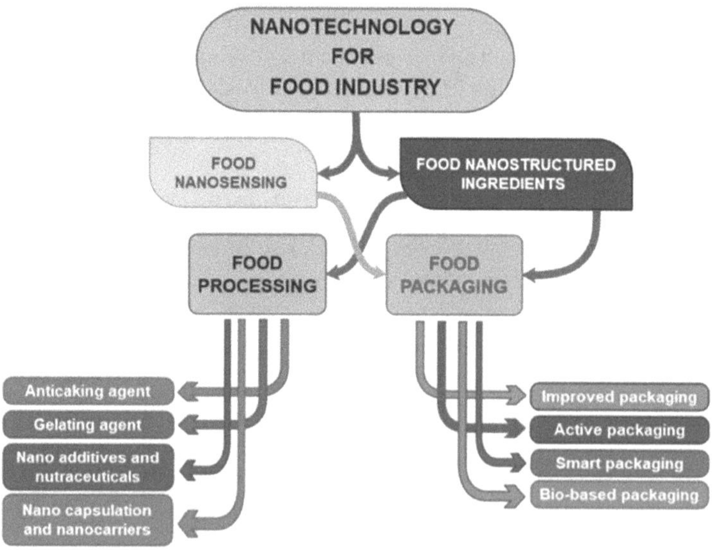

FIGURE 4.6 Applications of nanotechnology in diverse food industries.

quality and safety. Moreover, nanotechnology applications are expanding beyond food packaging to encompass diverse sectors like textiles, leather, and cosmetics, where they enhance functionality and sustainability.

4.6.6 Nanotechnology in Construction and Civil Engineering

Efficient construction practices are increasingly incorporating nanomaterials to advance sustainability in the industry. Nanoparticles are improving cement's strength, durability, and workability, with materials like silicon dioxide now produced at the nanoscale. Modern technologies such as vacuum insulation panels (VIPs) and phase change materials have emerged from these advancements, significantly boosting thermal insulation and energy efficiency in buildings. Nanomaterials in construction offer benefits like lightweight structures, enhanced durability, low maintenance needs, and improved insulation and fire resistance. Nano-SiO_2 in cement enhances compressive strength, pore size distribution, and durability, while advancements in steel and coatings through nanotechnology enhance properties like corrosion resistance and antibacterial capabilities. Nanotechnology is also fostering breakthroughs in materials such as carbon nanotubes for structural reinforcement and titanium dioxide for self-cleaning and germ-repellent surfaces.

4.6.7 Nanotechnology, Healthcare, and the Medical Industry

Nanomedicine, a significant offshoot of nanotechnology, has revolutionized diagnostics, treatment, disease prevention, and healthcare. Advancements in nano-genomics, tissue engineering, and gene therapy are driving progress in drug manufacturing, design, conjugation, and delivery. In bone regeneration, nanotechnology aids bone grafts and muscle restructuring through biomineralization and collagen coatings. Drug delivery is improved using nanorobots for precise targeting and nanoscale surgeries.

Oncology benefits from nano agents, molecular diagnostics, and targeted cancer treatments. Nanomedicine carriers like liposomes and gold nanoparticles enhance tumor imaging and immunotherapy. These advancements promise a precise and technologically advanced healthcare approach through interdisciplinary collaboration.

4.6.8 Nanotechnology Industries and the Environment

Nanotechnology is integral to sustainable development, revolutionizing industries and environmental applications. Its impact on chemistry, physics, biotechnology, and more directly enhances environmental sustainability. Nanomaterials are being increasingly applied in processing raw materials, manufacturing, treating contaminants, and managing soil and wastewater. They also enhance energy storage and hazardous waste management. Developed nations advocate for nanotechnology in addressing environmental issues, including water purification, detoxification of pollutants, and improving material durability. Current research focuses on remediation technologies, especially water treatment, due to global water scarcity. The scientific community must explore nanomaterials' properties for a sustainable future.

4.6.9 Nanotechnology in the Renewable Energy Industry: Focus on Solar Power

Renewable energy is essential for tackling environmental issues, making it a central component of the environmental sector. Nanotechnology significantly enhances the production of solar, hydrogen, biomass, geothermal, and tidal energy. Extensive research is focused on enhancing solar collectors with nanotechnology, driving innovations in various designs such as flat solar plates, direct absorption plates, parabolic troughs, wavy plates, and heat pipes. The use of nanofluids has become common for boosting the efficiency of these systems. However, additional studies are needed to refine the use of nanomaterials in solar panels and to assess both the cost-effectiveness and performance of traditional solar technologies compared to those utilizing nanotechnology.

4.7 CONCLUSION

The overview of cutting-edge computation within the context of the fourth industrial revolution shows how these developments have the potential to revolutionize a wide range of industries. Traditional business structures and operations are changing as a result of the integration of cutting-edge computing techniques in the domains of AI, ML, HPC, and quantum computing. This is resulting in enhanced productivity, efficiency, and creativity. Furthermore, the way organizations run is being completely transformed by the use of advanced computing tools in fields like supply chain optimization, tailored manufacturing, and predictive maintenance. Using these tools, businesses can quickly review enormous amounts of data, see patterns and trends, and make well-informed decisions that boost their competitiveness and growth potential. Industry 4.0's embrace of modern computing is not without its difficulties, though. For many businesses, challenges including employee upskilling, data security and privacy, and integrating new technology into legacy systems continue to be major obstacles. For Industry 4.0 technologies to be implemented successfully, several issues must be resolved. To sum up, Industry 4.0 is primarily driven by improved computing, which is also boosting industry competitiveness, efficiency, and innovation. Businesses that continue to adopt these technologies will be in a better position to take advantage of new development and success prospects and navigate the complexity of today's business environment.

4.8 FUTURE PROSPECTS

Industry 4.0 uses modern computer technologies such as big data analysis, AI, ML, and cloud computing to optimize industrial processes, improve product distribution, and boost overall efficiency. These improvements allow for the development of smart factories, which are equipped with sensors, embedded software, and robotics that collect and analyze data to help make smarter decisions. Combining these contemporary computer technologies yields automation, preventative maintenance, process improvement, and better customer responsiveness. The application of nanotechnology is widely recognized as a key driver of the future Fourth Industrial Revolution.

It entails changing materials at the nanoscale to produce novel materials and technologies with distinct properties and functions. The combination of nanotechnology with modern computers in Industry 4.0 has the potential to result in more intelligent and efficient manufacturing processes, as well as the creation of unique products with improved capabilities.

REFERENCES

1. L. Da Xu, E. L. Xu, and L. Li, Industry 4.0: state of the art and future trends, *International Journal of Production Research*, vol. 56, no. 8, pp. 2941–2962, 2018. https://doi.org/10.1080/00207543.2018.1444806
2. F. J. Folgado, D. Calderón, I. González, and A. J. Calderón, Review of Industry 4.0 from the perspective of automation and supervision systems: definitions, architectures and recent trends, *Electronics (Switzerland)*, vol. 13, no. 4, 2024. https://doi.org/10.3390/electronics13040782
3. A. Adel, Future of industry 5.0 in society: human-centric solutions, challenges and prospective research areas, *Journal of Cloud Computing*, vol. 11, no. 1, 2022. https://doi.org/10.1186/s13677-022-00314-5
4. F. Y. Abdelmajied, Industry 4.0 and its implications: concept, opportunities, and future directions. [Online]. Available: www.intechopen.com. https://doi.org/10.5772/intechopen.102520
5. A. Bilad, M. Zaim, and F. Zaim, Towards a systematic review on Industry 4.0: big data & internet of things, *ITM Web of Conferences*, vol. 46, p. 03004, 2022. https://doi.org/10.1051/itmconf/20224603004
6. M. K. Yadav, R. Kumar, R. K. Ratnesh, J. Singh, R. Chandra, A. Kumar, V. Vishnoi, G. Singh, and A. K. Singh, Revolutionizing technology with spintronics: devices and their transformative applications, *Materials Science & Engineering B*, vol. 303, p. 117293, 2024. https://doi.org/10.1016/j.mseb.2024.117293
7. S. Pathak, N. Kishore, G. Upadhyay, R. K. Ratnesh, and R. Mishra, A compact size planar microstrip-fed patch antenna with hexagonal DGS slot for WLAN application, recent trends in electronics and communication, LNEE, Springer, vol. 777, pp. 263–271, 2021. https://doi.org/10.1007/978-981-16-2761-3_25
8. E. Vyhmeister, and G. G. Castañe, When industry meets trustworthy AI: a systematic review of AI for Industry 5.0, 2022. https://doi.org/10.3389/frai.2024.1429186
9. O. Bongomin, G. Gilibrays Ocen, E. Oyondi Nganyi, A. Musinguzi, and T. Omara, Exponential disruptive technologies and the required skills of industry 4.0, *Journal of Engineering (United Kingdom)*, vol. 2020, 2020. https://doi.org/10.1155/2020/4280156
10. R. K. Ratnesh, M. K. Singh, and J. Singh, Enhancing ZnO/Si heterojunction photodetector performance for ultra high responsivity across wide spectral range, *Journal of Material Science: Materials in Electronics*, vol. 35, p. 756, 2024. https://doi.org/10.1007/s10854-024-12516-5
11. R. K. Ratnesh, M. K. Singha, V. Kumar, S. Singh, R. Chandra, M. Singh, and J. Singh, Mango leaves (Mangifera Indica) derived highly florescent green graphene quantum dot nanoprobes for enhanced on-off dual detection of cholesterol and Fe^{2+} ions based on molecular logic operation, *Biomaterials Science & Engineering, American Chemical Society*, vol. 7, pp. 4417–4426, 2024. https://doi.org/10.1021/acsabm.4c00292
12. B. Bhaskaran, B. Babu, J. D. Nallasivam, and S. Aravinth, Overview of Industry 4.0, 2023. [Online]. Available: https://www.researchgate.net/publication/372371011
13. V. Alcácer, and V. Cruz-Machado, Scanning the Industry 4.0: a literature review on technologies for manufacturing systems, *Engineering Science and Technology, an International Journal*, vol. 22, no. 3, pp. 899–919, 2019. https://doi.org/10.1016/j.jestch.2019.01.006

14. L. Duan, and L. Da Xu, Data analytics in Industry 4.0: a survey, *Information Systems Frontiers*, 2021. https://doi.org/10.1007/s10796-021-10190-0
15. R. Rai, M. K. Tiwari, D. Ivanov, and A. Dolgui, Machine learning in manufacturing and industry 4.0 applications, *International Journal of Production Research*, vol. 59, no. 16, pp. 4773–4778, 2021. https://doi.org/10.1080/00207543.2021.1956675
16. P. Garg, T. Chandra, R. Ahlawat, N. Mittal, R. K. Ratnesh, and S. K. Tripathi, Star Galaxy Image Classification Via Convolutional Neural Network, ICOSEC-2022, IEEE, 2022, pp. 1156–1168. https://doi.org/10.1109/ICOSEC54921.2022.9952065
17. G. Tsaramirsis, A. Kantaros, I. Al-Darraji, et al., A modern approach towards an Industry 4.0 model: from driving technologies to management, *Journal of Sensors*, vol. 2022, 2022. https://doi.org/10.1155/2022/5023011
18. J. Posada, I. Barandiaran, J. R. Sánchez. et al., Computer graphics and visual computing use cases for Industry 4.0 and Operator 4.0, *International Journal for Simulation and Multidisciplinary Design Optimization*, vol. 12, 2021. https://doi.org/10.1051/smdo/2021026
19. M. Javaid, A. Haleem, R. P. Singh, R. Suman, and E. S. Gonzalez, Understanding the adoption of Industry 4.0 technologies in improving environmental sustainability, *Sustainable Operations and Computers*, vol. 3, pp. 203–217, 2022. https://doi.org/10.1016/j.susoc.2022.01.008
20. M. Di Capua, A. Ciaramella, and A. De Prisco, Machine learning and computer vision for the automation of processes in advanced logistics: the integrated logistic platform (ILP) 4.0, *Procedia Computer Science*, pp. 326–338, 2022. https://doi.org/10.1016/j.procs.2022.12.228
21. S. Umbrello, Quantum technologies in Industry 4.0: navigating the ethical frontier with value-sensitive design, *Procedia Computer Science*, pp. 1654–1662, 2024. https://doi.org/10.1016/j.procs.2024.01.163
22. G. Lei, H. Yang, M. Li, and J. Wang, Advanced computing technology and standardization research, In *E3S Web of Conferences*, EDP Sciences, 2021. https://doi.org/10.1051/e3sconf/202123602026
23. International Advanced Research Centre for Powder Metallurgy and New Materials (ARCI). https://www.arci.res.in/.
24. S. Furber, Large-scale neuromorphic computing systems, *Journal of Neural Engineering*, vol. 13, no. 5, 2016. https://doi.org/10.1088/1741-2560/13/5/051001
25. S. R. Kothiyal, R. K. Ratnesh, and A. K. Shukla, Enhancing sensing with surface plasmons resonance: theoretical insights and design strategies in photonic crystal fiber, In *International Conference Artificial Intelligence and Information Technologies, (ICAIIT)-2024*, Taylor & Francis, 2023, pp. 279–285. https://doi.org/10.1201/9781003510833-46
26. H. Rehan, Revolutionizing America's cloud computing the pivotal role of AI in driving innovation and security, *Journal of Artificial Intelligence General Science (JAIGS)*, vol. 2, no. 1, pp. 239–240, 2024. https://doi.org/10.60087/jaigs.v2i1.p208
27. H. Sarker, AI-based modeling: techniques, applications and research issues towards automation, intelligent and smart systems, *SN Computer Science*, vol. 3, no. 2, 2022. https://doi.org/10.1007/s42979-022-01043-x
28. J. Oluwaseyi, and A. Luz, Seamless integration of ML with existing systems is crucial for successful adoption in banking. [Online]. Available: https://www.researchgate.net/publication/379534479
29. A. Gupta, L. V. Kale, F. Gioachin, et al., The who, what, why, and how of high performance computing in the cloud, In *Proceedings of the International Conference on Cloud Computing Technology and Science, CloudCom*, IEEE Computer Society, 2013, pp. 306–314. https://doi.org/10.1109/CloudCom.2013.47
30. S. Malik, K. Muhammad, and Y. Waheed, Nanotechnology: a revolution in modern industry, *Molecules*, vol. 28, no. 2, 2023. https://doi.org/10.3390/molecules28020661

31. S. K. Sahoo, S. S. Goswami, and R. Halder, Supplier selection in the age of Industry 4.0: a review on MCDM applications and trends, *Decision Making Advances*, vol. 2, no. 1, pp. 32–47, 2024. https://doi.org/10.31181/dma21202420
32. Garg, and R. K. Ratnesh, Design and simulation of GaAs/InP and Si/SiC heterojunction solar cells, *ICCCES-2022, Lecture Notes in Electrical Engineering (LNEE)*, Springer, 2022, pp. 867–875. https://doi.org/10.1007/978-981-19-7753-4_66
33. M. Di Capua, A. Ciaramella, and A. De Prisco, Machine learning and computer vision for the automation of processes in advanced logistics: the Integrated Logistic Platform (ILP) 4.0, In *Procedia Computer Science*, Elsevier B.V., 2022, pp. 326–338. https://doi.org/10.1016/j.procs.2022.12.228
34. S. Kolasani, *International Journal of Sustainable Development through AI, ML and IoT Unleashing Exponential Intelligence: Transforming Businesses through Advanced Technologies*, [Online]. Available: https://ijsdai.com/index.php/IJSDAI/index
35. R. Kumar, R. K. Ratnesh, J. Singh, A. Kumar, and R. Chandra, IoT-driven experimental framework for advancing electrical impedance tomography, *ECS Journal of Solid State Science and Technology*, vol. 13, no. 2, p. 027002, 2024. https://doi.org/10.1149/2162-8777/ad2331

5 Synergizing Nanotechnology with Biomedical and Biotechnology for Healthcare

Yashaswi Sharma, Priyanshu Rajput, Anamika Brahma, Tanya Goel, Aarushi Singh, Jitender Madan, Ratneshwar Kumar Ratnesh, Ramesh Chandra, and Snigdha Singh

5.1 INTRODUCTION

Nanotechnology is a dynamic and rapidly advancing discipline that involves manipulating matter at incredibly small scales, typically spanning from 1 to 100 nm.[1] At this scale, materials display exceptional properties that can markedly differ from those of their bulk counterparts.[2] Synthesizing nanomaterials involves a wide range of approaches and techniques tailored to achieve specific characteristics and applications.[3] Some of the discussed methods include gas phase evaporation, matrix isolation technique, metal vapour–solvent co-deposition, sol–gel processing, solvothermal and hydrothermal methods, and wet chemical reduction.

Nanotechnology has numerous industrial applications, including electronics, medicine, energy, environmental science, and materials science.

- Nanotechnology has revolutionized electronics by facilitating the fabrication of electronic devices that are smaller, faster, and more efficient, such as transistors and memory chips.[4,5]
- In medicine,[6] nanotechnology is revolutionizing drug delivery, diagnostics, imaging,[7] and regenerative medicine. Nanoparticles (NPs) can be functionalized to target particular tissues or cells within the body, thereby enhancing efficacy and reducing treatment side effects.
- Nanotechnology plays a crucial role in advancing energy technologies. Nanomaterials are being used for the development of more efficient solar cells,[8] enabling the development of energy storage devices[9] and catalytic systems for renewable energy production.[10,11]

DOI: 10.1201/9781003634553-5

In material sciences, nanotechnology allows for the creation of novel materials with better electrical, mechanical, and optical properties, finding applications in the automotive,[12] aerospace,[13] and construction industries, among others.[14]

Nanotechnology offers myriad promising applications in the biomedical industry, transforming multiple dimensions of healthcare.[15] It enables targeted delivery of drugs, early detection of diseases, tissue engineering, and medical monitoring at the molecular level.[16,17] NPs can be designed as specialized drug carriers to effectively treat cancer cells and tissues in the body, thereby decreasing the adverse effects of traditional medications and improving therapeutic outcomes. They can be engineered to transport therapeutic drugs to specific cancer cells while bypassing healthy cells, thereby minimizing side effects. They can also be utilized as contrast agents in imaging techniques such as magnetic resonance imaging (MRI), computed tomography scans, and ultrasound. For instance, FeO NPs served as contrast agents for MRI, magnetic particle imaging, and ultrasonic techniques; zinc oxide (ZnO) nanowires are employed for imaging cancer cells, and TiO2 NPs are useful in bone regeneration. Functionalized NPs possess specific targeting abilities for tissues or biomolecules, thus enhancing image resolution and sensitivity. Nanotechnology enables the advancement of biosensors with high sensitivity and selectivity for detecting biomarkers related to diseases.[18] Nanomaterials like quantum dots (QDs), carbon nanotubes (CNTs), and gold NPs (GNPs) can be integrated into sensor platforms for rapid and accurate diagnostics.[19] Nanotechnology is pivotal in tissue engineering and advancements in medicine as it furnishes scaffolds with nanostructured surfaces that mimic the extracellular matrix. These surfaces facilitate cell adhesion, proliferation, and differentiation, which are essential processes for tissue formation and regeneration.[20] NPs can also be engineered to selectively target cancer cells while sparing healthy tissues, thereby enhancing the efficiency and mitigating the adverse effects of conventional cancer treatments such as chemotherapy, radiation, and photodynamic therapy. NPs also function as efficient carriers for delivering nucleic acids (deoxyribonucleic acid (DNA) or ribonucleic acid (RNA)) into cells for gene editing, gene silencing, or gene therapy applications. They protect these nucleic acids from degradation and facilitate their uptake by target cancer cells.[21]

Three methods are generally utilized for synthesizing nanomaterials: physical, chemical, and green approaches (Figure 5.1). The physical methods of synthesizing NPs include mechanical mining, lithography, laser ablation, and electrospinning. These methods are suitable when purity is the top priority. Additionally, physically synthesized NPs provide high yield, homogeneity, minimal chemical contamination, and controllable particle sizes. Some important physical methods to synthesize nanomaterials are shown in Figure 5.2. However, these methods require extremely high energy, involve complex equipment, demand high maintenance, and often result in environmental toxicity due to the formation of waste by-products during the process. Secondly, numerous chemical methods exist (Figure 5.3), such as the sol–gel method, chemical vapour deposition (CVD), hydrothermal methods, solvent-based synthesis, and others. Chemical methods enable the creation of a diverse array of nanomaterials, offering meticulous control over their characteristics, thus rendering them indispensable across multiple fields such as electronics, catalysis, biomedical

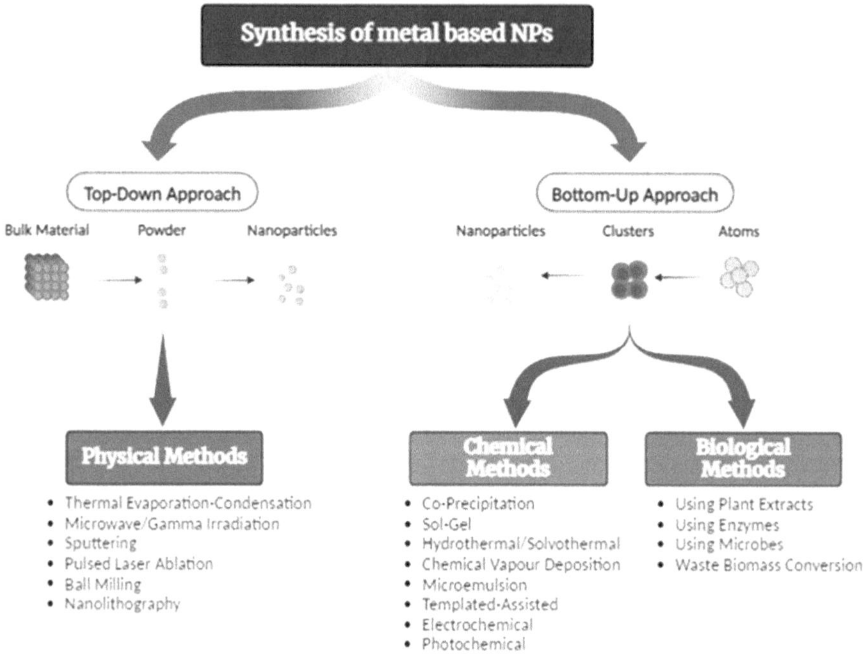

FIGURE 5.1 Different methods of synthesizing metal-based NPs.

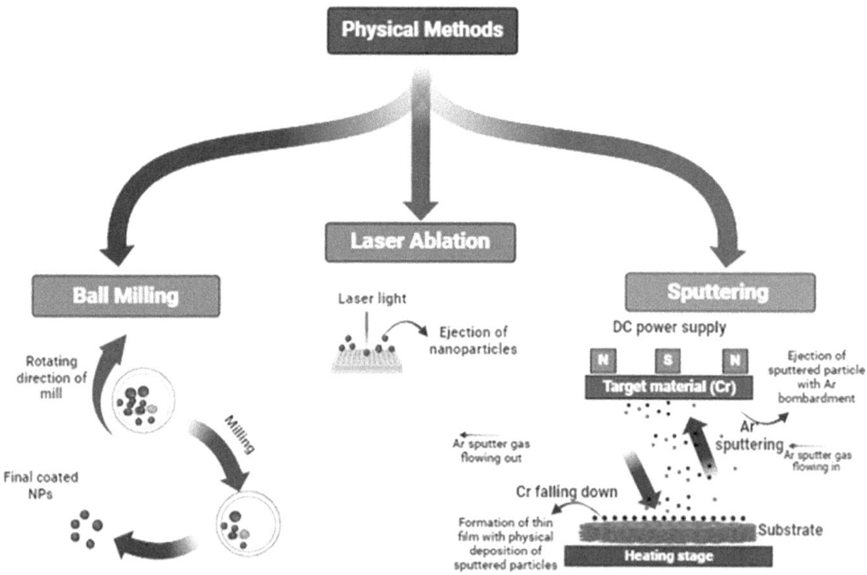

FIGURE 5.2 Some physical methods of synthesizing metal-based NPs.

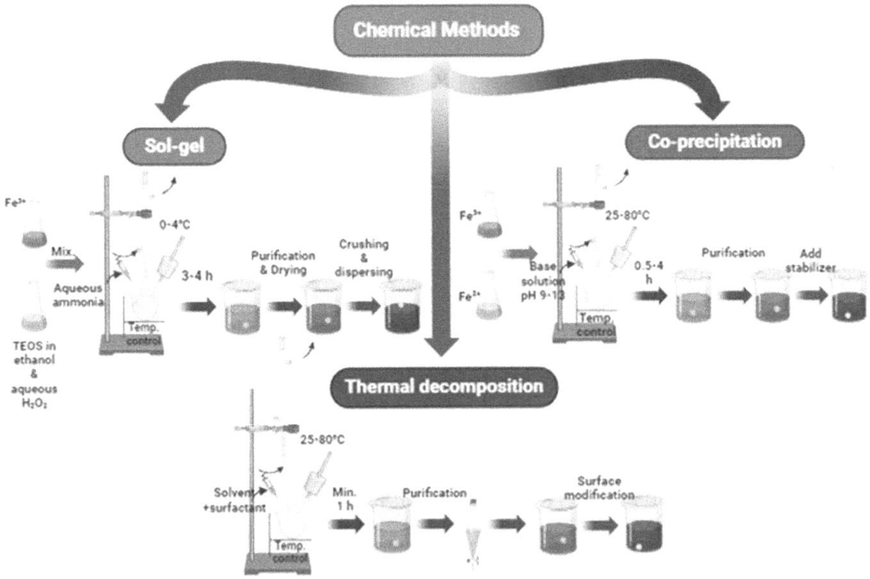

FIGURE 5.3 Some chemical methods of synthesizing metal-based NPs.

engineering, and energy conversion/storage. Despite the numerous advantages that chemical synthesis offers, there are certain drawbacks that make these techniques less desirable in the modern era. Some disadvantages include difficulty in controlling size and shape, contamination from solvents, surfactants, stabilizers, etc., intensive energy and resource requirements, lengthy synthesis (up to days), and safety concerns. Considering the obstacles, biological or green synthesis is the most preferred approach. Several biological elements from the environment are used in this process, including enzymes, bacterial strains, and plant extracts from roots, leaves, flowers, bark, stems, among others.

Green methods have a significant advantage over other methods of nanomaterial synthesis. They are inexpensive, safer for the environment, non-toxic, simple to synthesize from biodegradable materials, reproducible, easy to manage in terms of design and size, and offer versatility and diversity in the development of NPs. This is the primary rationale for why the green approach is frequently accepted and utilized for creating NPs with any desired characteristics.

5.1.1 Physical Methods of Synthesis

5.1.1.1 Thermal Evaporation–Condensation

This is a top–down approach used to synthesize metal and alloy NPs. It involves the thermal evaporation of the target in an electron beam evaporation device. The challenge with this method is that non-uniform heating leads to the formation of non-homogeneous NPs.

5.1.1.2 Microwave/Gamma Irradiation

In this method, microwave (MW) radiation is used to heat and evaporate the target material. The difference lies in the fact that MW/gamma radiation provides uniform heating, leading to the generation of homogeneous NPs. Hasanpoor et al. synthesized ZnO NPs (zinc oxide NPs) using MW irradiation.[22]

5.1.1.3 Sputtering

This is also a top–down method of synthesizing NPs, where high-energy particles like Ar ions bombard a solid surface, causing surface erosion of the solid and formation of NPs. Ayyub et al. studied the production of NPs composed of metals and metal oxides. They examined how size and orientation are mainly affected by sputtering power, temperature, pressure, nature, and rate of flow of the sputtering gas.[23]

5.1.1.4 Pulsed Laser Ablation Method

In this method, a pulsed laser beam is directed on the target material to ablate NPs. The properties of synthesized NPs depend upon the properties of the laser used during the process.

5.1.1.5 Mechanical Method

The mechanical method, also called the ball milling method, is a top–down approach to synthesize NPs using mechanical energy. It involves crushing solids using steel, SiC, or WC balls contained in stainless steel. To achieve uniformity, the crushing steps last for about 150 hours, or roughly 6.5 days. Although it is an effortless and economical process, the shape and purity of the synthesized NPs are challenging to control.

5.1.1.6 Nanolithography Method

This technique mainly involves thin film deposition, self-assembly and self-organization. In this method, a beam of electrons, ions or light is focused on the target material to produce the NPs.[24]

5.1.2 Chemical Methods of Synthesis

5.1.2.1 Precipitation

In the precipitation method, soluble precursors are mixed in a solvent to form a supersaturated solution. Nucleation and NP formation are triggered by the addition of a precipitating agent. Benefits include simplicity, scalability, and the ability to generate a broad range of nanomaterials.

5.1.2.2 Sol–Gel Synthesis

Sol–gel synthesis involves the conversion of a precursor solution into a solid network via hydrolysis and condensation reactions. This technique offers good control over the morphology, particle size, and surface properties of metal oxides such as silica, titania, and alumina NPs.[25]

5.1.2.3 Hydrothermal and Solvothermal Method

Hydrothermal and solvothermal synthesis methods involve the reaction among precursors in organic and aqueous solvents at high pressure and temperature. Such conditions promote brisk nucleation and crystallization, resulting in the formation of nanomaterials with controllable size, shape, and crystallinity.

5.1.2.4 CVD Method

CVD techniques entail depositing QDs vapour-phase reactants onto a substrate, where they undergo reactions to generate thin films and coatings. CVD is commonly used in semiconductor nanomaterials, such as graphene, CNT, and semiconductor QDs, with accurate control over composition, film thickness, and crystal structure.[26]

5.1.2.5 Microemulsion Synthesis

Microemulsion synthesis utilizes surfactant-stabilized emulsions to facilitate the growth of NPs. This method allows control over the size, shape, and surface properties of NPs. It is suitable for producing NPs with precise size distributions and high uniformity.

5.1.2.6 Template-Assisted Synthesis

Template-assisted synthesis involves the use of templates, such as nanoporous membranes or molecular scaffolds, that control the shape, size, and structure of synthesized NPs. This technique enables the development of complex nanostructures, including nanowires, nanotubes, and hierarchical assemblies, with precise dominance over dimensions and properties.

5.1.2.7 Electrochemical Synthesis

Electrochemical methods involve the reduction or oxidation of metals at electrode surfaces to generate NPs. This approach allows exceptional control over particle size, morphology, and composition and is extensively applied for the synthesis of metal and their metal oxide NPs, as well as nanomaterials for energy storage and conversion applications.

5.1.2.8 Photochemical Synthesis

Photochemical methods utilize light energy to initiate chemical reactions for nanomaterials synthesis. Photocatalytic processes, such as photoreduction and photodeposition, enable the preparation of semiconductor NPs and nanocomposites with controlled properties and enhanced photocatalytic activity.

5.1.3 Green Methods of Synthesis

5.1.3.1 Plant-Mediated Synthesis

Plant extracts, such as those from leaves, stems, roots, or fruits, are also used as stabilizing and reducing agents in the development of NPs. Phytochemicals found in the extracts, such as flavonoids, polyphenols, and terpenoids, facilitate the reduction of metal ions to form NPs. This is an eco-friendly, cost-effective method that enables the development of metal and metal oxide NPs.

5.1.3.2 Microbial Synthesis

Microorganisms like fungi, algae, bacteria, and yeast facilitate the development of NPs through intra- or extracellular reduction processes. Microbes produce enzymes and metabolites that can easily reduce metals present as ions and facilitate NP formation. Numerous benefits of microbial synthesis include increased yield, more favourable reaction conditions, and the potential for large-scale production.

5.1.3.3 Bioinspired Synthesis

To create NPs, bioinspired synthesis strategies imitate the inherent mechanisms and processes found in living organisms. Biomolecules such as proteins, peptides, DNA, or RNA can act as templates, scaffolds, or directing agents in the development of NPs. These biomolecules enable the creation of complex nanostructures with distinctive features and offer fine control over size, shape, and assembly.

5.1.3.4 Enzymatic Synthesis

Enzymatic synthesis methods utilize enzymes as biocatalysts to reduce metal ions and aid the generation of NPs. Enzymatic synthesis offers advantages such as excellent specificity, moderate reaction conditions, and biocompatibility. This method is preferably applied in biomedicine, catalysis, and environmental remediation.

5.1.3.5 Waste Biomass Conversion

Waste biomass materials, such as agricultural residues, food waste, or industrial byproducts, can be utilized as precursor sources for NP synthesis. These materials consist of organic compounds that serve as carbon sources, reducing agents or for the formation of carbon-based nanomaterials, like CNTs, graphene, or biochar. Waste biomass conversion not only reduces waste generation but also adds value to underutilized resources.

Green synthesis, which includes the use of different bioactive agents like microorganisms, plant materials, enzymes, and biomolecules, (Figure 5.4) provides a sustainable approach to NP synthesis. These methods rely on the use of natural starting materials, simpler reaction conditions, and almost no environmental toxicity. This makes the green approach superior to traditional methods and provides solutions to the limitations of pre-existing methods of nanomaterial synthesis

5.2 CHARACTERIZATION OF NPS

The subsequent step after synthesis is characterization. It involves the use of various techniques to study different physical and chemical properties, such as size, shape, composition, and functionality of the NPs. The techniques include light scattering, electron microscopy, and spectroscopic methods of analysis. Some commonly used characterization techniques and their uses are listed in Table 5.1.

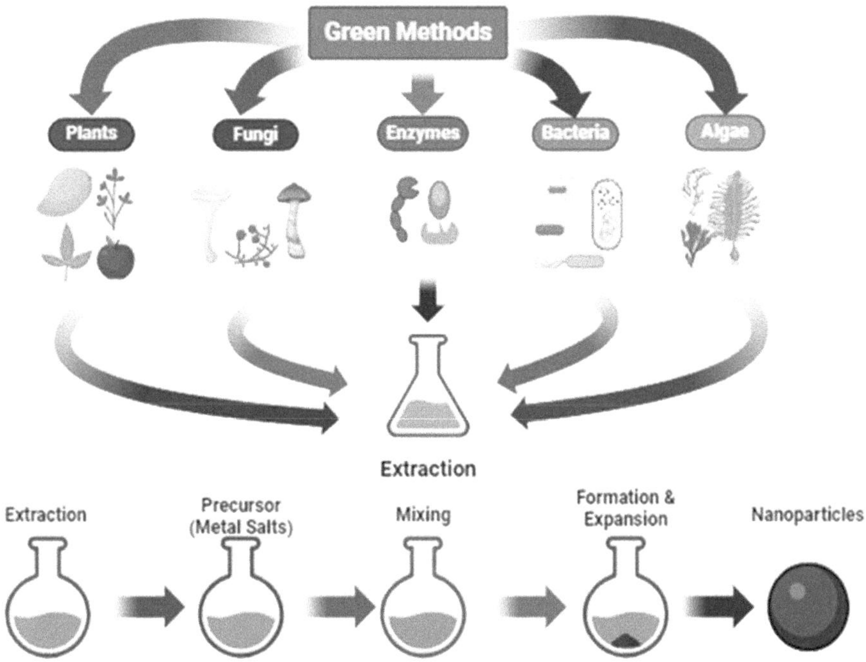

FIGURE 5.4 Some green methods of synthesizing metal-based nanoparticles.

TABLE 5.1
Nanoparticles' Characterization Techniques

S.No.	Characterization Technique	Property Analysed
1	Static light scattering	Weight, size, morphological structure
2	Dynamic light scattering	Size
3	Scanning electron microscopy	Size, shape, texture
4	Transmission electron microscopy	Size, shape, composition, distribution
5	Powder X-ray diffraction	Crystal structure, composition
6	UV–visible	Size, shape, optical properties
7	Fourier transform infrared spectroscopy	Composition, structure

5.2.1 Static Light Scattering

It is a technique mainly used for the determination of the weight and size of molecules. Static light scattering involves the interaction of particles with the electromagnetic field of light, leading to the scattering of light in different directions.

5.2.2 Dynamic Light Scattering

This technique is applied to determine the hydrodynamic size of NPs. In this method, the time scale of Brownian motion is measured by illuminating the sample with a beam of laser light, and the waves of scattered light exhibit constructive and destructive interference due to the random diffusion of particles.

5.2.3 Scanning Electron Microscopy (SEM)

This technique is used to measure the size, shape, and texture of NPs. The advantage is that it only requires a small quantity of the sample. In SEM, scanning across the sample is performed using a fine beam of electrons.

5.2.4 Transmission Electron Microscopy

It is another effective method for analysing the size, shape, and distribution of NPs. Compared to SEM, transmission electron microscopy has a higher resolution and uses transmitted electrons for detection, rather than secondary and backscattered electrons as in SEM.

5.2.5 Powder X-Ray Diffraction

Powder X-ray diffraction is necessary to provide data about the crystal phase, lattice constant, orientation, and average size of the particles. It is helpful in the synthesis of bimetallic NPs, confirming the alloy structure of the NPs.

5.2.6 UV–Visible Spectroscopy

The optical analysis of NPs is conducted using this technique. It also provides information about the electronic structure of the particles by relating the absorption bands to the aspect ratio and diameter of metal NPs. This technique involves the absorption of UV or visible light by the sample, resulting in the production of distinct spectra.

5.2.7 Fourier-Transform Infrared Spectroscopy

This technique is used to analyse the composition and structure of molecules. The infrared radiation influences the vibrational and rotational modes of the materials, which are measured using Fourier transform infrared spectroscopy.

5.3 NANOTECHNOLOGY IN BIOTECHNOLOGY

Nanotechnology is applied in biotechnology to create nanodevices for biological studies at the nanoscale[27] or to explore various ways of studying and detecting biological systems.[28,29] The nanodevices can detect differences at the single-cell level.

The NPs are also employed in chemotherapy,[30] immunotherapy,[31] gene therapy,[32,33] and more.[34] Furthermore, nanotechnology is used for selectively identifying illnesses in the body or specific cells.[35] Here are some key applications of nanotechnology in biotechnology:

5.3.1 Biosensors

Biosensors are analytical devices designed to detect and quantify biological or biochemical reactions within living systems. They operate by generating and detecting electronic signals triggered by the interaction between the targeted analyte and a biological recognition substance. Nanotechnology-based biosensors, or nanosensors, are developed to enhance the capabilities of biosensors and enable the rapid and precise measurement of signals at the nanoscale.[36] They are employed for sensing water quality, determining drug content in foods, and detecting pesticides and other contaminants in river waters. Some nanomaterials commonly applied in nanobiosensors include CNTs, graphene, QDs, and chitosan, as shown in Figure 5.5. For example, CNTs are extensively utilized in biosensing, tissue engineering, diagnostics, and drug delivery. Their hollow shape allows them to act as enhanced carriers for drugs and other biomolecules. They've been utilized to detect sugars, proteins, enzymes, and cancer cells. Graphene, on the other hand, finds application in the advancement of biosensors for the detection of analytes, ions, biomolecules, and neurotransmitters. Their availability, along with their exceptionally high sensitivity and specificity,

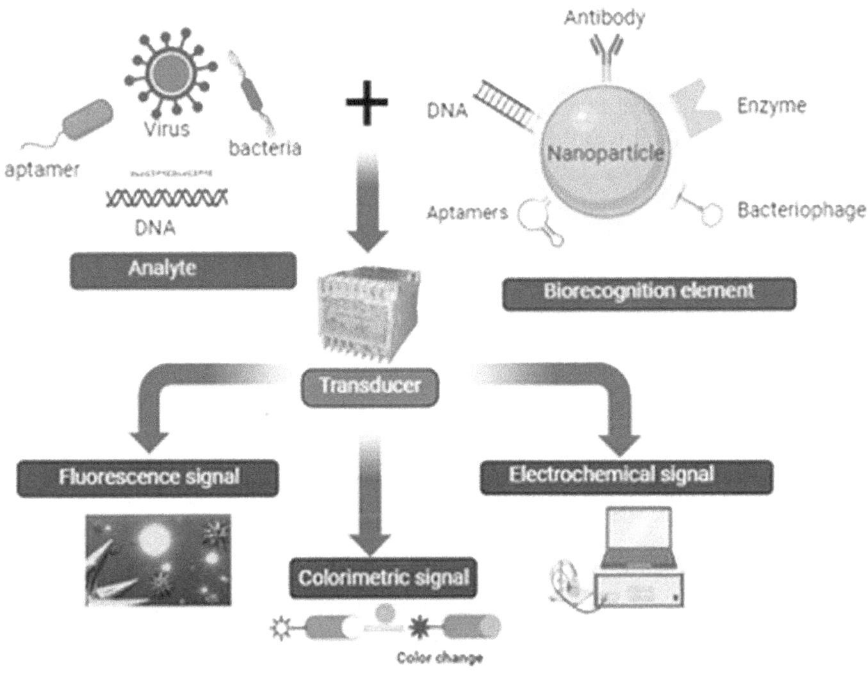

FIGURE 5.5 Biosensing application of metal-based nanoparticles.

makes them useful for the identification of cancer cells. As discussed in section 5 (b), chitosan-based nanomaterials are highly specific for their use and harmless to the environment, making them useful in biosensing and diagnostics applications.[37]

5.3.2 Nanofluidic Devices

Microfluidic devices are engineered to manipulate and control tiny volumes of fluids at the microscale level. These devices have the ability to perform multiple operations concurrently within the same platform. With the development of nanoelectrochemical systems and nanofluidics, a deeper field within microfluidics has emerged, offering more advanced capabilities. These advancements include the ability to stretch DNA for detecting variations in single sequences, mapping genomes, sequencing DNA/RNA, and separating biopolymers.[38]

5.3.3 Gene Therapy

In recent years, nanotechnology has sparked a revolution in the healthcare industry. Gene therapy and nanobiotechnology typically work together to develop novel and more effective therapeutic treatments for a range of diseases.[39] Gene therapy entails introducing DNA molecules into the patient's cells to substitute the altered genes. Nanobiotechnology facilitates the rapid and selective delivery of particular genes to the desired cells, as highlighted in Figure 5.6. Moreover, it prevents the degradation of DNA molecules and facilitates their penetration into the cell membrane.[40] In gene therapy, polymeric,[41] and micellar[42] NPs are utilized due to their ability to self-assemble. Many diseases like cancer, tumours, and genetic disorders can be effectively treated through nanotechnology in gene therapy. For example, in Alzheimer's disease treatment, siRNA delivery is performed using nanoscale delivery systems which improves the efficacy of traditional gene therapy methods.[43]

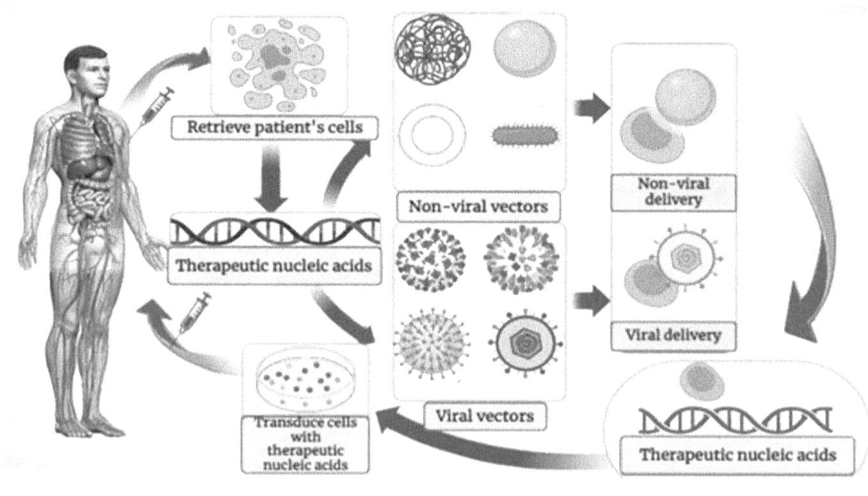

FIGURE 5.6 Gene therapy application of metal-based nanoparticles.

5.3.4 Nano Diagnostics

There is a constant and urgent need to improve traditional diagnostic techniques to overcome new challenges and enhance the efficacy of existing treatment methods. Nanotechnology can effectively be used to design and develop a range of methods for the synthesis, delivery, and detection of biomolecules. There are three categories in nano diagnostics: 'chip-based devices', 'DNA nanomachines' (DNA NMs).[44,45] Chip-based devices show very high sensitivity and provide amplification-free diagnosis. DNA nanomachines, with their random structures and controllable motion behaviour, encompass DNA tweezers,[46] walkers,[47] and robots[48] that exhibit high potential in diagnostics. Random DNA NMs are applied in biorecognition and signal transduction.[49] In addition, nanotechnology is also applied in other areas like immune response modulation and drug delivery.[50]

5.4 MECHANISM OF THE ACTION OF NP INVOLVED IN BIOMEDICAL APPLICATIONS

Every activity that the NPs engage in follows a specific mechanism. Several theories are proposed for the activities of distant NPs, despite the fact that the precise mechanism of action is not well understood. The general mode of action involves penetration into the membrane, rupturing the target's cell wall or membrane and other essential components present in the target cell, ultimately causing cell death. NPs are capable of generating reactive oxygen species (ROS) inside the cell, causing oxidative stress to the cell structure, denaturation of DNA, and lipid peroxidation, eventually resulting in cell death. ROS are reactive molecular species that have a tendency to oxidize biomolecules and their cellular structures.[51] The action mechanisms of some activities exhibited by NPs are briefly discussed below:

5.4.1 Anticancer

NPs can cause cancer cells to undergo apoptosis through a series of mechanisms. ROS-mediated apoptosis in cancer cells is the most commonly employed method. Other mechanisms include upregulation and downregulation of proteins, transcription inhibition, and immunological interventions.[52]

The primary mechanism of action for NPs is ROS-induced apoptosis, which is the most extensively studied mechanism to date. Usually, it acts like a double-edged sword, having both pro-apoptotic and anti-apoptotic effects. Anti-apoptotic effects include cell promotion, invasion, proliferation, and metastasis, whereas pro-apoptotic effects contribute to cell cycle arrest, apoptosis, and necrosis.[53] Proteins like the agile histidine triad have the capability to induce ROS-mediated apoptosis.[54] NPs are more effective apoptotic inducers due to their nanoscale size and higher reactivity resulting from a larger surface area, which results in increased ROS formation. This ROS formation leads to increased oxidative stress, followed by DNA and protein damage, ultimately leading to cell death.[55]

Upregulation and down regulation of proteins occur when cells mimic a cellular response to a stimulus. Under oxidative stress, protein regulation increases, leading to

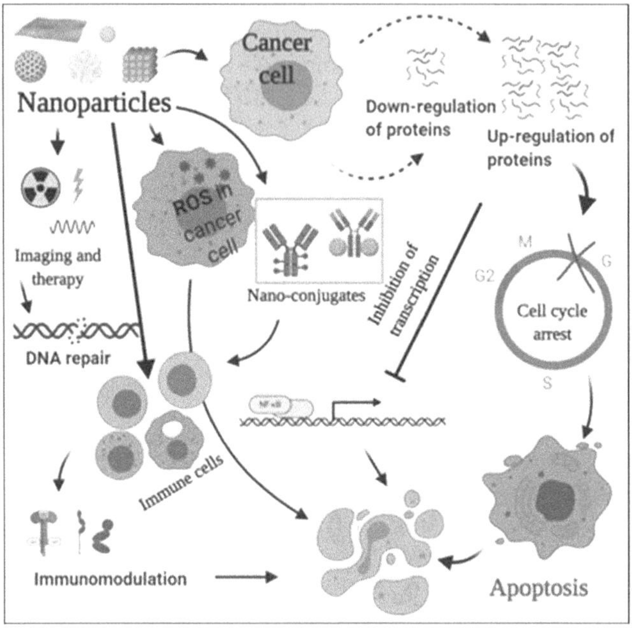

FIGURE 5.7 Anticancer mechanism of Action of NPs.[58]

the remodelling of metabolic and signalling pathways. This impacts the progression of the cell cycle and proliferation, ultimately resulting in cell death and suppression of the tumour.[56] NPs enhance the efficacy of protein regulation and have emerged as potential anticancer agents. For example, CuO NPs downregulate the expression of 'Bcl-2' and 'Bcl-xL' anti-apoptotic proteins in the human colon adenocarcinoma cell line 'HT-29', leading to the death of the targeted cell.[57] The anticancer mechanism of action of NPs, as summarized by Mundekkad et al., is shown in Figure 5.7.

5.4.2 Antimicrobial

The tiny size of NPs helps them adhere to the cell wall surface of microbes, enabling them to penetrate through the cell membrane, disrupting it and ultimately killing the cell.[59] Some metallic NPs, like silver (Ag) NPs, release free radicals that result in microbial cell death by destroying the cell membrane integrity and rendering it more porous.51

The antibacterial mechanism involves the release of ions from NPs, which inactivate vital enzymes through interaction with thiol groups. They inhibit several cellular functions, thereby causing harm to the cell. The generation of ROS occurs within the cell, ultimately leading to self-destruction of the microbial cell.

Apart from this, the metal NPs, such as Ag, which are soft acids, tend to react with the base components like sulphur (S) and phosphorus (P) present in DNA. This interaction interrupts the DNA structure and results in targeted cell death. The surfaces of microbial membranes, such as those of viruses, bacteria, and fungi, usually carry a negative charge, whereas NPs have a positive surface charge. The interaction between these oppositely charged surfaces leads to protein denaturation as metal NPs bind with the functional groups of the proteins.[60]

5.4.3 Antioxidant

NPs also function as antioxidants either by hydrogen or electron transfer. Oxidative molecules are reduced due to the electron transfer, which involves the donation of electrons from antioxidant species. However, in hydrogen transfer, hydrogen atoms scavenge free radicals.[61] Diseases associated with imbalances in ROS and defences against antioxidants are mainly caused by oxidative stress. The antioxidant enzyme-mimetic mechanism explains the antioxidant effects. For instance, to remove superoxide and peroxynitrate, superoxide dismutase imitates the functions of agents involved in catalysis.[62]

During the natural oxidative processes, ROS are generally created by the breaking of covalent bonds in molecules.[63] The generation of different ROS, derived from oxygen molecules through electronic excitation and redox reactions, induces damage to the molecules.[64] Inorganic NPs are able to scavenge ROS like hydrogen peroxide efficiently. For example, Ce NPs can remove the OH radicals that hydrogen peroxide produces.[65] Se NPs also act as effective antioxidants by scavenging free radicals.[66]

5.5 SILVER-BASED NPS

Since ancient times, people have utilized AgNPs as antimicrobial agents for their antibacterial properties. Ancient civilizations, such as the Greeks, Romans, and Egyptians, employed silver-based utensils for water storage and other liquids to prevent microbial growth. Additionally, it was used in wound dressings and for various medicinal purposes because of its perceived antibacterial properties.[67,68] During the Middle Ages, silver coins were often placed in milk and other perishable liquids to delay spoilage. This practice began after discovering that silver halted the growth of bacteria and other microorganisms. With advancements in nanotechnology, researchers began exploring the antimicrobial properties of silver NPs. Compared to bulk silver, AgNPs have an increased surface area-to-volume ratio, which enhances their antimicrobial efficacy. Additionally, they exhibit antimicrobial properties against a wide range of microbes such as bacteria, fungi, and viruses because they disrupt microbial cell membranes, inhibit enzyme activity, and induce oxidative stress. Figure 5.8 shows the different types of AgNPs synthesized by researchers so far.

Research on creating silver NPs and examining their potential applications is an active area of study. For several years, there has been a surge in research focused on synthesizing AgNPs for their application in numerous fields (Figure 5.9), including medicine, biotechnology, environmental science, and consumer products. The application of AgNPs is practiced in wound dressings, textiles, medical devices, cosmetics, and household items because of their antimicrobial properties. This chapter will cover the manufacture of AgNPs by both green and chemical synthesis methods and their use in biological research.

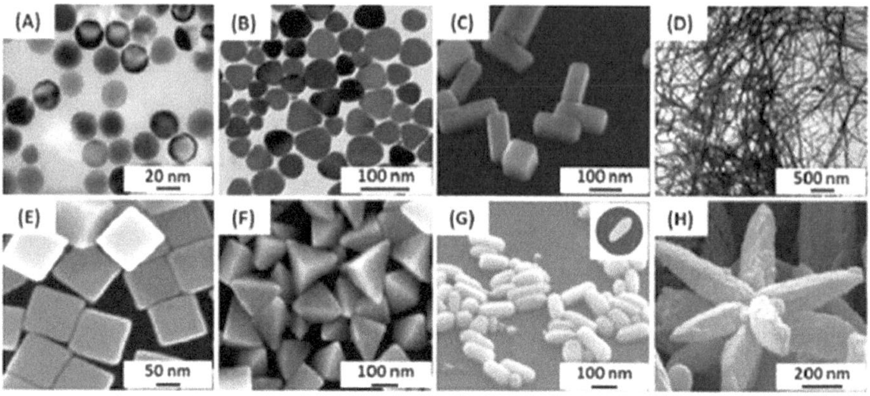

FIGURE 5.8 Different shapes of AgNPs TEM images: (A) nanospheres, (B) nanoprisms,[69] (C) nanobars,[70] and (D) nanowires.[71] SEM images of (E) nanocubes, (F) pyramids,[72] (G) nanorice, and (H) nanoflowers.[73,74]

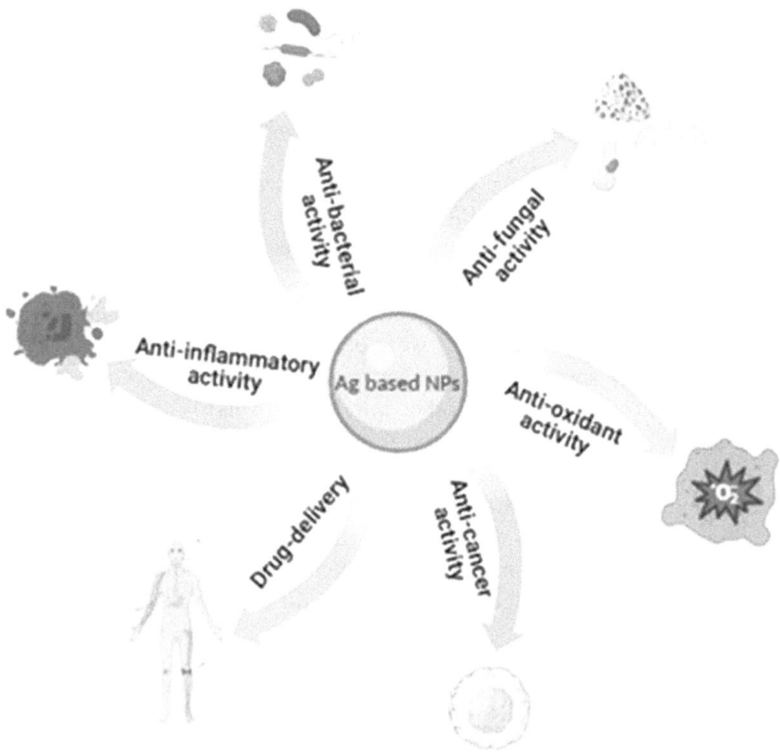

FIGURE 5.9 Biomedical applications of Ag-based NPs.

TABLE 5.2
Chemical Methods for Synthesis of Silver-based Nanoparticles

S.No.	Nanocomposites	Biological Applications	Chemical Method	References
1.	Ag/Bhm NC	Scolicidal and Apoptotic effects	Precipitation method	Bagheriv and et al.[76]
2.	AgNPs	Antioxidant, antibacterial, antifungal, enzyme inhibitor, nanozyme activity	Photochemical reduction	Sajjad et al.[77]
3.	AgNP-gemcitabine	Nanocarrier for anticancer drugs	Chemical reduction using borohydride and PVP	Karuppaiah et al.[78]
4.	AgNP-methotrexate	Anticancer	Chemical reduction using borohydride and citrate	Rozalen et al.[79]
5.	$AgFeO_2$	Antibacterial and antioxidant	Chemical reduction using sodium borohydride	Muthukumar et al.[80]

5.5.1 Chemical Synthesis of AgNPs

The chemical methods for synthesizing AgNPs include chemical and photochemical reductions, MW and gamma ray irradiation, electrochemical methods, and laser ablation[75] (Table 5.2).

In recent years, there has not been much work on synthesizing NPs *via* chemical synthesis due to obvious reasons of environmental toxicity and other disadvantages such as high cost, time-consuming procedures, and complex reaction conditions.[81] Still, here we have discussed some key research findings. Recently, Kumar et al. published their work on synthesizing different AgNPs and nanocomposites and studied their apoptotic and scolicidal activity on *Echinococcus granulosus* protoscoleces. Their study showed that the Ag/Bhm nanocomposite exhibited scolicidal and apoptotic effects against *E. granulosus* and should be considered for the treatment of *E. granulosus*-caused parasitic zoonotic infections such as cystic hydatid disease. They synthesized Ag/Bhm NCs using $AgNO_3$, Boehmite mineral, polyethylene glycol and triethanolamine.[75] Park et al. in 2022 synthesized AgNPs via photochemical reduction. The synthesized NPs exhibited antioxidant, antibacterial, antifungal, enzyme-inhibiting, and nanozyme activities. Before that, several researchers synthesized different AgNCs by combining Ag with other metals or drugs to enhance the characteristics of AgNPs alone.[76] Karuppaiah et al. (2020) conjugated AgNPs with the anticancer drug 'gemcitabine' for breast cancer. They concluded that synergizing AgNPs with gemcitabine enhanced the anticancer effect by serving as a nanocarrier for the drug.[79] Similarly, Rozalen et al. (2020) synthesized AgNPs and combined them with another anticancer drug 'methotrexate' enhancing its anticancer effect against colon and lung cancer cells.[80] Muthukumar et al. (2020) synthesized Silver Ferrite NPs ($AgFeO_2$) by chemical reduction. The obtained $AgFeO_2$ NPs showed antibacterial and antioxidative properties.[81]

As discussed before, the chemical synthesis of NPs has many disadvantages; hence, green methods are preferred over chemical or physical ones. Chemically synthesized AgNPs have a very short lifespan, byproducts are formed during synthesis, and the process is costly, consuming high power and requiring higher maintenance. Consequently, scientists began to look for other alternatives and discovered greener methods for nanomaterial synthesis. Most of the issues and challenges were resolved by green synthesis, which is now the most commonly used and preferred route for creating nanomaterials.

5.5.2 Green Methods to Synthesize Silver NPs

The biological (green) synthesis of nanomaterials involves the use of living organisms such as fungi, bacterial strains, plant extracts, and algae, which serve as reducing, capping, or stabilizing agents, reducing Ag from a +1 to a zero-oxidation state[82] (Table 5.3).

Recently, the green or biological synthesis of NPs has become the most commonly used. The reasons are quite self-explanatory: there is no use of any toxic chemicals, and it is biodegradability and easily accessible and produces better products with better properties. The NPs synthesized from green methods exhibit several biological properties. For example, Langer et al. produced AgNPs using '*Lallemantia royleana*' leaf extract which showed antimicrobial, antioxidant, anti-inflammatory, anti-arthritic, and cytotoxic properties against several microbes.[82] Earlier they synthesized AgNPs from *Astragalus tribuloides Delile's* root extract and discovered that the synthesized NPs had antibacterial, antioxidant, and anti-inflammatory activities. Plucinski et al. synthesized AgNPs using *Curcumin* root's aqueous extract which showed anti-inflammatory properties.[85] As per Li et al. synthesized AgNPs using the plant extract of *Morinda lucida*, which exhibited anti-oxidant and antibacterial activity.[86] During 2020, many researchers worked in this field to produce different AgNPs using different green sources. Li et al. used leaf extract of *Azadirachta indica* to perform AgNPs biosynthesis in a very simple way and studied their antibacterial properties against *S. aureus and E. coli*, as well as their cytotoxicity on leukemic cells.[90] Mahamuni-Badiger et al. synthesized AgNPs from *Gleichenia pectinata (Willd.) C. Presl* plant extract and showed that the designed AgNPs exhibited effective antibacterial and antifungal activities against multidrug-resistant pathogens.[91]

In addition to using AgNPs alone, researchers have developed AgNCs and/or combined AgNPs with other metals, polymers, or extracts of plants or bacteria to enhance their existing properties. In 2023, Sivaram et al. synthesized phyto-functionalized AgNPs from the bark extracts of *Abies alba* and *Pinus sylvestris*, which resulted in enhanced antibacterial and antioxidant properties of AgNPs.[84] Before this, Cai et al. engineered phyto-fabricated AgNPs from the leaf extract of *Otostegia persica (Burm.) Boiss.* and witnessed an enhancement of the antibacterial, anti-oxidant, anti-inflammatory, and antifungal activities of NPs.[87] Earlier in 2019, Horky et al. synthesized AgNPs from *Cichorium intybus* aqueous extract. Human breast cancer 'MCF-7' demonstrated anticancer properties for the developed NPs.[92] Apart from just phyto fabricated AgNPs, scientists also developed AgNPs in combination with other elements. For example, in 2020 Safdar et al. produced $AgFeO_2$ NPs using '*Amaranthus blitum*' leaves extracts. The resulting NPs showed antioxidant and antibacterial activities.[79] In the same year, Kothyal et al. synthesized AgCl-NPs

TABLE 5.3
Green Methods of Synthesis of Silver-Based Nanoparticles

S.No.	Nanocomposites	Biological Applications	Green Method	References
1.	AgNPs	Antioxidant, antimicrobial, anti-inflammatory, anti-arthritic, antioxidant, antibacterial, and anti-inflammatory	*Lallemantia royleana* leaf extract *Astragalus tribuloides Delile.* root extract	Sharifi-Rad et al.[83] Sharifi-Rad et al.[84]
2.	Phyto-functionalized AgNPs	Antioxidant and antibacterial	*Abies alba* and *Pinus sylvestris* bark extracts	Macovei et al.[85]
3.	AgNPs	Anti-inflammatory	aqueous extract of *curcumin* root	Khashan et al.[86]
4.	AgNPs	Antioxidant and antibacterial	plant extract of *Morinda lucida*	Labulo et al.[87]
5.	Phytofabricated AgNPs	Antioxidant, antibacterial, antifungal, and anti-inflammatory activities	*Otostegia persica (Burm.) Boiss.* leaf extract	Pathak et al.[88]
6.	AgNPs	Antibacterial	*Bacillus subtilis*	Alsamhary[89]
7.	$AgFeO_2$	Antibacterial and antioxidant	*Amaranthus blitum* leaves extract	Safdar et al.[79]
8.	AgCl-NPs	Antibacterial, antifungal, and antioxidant	*Pulicaria vulgaris Gaertn.* aerial part extract	Sharifi-Rad et al.[90]
9.	AgNPs	Antibacterial	Leaf extract of *Azadirachta indica*	Asimuddin et al.[91]
10.	AgNPs	Antibacterial and antifungal	Plant extract of *Gleichenia pectinata (Willd.) C. Presl.*	Femi-Adepoju et al.[92]
11.	Fabricated AgNPs	Anticancer activity	*Cichorium intybus* aqueous leaf extract	Behboodi et al.[93]
12.	AgNPs	Antibacterial against multidrug-resistant bacteria	Strain of *Streptomyces* sp.	Shirley et al.[94]

by utilizing aerial part extracts of '*Pulicaria vulgaris*', and the resulting NPs exhibited excellent antibacterial, antifungal, and antioxidant activities that can be useful for treating several illnesses.[89] Researchers have also tried to explore other alternatives apart from plant extracts. Elahi used *Bacillus subtilis,* a bacterium, to synthesize AgNPs that exhibited antibacterial activities.[88] In 2010, when nanotechnology was still not well explored, Uikey et al. synthesized AgNPs using a unique strain of *Streptomyces sp.* and examined the antibacterial potency against several bacterial strains.[93]

5.5.3 Challenges and Concerns

Despite their effective antimicrobial properties, there are concerns about the potential cytotoxic effects on the environment. Since nanosilver has antibacterial properties, it is used extensively. Its widespread usage causes more harm to the environment compared to other metal NPs. Silver in nano form is more toxic to prokaryotic cells than its ionic form, making it an excellent antimicrobial agent. As a result, it has been excessively used in recent years, causing ecosystem and environmental exploitation. The characterization of AgNPs is very necessary before the beginning of any research or trials, as agglomeration occurs when they are diluted in a medium with high ionic strength. This agglomeration is often responsible for the reduced rate of degradation of AgNPs. Research efforts are ongoing to address these challenges and optimize the synthesis and applications of silver NPs for safe and sustainable use.

5.6 COPPER-BASED NPs

In recent years, there has been an increasing importance of metal NPs and their applications in several fields, including catalysis, conversion, biomedicine, energy storage, and drug delivery. Metal NPs of various metals, including silver, zinc, iron, gold, nickel, and titanium, have shown promising antibacterial, anticancer, anti-inflammatory, and antimicrobial activities. Consequently, due to their economic viability, these metal NPs have various uses in the pharmaceutical and medical industries. Because of their superior antimicrobial activity, extended shelf life, high stability, and exceptional quantum efficiency, copper-based NPs have attracted significant attention in the field of biology, compared to other metal NPs. Figure 5.10 highlights various biomedical applications exhibited by copper NPs.

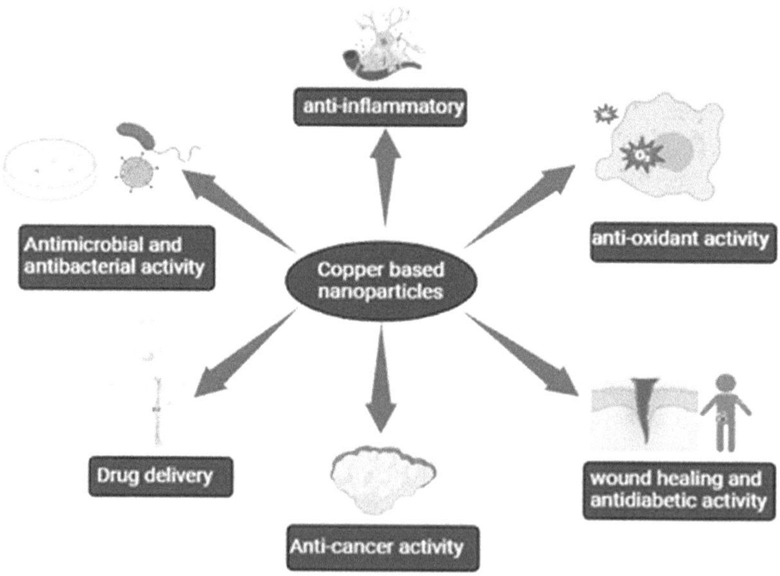

FIGURE 5.10 Biomedical applications of Cu-based nanoparticles.

5.6.1 Chemical Methods to Synthesize Copper NPs

The chemical methods for preparing CuNPs (starch-coated copper NPs) include the sol–gel method, chemical reduction, CVD, aqueous reduction method, and bovine serum albumin (BSA) (Table 5.4).

Cu NPs have been prepared using argon plasma reduction technology. For brown Cu NPs, the formed NPs contain Cu in a zero-oxidation state, while a combination of CuO and Cu(OH)2 is present for green Cu NPs. Both brown and green Cu NPs exhibit biocompatibility and display antibacterial effects. Nanofiber scaffolds have been synthesized using polycaprolactone and polylactic acid as matrices, with Cu NPs and curcumin incorporated using electrospinning, resulting in a hybrid nanocomposite scaffold. This scaffold shows various properties, including drug delivery, antibacterial activity, and wound dressing because of its porosity and increased surface-to-volume ratio. CuO nanoarrays have been synthesized using MW–solvothermal technology, and their biocompatibility has been assessed through haemolytic assays. Additionally, their bioactivity, including antioxidant and anti-inflammatory properties, has been determined. The results showcased a strong anti-inflammatory effect of CuO nanoarrays with low haemolytic potential. According to recent research, ascorbic acid is used for the preparation of Cu-SNPs, where starch serves as a capping agent, thereby preventing the NPs from oxidizing. As a result, Cu-SNPs show potential antimicrobial and anticancer effects.[95] Kumar et al.[94] have used the aqueous reduction method to prepare BSA-coated Cu nanocomposites. BSA provides a hydrophilic coating to the NPs, thereby improving their solubility, stability,

TABLE 5.4
Chemical Method of Synthesis of Copper-Based nanoparticles

S.No.	Nanocomposites	Biological Applications	Synthetic Method	References
1.	BSA @Cu NPs	Anticancer activity	Aqueous reduction method	Amatya et al.[95]
2.	Cu@S NPs	Cytotoxic, apoptotic, and necrotic effects on pancreatic cancer cell line 'Capan-1'	Ascorbic acid and starch as a capping agent	Ilbasmis-Tamer et al.[96]
3.	$Fe_3O_4/SiO_2/$ APTS(~NH_2)-Schiff base/Cu(II)	DNA and topoisomerase-II molecular targets with compounds and anticancer medications	Chemical reduction	Eshaghi-Malekshah et al.[97]
4.	DS-cappedCuInSe2/ ZnS nanosuspensions	Bioimaging and cytotoxicity	Mechanochemical synthesis	Dutkova et al.[98]
5.	Graphene–copper nanoparticles	Antibacterial activity	Chemical vapour deposition method	Vasilopoulos et al.[99]
6.	CuO NPs	Antimicrobial activity	Sol–gel method	Swamy et al.[100]

and biocompatibility, as well as extending the NPs' plasma half-life. The BSA@ Cu NPs demonstrated good photothermal stability and light-to-heat conversion efficiency, along with exceptional anticancer activity. Organ silane-modified Fe3O4/ SiO2 NPs functionalized with a copper–Schiff base complex show potential for delivering cancer-fighting drugs. Dutkova et al. synthesized CuInSe2/ZnS using a mechanochemical approach with sodium dodecyl sulphate as a stabilizing agent, which showed potential biomedical applications such as bioimaging and cytotoxicity. Biocompatible graphene-wrapped Cu nanocomposites anchored on silicon wafers were developed by Vaileios Vasilopoulos et al. using the electrodeposition of copper on Si wafers followed by CVD of graphene, demonstrating antibacterial properties. Swamy et al. synthesized Cu, Ni, and Zn metal–oxide NPs via the sol–gel technique, with CuO NPs exhibiting the highest antimicrobial properties among all. Meanwhile, ZnO and NiO showed better antifungal activity. The research underscores the significance of metal oxide NPs in nanomaterials research, particularly in electronics, optics, energy storage, and catalysis. The size and surface-to-volume ratio of metal oxide NPs are primarily responsible for their antibacterial characteristics.

5.6.2 Green Methods to Synthesize Copper NPs

The green method of Cu NPs synthesis has garnered tremendous attention in recent years. Several eco-friendly synthesis methods for Cu NPs have been acknowledged, employing bottom–up synthesis approaches. The green-mediated production of Cu NPs is particularly advantageous in biomedicine because it employs non-toxic solvents and stabilizing agents. Different physical and chemical techniques, such as sonication, thermal decomposition, and chemical and electrochemical reduction, are utilized in fabricating NPs of metals or metal oxides. However, these methods have many drawbacks, such as the requirement for greater energy, toxic materials, and solvents. To address these issues, researchers have investigated a plant-based method called 'green synthesis' for the manufacturing of NPs. This offers great advantages, including reduced toxicity, cost-effectiveness, shorter production time, and eco-friendliness. Plants contain various metabolites, including reducing sugars, amino acids, tannins, saponins, alkaloids, steroids, flavonoids, phenolic glycosides, and anthraquinones, which act as cappers, reducers, and stabilizers while metal ions are being transformed into metal NPs, thereby imparting desirable properties (Table 5.5).

Pratibha et al. prepared Cu NPs using leaves of *Flueggea leucopyrus*, a wild plant that stabilizes the prepared NPs and also acts as a capping agent, aiding in the prevention of agglomeration. The synthetic method used is environmentally friendly and has great biological properties, such as antibacterial activity and alpha-amylase inhibition. The dose-dependent inhibition of these NPs shows that these particles have prospective applications as antioxidants, antidiabetic medications, and agents for bacterial resistance, respectively. Additionally, Khodashenas et al. prepared Cu-NPs using *A. scholaris*, a medicinal plant with various bioactive components in the leaf extracts. Comparing the solvent extracts, it was found that ethyl acetate and chloroform contained the most bioactive compounds. Furthermore, it was discovered that the methanol extract exhibited the highest antimicrobial activity, leading to its

TABLE 5.5
Biological Method of Synthesizing Copper-Based Nanoparticles

S.No.	Nanocomposites	Biological Applications	Green Method	References
1	Cu NPs	Antibacterial, antioxidant, and antidiabetic	*Flueggea leucopyrus* wild leaf	Pratibha et al.[101]
2	Cu NPs	Antimicrobial studies	*Alstonia scholaris* leaves	Labaran et al.[102]
3	CuO NPs	Recyclability and catalytic activity for biomedical application	*Bougainvillea* plant flower extract	Natrayan et al.[103]
4	CuO NPs	Antibacterial, antioxidant, and anticancer	Floating fern *Salvinia cucullata Roxb.*	Santhosh et al.[104]
5	CuO NPs	Antimicrobial	*Allium cepa L.*	Kenneth et al.[105]
6	ZnO-CuO	Antifungal activity	Aspergillus fumigatus	Gaber et al.[106]
7	CuO NPs	Antibacterial and anticancer	Dried fruit of *Amomum subulatum*	Dhir et al.[107]
8	AS@CuO NPs	Molluscicidal activity	*Annona squamosal* seed extract	Hussain et al.[108]
9	CuO NPs	Antibacterial and antioxidant activity	*Cymbopogon citratus* plant extract	Jabeen et al.[109]
10	Fe@Cu NPs	Antioxidant, antidiabetic, anticholinergic	*Strobilanthes cordifolia* leaf extract	Kirubakaran et al.[110]
11	Cu NPs	Antibacterial, antidiabetic, anti-inflammatory	*Emilia sonchifolia* leaf extract-mediated	Shankara Narayanan et al.[111]
12	CuO NPs	Antioxidant, antimicrobial	*Phormidium* sp.	Asif et al.[112]
13	CuO NPs	Antimicrobial	*Asparagus aethiopicus* and *Asparagus densiflorus*	Stephen et al.[113]
14	Cu NP	Cancer treatment	Himalayan *Hippophae rhamnoides L* extract plant	Ratnesh et al.[114]
15	CuS-PNIPAm NPs	Photothermal treatment of infected skin	By forming aggregates with bacterial cells	Wang et al.[115]
16	Cu NPs	Anticancer, antimicrobial, antioxidant	*Phragmanthera austroarabica* extract	Alahdal et al.[116]
17	CuO NPs	Anticancer activity	*Spirulina platensis*	Doman et al.[117]
18	CuO NPs	Antibacterial antioxidant	*Cotoneaster* plant extract	Isikel et al.[118]
19	Fenugreek@ CuNP	Antibacterial activity against aquaculture pathogens and anticancer potency against human fibroblast	*Trigonellafoenum graecum*	Ghuglot et al.[119]
20	Cu NPs	Antioxidant, antimicrobial, and anticancer activities	Copper nanoparticles synthesis through use of starch-based nanocomposite	Hasanin et al.[120]
21	Cu NPs	Anticancerous, antibacterial	*Aspergillus niger stains STA9*	Noor et al.[121]

(Continued)

TABLE 5.5 (*Continued*)
Biological Method of Synthesizing Copper-Based Nanoparticles

S.No.	Nanocomposites	Biological Applications	Green Method	References
22	CuO NPs	Antibacterial activity	*Aloe vera* extract	Hemalatha et al.[122]
23	CuO NPs	Morphology-dependent antibacterial activity	*Aloe vera* leaf extract	Tavakoli et al.[123]
24	Cu NPs	Antimicrobial and antioxidant	*S. lavandulifolia* extract	Hemmati et al.[124]
25	Polyvinyl alcohol/ CuO	Applications in antibacterial and wound dressing	*Momordica charantia* extract	Sarwar et al.[125]

common utilization in the preparation of Cu NPs. Notably, this method demonstrated versatile applications in both biomedical and environmental fields. Cu NPs have also been produced using the leaf extract of '*Emilia sonchifolia*', where it serves as both a reducing and capping agent.[110] Additionally, when comparing green Cu NPs with the leaf extract, it was found that the Cu NPs exhibit stronger anti-inflammatory, antibacterial, and antidiabetic properties. Cu NPs have been prepared using the extract of '*Hippophae rhamnoides L*', a Himalayan plant that exhibited anticancer properties. NPs from the extract of '*Phragmanthera austroarabica*' showed sensing applications for Cr (IV), degradation of various organic dyes including Congo red and methylene blue (MB), and exhibited antioxidant, antifungal, antibacterial activity, and cytotoxicity against human breast cancer cells.[115] Notably, *Aspergillus niger* stain *STA9* has been employed in the synthesis of copper NPs that showed significant cytotoxicity against human hepatocellular carcinoma cells, along with antidiabetic effects and antimicrobial properties. Cu NPs prepared with the aqueous extract of '*Stachys lavandulifolia vahl*' flowers act as potent agents against microbial infections. Rozalen et al.[116] employed an aqueous solution of blue–green algae, specifically '*Spirulina platensis*', as both a reducing and capping agent, which was found to possess cytotoxicity against various cell lines, thus proving it to be a potential anticancer agent. The '*Strobilanthes cordifolia*' plant, belonging to the Acanthaceae family and known for its diverse biological properties, has been used for synthesizing Cu NPs. These NPs exhibit antibacterial activity against both *Staphylococcus aureus* and *Escherichia coli*, in addition to showing effective antioxidant and antidiabetic activities.

Lee et al.[102] developed stable CuO NPs using an aqueous extract of *Bougainvillea* flowers, which serve as stabilizers, capping agents, and reducing agents, having a negative zeta potential. Green synthesis provides an alternative to conventional methods, which often involve hazardous chemicals and high temperatures. Sirisha et al.[103] have also reported the synthesis of CuO NPs using the aqueous extract of *Salvinia cucullata*, which shows inhibitory effects against various bacterial strains, anticancer activity, and significant cytotoxicity on the 'MCF-7' breast cancer cell line. The CuO NPs synthesized using the aquatic weed *S. cucullata* are a potential natural resource for the development of antibiotics and anticancer drugs.[106] Sajjad et al.[104]

have prepared CuO NPs using the aqueous extract of *Allium cepa* (onion), showing strong antimicrobial activities. Mundekkad et al.[106] introduced another approach using the dried fruit of *Amomum subulatum* as a stabilizing, reducing, and capping agent. Additionally, it displays promising antibacterial activity and significant cytotoxicity against human cervical and breast cancer cells. Incorporating *Annona squamosa* seed extract onto CuO NPs shows a molluscicidal effect on *'Biophalarial alexandrina'* snails. The toxic profile of green CuO NPs produced by *A. squamosa* might harm the biological activity of *B. alexandrina* snails, suggesting that this compound could serve as a foundation for molluscicidal applications.[107] AS@CuO NPs show harmful effects on the digestive and reproductive systems of *B. alexandrina*, leading to severe consequences for their overall health and feeding behaviour. Hence, *B. alexandrina* has established itself as a useful biomarker for detecting nanomaterial contamination. A novel method for the synthesis of CuO NPs has been introduced using *Cymbopogon citratus* plant extract.[108] It notably exhibits remarkable antioxidant activity and potential antibacterial properties against *Escherichia coli* and *Staphylococcus aureus*, as well as promising anticancer properties by demonstrating cytotoxicity against breast and rhabdomyosarcoma cell lines while maintaining the viability of healthy cells.

Additionally, CuO NPs have been prepared using cyanobacteria-derived methods, which showed antioxidant and anti-inflammatory activity, along with toxicity against bacterial and fungal strains, and displayed synergistic interactions with combined drugs. CuO NPs have been synthesized by Stephens' group using an aqueous extract of shed dried parts of the plants *Asparagus aethiopicus L.* and *Asparagus densiflorus jessop,* inhibiting the growth of *Pseudomonas aeruginosa* bacteria. *Cotoneaster* plant extract has also been used for CuO NPs synthesis, which exhibits remarkable antibacterial activity against *Escherichia coli* and *Staphylococcus aureus* and shows antioxidant properties. Aloe vera extract has also been used for the preparation of CuO NPs and displayed considerable antibacterial effectiveness against both Gram-positive and Gram-negative bacteria, along with antimicrobial properties demonstrating inhibitory effects, particularly against *Staphylococcus aureus.* Tavakoli et al. showed that different morphologies of CuO NPs influence the efficiency, with the enhanced surface area and rod-shaped particles exhibiting the highest antibacterial activity. A tricomponent electrospun nanocomposite has been prepared by Alsamhary et al.[124] containing PVA, CuO NPs, and *Momordica charantia* extract, which is found to have potential applications as a wound dressing as it possesses antimicrobial properties. The *Momordica charantia* extract within the nanocomposite reduces the toxicity of CuO NPs and hence imparts antioxidant properties to the nanofibers.

CuO-ZnO NPs were synthesized through a greener method by employing *Aspergillus fumigatus.* Here, the fungal extract acts as a stabilizing and reducing agent, converting metallic ions to bimetallic CuO-ZnO NPs and thus maintaining their colloidal form. Additionally, it is known to exhibit promising antifungal activity against *F. oxysporum*, showing potential in monitoring fusarium wilt infection. CuS-PNIPAm (polyisopropylacrylamide)[114] NPs have been prepared to attain a biocompatible photothermal therapy agent with the capability to actively capture and kill bacteria by modifying CuS NPs with 4sPNIPAm (four-arm star thermo-sensitive polyisopropylacrylamide). It induces toxicity through photothermal properties, offering a viable method for efficiently ensnaring and eradicating microorganisms, thereby potentially benefiting the treatment of chronic wounds. Sharifi-Rad et al. have synthesized stable fenugreek@Cu NPs through a rapid and eco-friendly route using fenugreek as a stabilizing and reducing agent, which shows remarkable antibacterial activity against *Vibrio* species.[118]

5.6.3 Conclusion

Mechanical, chemical, and green methods have all been employed for the preparation of Cu and CuO NPs. The green methods of synthesis include natural sources involving plant and fungal extracts, which provide a great alternative to traditional chemical routes that involve hazardous reducing and stabilizing agents. These NPs exhibit a comprehensive range of applications, including antibacterial, antimicrobial, anticancer, and antidiabetic activities, positioning them as promising candidates in biomedical therapeutics. The incorporation of Cu NPs into various nanocomposites further enhances their properties, enabling their application in drug delivery, wound healing, and other biomedical activities.

5.7 IRON-BASED NPs

In recent decades, there has been a consistent increase in the usage of iron-based NPs in numerous biomedical fields, including drug delivery, gene delivery, hyperthermia, and cell labelling. Some methods to synthesize iron-based NPs and their biomedical applications are shown in Figure 5.11. Iron NPs with favourable characteristics include their high magnetic susceptibility, high saturation, non-toxicity, and chemical stability. Iron NPs can be found in two specific nanosystems: magnetic NPs and bimetallic NPs. Magnetic NPs like Fe3O4, Mn-Fe2O4, and Co-Fe2O4, and bimetallic NPs include Fe-Pt, Fe-Ni, and Fe-Co. Magnetic NPs are the most researched and recently used materials in biomedical applications due to their distinctive chemical, biological, and magnetic qualities such as magnetization, biocompatibility, non-toxicity, and chemical stability.[126]

Furthermore, iron oxide NPs can exist in several oxidation states, such as Fe(II) oxides (FeO), Fe(III) oxides (Fe2O3), and Fe(II, III) oxides (Fe3O4). Bimetallic NPs are also known to possess suitable magnetic properties, including high Curie temperature and superparamagnetism. Studies have revealed that Fe-Pt NPs possess distinct chemical and mechanical characteristics, such as high X-ray absorption, superparamagnetization, a high Curie temperature, and chemical stability. Consequently, they are typically employed as MRI contrast agents. Additionally, Fe-Co NPs are used in various biomedical applications as MRI contrast agents and for magnetic hyperthermia because of their high saturation magnetization.[127]

5.7.1 Chemical Methods of Synthesizing Iron-Based NPs

Numerous physical and chemical techniques, such as hydrothermal, solvothermal, sol–gel, co-precipitation, thermal decomposition, laser ablation, and micro-emulsion methods, have been attempted to produce iron-based NPs. Chemically generating

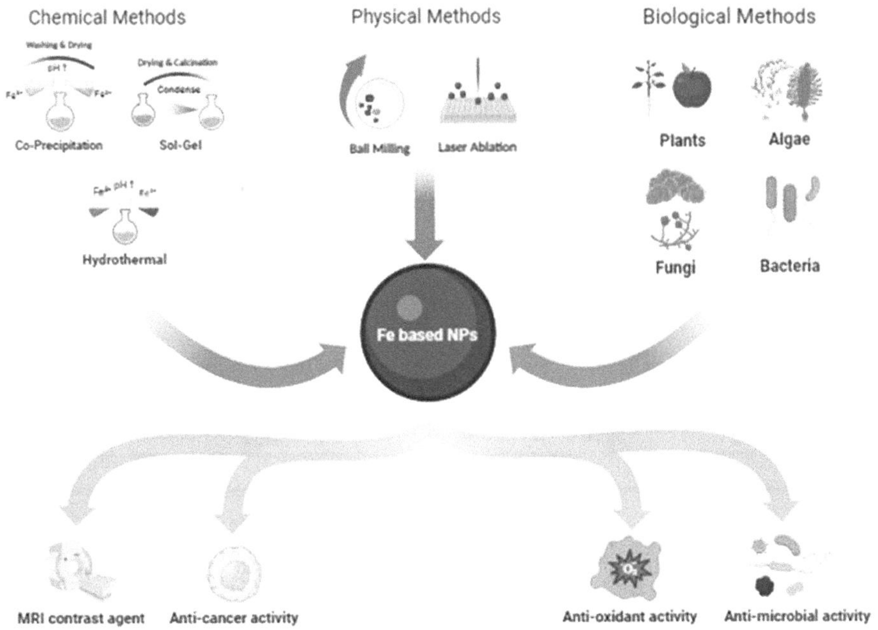

FIGURE 5.11 Methods of synthesizing iron-based NPs and their biomedical applications.

TABLE 5.6
Chemical Methods of Synthesizing Iron-Based Nanoparticles

S.No.	Nanocomposites	Biological Applications	Chemical Method	References
1.	$Fe_3O_4@SiO_2$	Absorb methyl orange from solution	Sol–gel method	Li et al.[128]
2.	$Fe_3O_4@MnO_2$	Rapid degradation of methylene blue	Two-step hydrothermal method	Gao et al.[129]
3.	Fe_3O_4@carbon/graphene oxide	To remove different harmful heavy metal ions from wastewater	Hydrothermal method	Wu et al.[130]

iron-based NPs has several disadvantages. These include liberating hazardous byproducts, using potentially harmful chemicals, contaminating the final product with precursor chemicals, and making the final product highly reactive and prone to form aggregates that lose reactivity. These limitations restrict their use within biological systems due to their limited biocompatibility. Moreover, typical hydrothermal and thermal decomposition methods involved in the production of iron-based NPs necessitate high pressure, temperatures, and large amounts of expensive and dangerous organic solvents (Table 5.6).

Femi-Adepoju et al.[127] $Fe_3O_4@SiO_2$ magnetic core–shell NPs were constructed using the sol–gel technique. The textile, leather, printing, and cosmetic industries require a lot of organizational dye. These industries also release ample amounts of dye wastewater annually, which eventually enters the ecosystem and exerts adverse effects on the environment, posing a significant risk to both human health and aquatic ecosystems. Methyl orange (MO) is a common anionic dye. Due to its complicated structure and difficulty in biodegrading, MO requires appropriate wastewater treatment methods. In this research, MO was absorbed from the solution using the SiO2-coated core–shell structured magnetic nanocomposite. Behboodi et al.[128] prepared the Fe_3O_4@MnO_2 core–shell structure catalyst in the shape of a ball using a two-step hydrothermal method. A widely recognized cationic and thiophene dye, MB, is an example of an aromatic heterocyclic alkaline dye. The textile, paint, plastic, printing, medical, pharmaceutical, and paper sectors all exploit MB because of its strong water solubility and stability. Extreme usage of MB dye can lead to respiratory illnesses, vomiting, skin irritation, mental disorders, and other related issues. MB dye is non-biodegradable and toxic. Fe3O4@MnO2 was employed for peroxymonosulphate (PMS) activation to break down the MB target pollutant.[127] Shirley et al. successfully synthesized the magnetic Fe_3O_4@C-GO (Fe_3O_4@carbon/graphene oxide) nanocomposite, using a simple one-pot hydrothermal technique to effectively remove Cu(II) from aqueous solutions.[129] The nanocomposite is composed of GO and Fe_3O_4 coated with ultra-thin carbon. The Fe_3O_4@C-GO nanocomposite's shape, surface area, magnetic properties, and porosity were also examined and it was observed that the as-prepared Fe_3O_4@C-GO nanocomposite possessed superparamagnetism, a large specific surface area, and functional groups containing oxygen. This study offers an innovative perspective on the production of a Fe_3O_4@C-GO magnetic nanocomposite with competitive traits that is both practical and environmentally favourable. In the future, it is expected that this novel Fe_3O_4@C-GO nanocomposite will be recyclable when applied as a significant magnetic adsorbent to extract various harmful heavy metal ions from wastewater or effluents.[128]

5.7.2 Green Methods of Synthesizing Iron-Based NPs

This process has gained attention as a safer, eco-friendly, and sustainable way to synthesize metal-based NPs, including iron-based NPs.[131] Because of their magnetic characteristics, FeNPs have drawn special attention. To generate iron-based NPs, biomaterials from plants, bacteria, fungi, and algae can be engaged in green synthesis. Moreover, it can function as a capping, stabilizing, fabricating, and reducing agent during the environmentally friendly production of NPs. Recently, in 2024, Tamer et al. synthesized FeO NPs using *Ziziphus oenoplia* fruit extract that showed promising antibacterial, antioxidant, larvicidal and catalytic degradation activity. For biological and environmental applications, the study intends to create a unique, eco-friendly, safe, economical, single-step method for producing high-purity and stable iron-based NPs from ZOF aqueous extract131 (Table 5.7).

Malekshah et al. successfully synthesized bimetallic Ag-Fe NPs using *clove bud* extract to co-reduce metal salts, which exhibited antioxidant, antimicrobial, and dye adsorption behaviour. Indian Ayurvedic medicine, Western herbalism, Chinese medicine, and dentistry all employ 'cloves' (dried flower buds of a tree). The essential oil of cloves is used as an anodyne (painkiller) for dental crises.[132] Dutková et al.'s work in 2023 used the food-origin bacterium '*Enterococcus faecalis*' isolated to

TABLE 5.7
Green Methods of Synthesizing Iron-Based Nanoparticles

S.No.	Nanocomposites	Biological Applications	Green Method	References
1.	FeONPs	Antioxidant, antibacterial, catalytic degradation, larvicidal activity	*Ziziphus oenoplia* fruit	Selvam et al. [132]
2.	Ag-Fe NPs	Antioxidant, antimicrobial, dye removal	Clove (*Syzygium aromaticum*) buds	Garg et al. [133]
3.	FeONPs	Antibiofilm	*Enterococcus faecalis*	Ali and Alkhafaji[134]
4.	FeONPs	Antifungal, anthelmintic, and antibacterial	Neem (*Azadirachta indica*) leaves	Zambri et al. [135]
5.	FeONPs	Decolourization of dye	Apple (*Malus pumila*) peels	Ting et al. [136]
6.	Fe NPs	Antibacterial, antioxidant, and catalytic agent	Green tea (*Camellia sinensis*) leaves	Gupta et al. [137]
7.	Fe NPs	Anti-inflammatory, antimicrobial, antioxidant, antidiabetic	Curry (*Murraya koenigii*) leaves	Sivakami et al. [138]
8.	FeONPs	Anticancer activity	*Chaetomium cupreum*	Wani et al. [139]
9.	Fe NPs	Memory-enhancing agent	*Allium cepa* peel	AL-Salman et al. [140]
10.	FeONPs	Antibacterial, antibiofilm, anti-inflammatory, antioxidant activity	*Eucalyptus camaldulensis*	Zia et al. [141]

produce iron oxide NPs in an easy, affordable, and ecologically acceptable manner.[133] Vasilopoulos et al.[134] successfully synthesized iron oxide NPs by utilizing a non-toxic leaf extract of the neem tree, which belongs to the Meliaceae family. They followed a straightforward one-pot, two-step co-precipitation approach utilizing *Azadirachta indica* (neem leaves) extract as the green reducing agent. Neem tree products are known to have anthelmintic, antidiabetic, antiviral, antifungal, antibacterial, contraceptive, and sedative effects and therefore have been utilized in India for over 2000 years for their therapeutic benefits. Swamy et al.[135] studied iron-based NPs using *apple peel* extract. Apple peels have higher polyphenolic and flavonoid components than the pulp, making them a significant source of antioxidants and reducing agents, and appealing substitutes for metal-based NP manufacturing. This work suggests that apple peel extract can be used to make iron-based NPs that aid in the removal of malachite green dye. Pratibha et al.[136] synthesized a safer and eco-friendly process for producing iron-based NPs in an aqueous medium by employing *green tea leaf* extract as the bio-reductant. They demonstrate antibacterial, antioxidant, and catalytic properties. Labaran et al.[137] synthesized iron-based NPs using *curry leaf (Murraya koenigii)* extract. Green synthesis is an economical, ecologically favourable process that prevents the creation of potentially dangerous byproducts. When the fluid changes to a brownish-black colour during the synthesis process, it indicates that iron-based NPs have formed. The phytochemicals in the extract of *Murraya koenigii* leaves acted as stabilizers and capping agents during the production of iron-based NPs. *Curry leaves*, which are high in folic acid, help improve blood circulation, reduce inflammation, and aid in the absorption of iron. They also possess antibacterial, antidiabetic, anti-inflammatory, antioxidant, and cholesterol-lowering properties, in addition to slowing the pace of premature greying. This tree is indigenous to India and is tropical to subtropical in nature. In Indian cooking, especially on the west and south coasts, the leaves are highly prized as a spice and are also utilized in Ayurvedic medicine as a herb. Natryan et al.[138] investigated iron-based NPs using the fungus *Chaetomium cupreum*. The investigation focuses on the in vitro anticancer activity of iron oxide NPs and their green production from *Chaetomium cupreum* extract. Several biological activities are present in the *Chaetomium* cupreum ethyl acetate extract that is utilized in the production of Fe_2O_3 NPs through an easy, cheap, and eco-friendly green method. Santosh et al.[139] investigated the effective synthesis of Fe_3O_4 NPs using the fruit extract of Iraqi onion peels (*Allium cepa*), which are biocompatible and therefore safer to use in medical applications. The capacity of Fe_3O_4 NPs to produce ROS, which inflict oxidative stress and bacterial cell destruction, is responsible for their antibacterial activity. Bacterial cells can die as a result of lipid peroxidation, protein oxidation, DNA damage, and membrane damage caused by the ROS produced by Fe_3O_4 NPs.[142] Kenneth et al. studied iron oxide NPs by using the plant extract of *Eucalyptus camaldulensis* (river red gum). These NPs demonstrated the largest zone of inhibition against foodborne pathogens after 120 hours of incubation. They also showed the highest percentage of anti-inflammatory and antioxidant activities, as well as maximum biofilm suppression. Therefore, iron oxide NPs studied by Eucalyptus are deemed powerful, efficient substitutes for state-of-the-art therapeutic medicines against foodborne bacteria.[140]

5.7.3 Conclusion and Concerns

Currently, it has been determined that due to the distinctive qualities of iron-based NPs, they exhibit extremely promising potential in numerous fields, such as medicine administration, environmental cleanup, catalysis, and more. However, there is a steady stream of emerging research in the relevant domains to achieve a more profound and comprehensive understanding of their behaviour in various situations. NP toxicity and impact on ecosystems are additional issues that need to be considered.

5.8 MISCELLANEOUS NPs

Miscellaneous NPs include

a. Metal-based NPs such as gold,[143] zinc,[144] titanium,[145,146] and platinum[147].
b. Polymeric/dendrimeric NPs such as chitosan,[148] BSA,[149] and human serum albumin[150]
c. Lipid-based NPs[151]
d. Micellar NPs[152]
e. Polysaccharide NPs[153]

There are many other material-based NPs synthesized for specific purposes. For example, GNPs are majorly used as biosensors[154] and in cancer nanotechnology.[155] Recently, different forms of GNPs, like spherical and nano-rod, have been developed to utilize GNPs in different fields.[156] Additionally, GNPs are applied in drug delivery systems for cancer treatments.[157,158] Zinc-based nanomaterials are used for their unique properties (variable morphology, high surface-to-volume ratio, and antimicrobial properties). They are mainly applied for their use in biofilm inhibition,[159] as antibacterial agents[160] and in agriculture.[161] The polymer-based NPs are very specifically used because of their properties like large surface area and branched structures. For example, chitosan-based NPs are specifically employed in drug delivery[162,163] or protein delivery.[164] BSA NPs are also utilized in drug delivery systems, where they are adsorbed with anticancer drugs.[165,166] Lipid-based NPs find applications in drug delivery[167] as well as nucleic acid delivery systems.[168] The micellar NPs are also used as nanocarriers in drug delivery, either by fabricating or coating with the drug.[169,170] Lastly, both natural and synthetic polysaccharides are employed in drug delivery system[171] and as antibacterials.[172] However, they are typically not used alone; fabrication is often necessary.[173] These miscellaneous NPs are developed solely to enhance the activities of existing drugs and treatments, and their applications are limited.

5.9 SUMMARY

The study of particles at the nanoscale, usually falling in the range of 1–100 nm, is the focus of the emerging scientific discipline known as nanotechnology. Electronics, healthcare, energy, environmental science, and materials science are just a few of the industries where nanotechnology finds widespread application. Nanotechnology offers promising remedies in medicine, encompassing targeted drug delivery, imaging, and diagnostics. Nanocomposites can be created through three basic techniques: physical, chemical, and green methods. Each method possesses its own advantages and disadvantages. Physical methods, such as mechanical mining and laser ablation, produce NPs with high purity, yield, and homogeneity. However, they require a lot of energy and equipment and can cause environmental toxicity. Chemical methods, like sol–gel and hydrothermal synthesis, offer control over NP properties but encounter challenges in size and shape control. Additionally, they involve the use of contaminating solvents and are energy-intensive. The green synthesis method incorporates the use of plant extracts, enzymes, and bacteria. It is cost-effective, environmentally friendly, and simple, producing biodegradable NPs with controlled size. This is why the green approach is widely favoured for NP synthesis. Silver NPs show antimicrobial properties and are useful in wound dressings, textiles, and cosmetics. Similarly, copper NPs exhibit antibacterial, antioxidant, and anticancer activities and can also be synthesized chemically or through green methods. Iron NPs find applications in drug delivery, hyperthermia, and imaging due to their magnetic properties. Additionally, NPs of other materials like gold, zinc, titanium, platinum, and polymeric NPs find numerous applications based on their specific characteristics. However, common concerns exist regarding the toxicity and environmental impacts of NPs, which require further research.

ACKNOWLEDGEMENTS

The authors are thankful to the Department of Chemistry, University of Delhi, for providing support and facilities. S.S. and R.C. are thankful to DST Indo-Russia: INT/RUS/RFBR/389 for providing financial assistance. T.G. is grateful for the DST-INSPIRE fellowship (IF200339). S.S. and R.K.R are also thankful to SERB-TARE: TAR/2022/000618 for their assistance.

CONFLICT OF INTEREST

The authors declare no conflicts of interests in publication of this work.

NOTES

1 N. Baig, I. Kammakakam, and W. Falath, *Mater. Adv.* **2**/6, 1821–1871 (2021). https://doi.org/10.1039/D0MA00807A
2 R. A. de Jesus, G. C. de Assis, R. J. de Oliveira, J. A. S. Costa, C. M. P. da Silva, H. M. Iqbal, and L. F. R. Ferreira, *Nano Struct. Nano Objects* **37**, 101071 (2024). https://doi.org/10.1016/j.nanoso.2023.101071
3 M. Tulinski, and M. Jurczyk, *Metrol. Stand. Nanotechnol. Protoc. Ind. Innov.* 75–98 (2017). https://doi.org/10.1002/9783527800308.ch4
4 J. SenGupta, *Handbook of Nanomaterials for Manufacturing Applications* (Elsevier, 2020, pp. 421–450). https://doi.org/10.1016/b978-0-12-821381-0.00017-x
5 A. Sharma, D. Manocha, A. Arora, A. Kumar, C. M. Srivastava, V. Rawat, H. Kim, M. Verma, and G. K. Rao, *Green Nanomaterials for Industrial Applications* (Elsevier, 2022, pp. 397–421). https://doi.org/10.1016/b978-0-12-823296-5.00016-2
6 F. A. Khan, and F. Khan, *Applications of Nanomaterials in Human Health* (Springer, Singapore, 2020, pp. 15–21). https://doi.org/10.1007/978-981-15-4802-4
7 S. Siddique, and J. C. Chow, *Nanomaterials (Basel)* **10**/9, 1700 (2020). https://doi.org/10.3390/nano10091700
8 V. K. Sethi, M. Pandey, and P. Shukla, *Int. J. Chem. Eng. Appl.* **2**/2, 77–80 (2011). https://doi.org/10.7763/IJCEA.2011.V2.79
9 S. De, and R. Madhuri, *Handbook of Functionalized Nanomaterials for Industrial Applications* (Elsevier, 2020, pp. 269–296). https://doi.org/10.1016/b978-0-12-816787-8.00011-9
10 A. Gupta, *Applications of Nanotechnology in Energy Industry Currently Employed & Future Prospects* (Doctoral dissertation, College of Management and Economics Studies, UPES, Dehradun, 2013).
11 M. Khalil, B. M. Jan, C. W. Tong, and M. A. Berawi, *Appl. Energy* **191**, 287–310 (2017). https://doi.org/10.1016/j.apenergy.2017.01.074
12 A. S. Malani, A. D. Chaudhari, and R. U. Sambhe, *Int. J. Mech. Mechatron. Eng.* **10**/1, 36–40 (2015).
13 R. D. Ramesh, A. Santhosh, and S. R. N. A. Syamala, Implementation of nanotechnology in the aerospace and aviation industry, in *Smart Nanotechnology with Applications* (CRC Press, 2020, pp. 51–69). https://doi.org/10.1201/9781003097532-4
14 V. P. Sharma, U. Sharma, M. Chattopadhyay, and V. N. Shukla, *Mater. Today Proc.* **5**/2, 6376–6380 (2018). https://doi.org/10.1016/j.matpr.2017.12.248
15 H. Chugh, D. Sood, I. Chandra, V. Tomar, G. Dhawan, and R. Chandra, *Artif. Cells Nanomed. Biotechnol.* **46**/supl, 1210–1220 (2018). https://doi.org/10.1080/21691401.2018.1449118

16 H. Tiwari, P. Gupta, A. Verma, S. Singh et al., *ACS Chem. Health Saf.* **31**/2, 153–161 (2024). https://doi.org/10.1021/acs.chas.3c00104
17 R. K. Ratnesh, M. K. Singh, J. Singh, Enhancing ZnO/Si heterojunction photodetector performance for ultra high responsivity across wide spectral range, *J. Mat. Sci.: Mat. Electron.*, **35**, 756 (2024). https://doi.org/10.1007/s10854-024-12516-5
18 M. Ramesh, R. Janani, C. Deepa, and L. Rajeshkumar, *Biosensors* **13**/1, 40 (2023). https://doi.org/10.3390/bios13010040
19 B. Mekuye, and B. Abera, *Nano Select.* **4**/8, 486–501 (2023). https://doi.org/10.1002/nano.202300038
20 F. Habibzadeh, S. M. Sadraei, R. Mansoori, N. P. S. Singh Chauhan, and G. Sargazi, *Heliyon* **8**/12, e12193 (2022). https://doi.org/10.1016/j.heliyon.2022.e12193
21 B. B. Mendes, J. Conniot, A. Avital, D. Yao, X. Jiang, X. Zhou, N. Sharf-Pauker, Y. Xiao, O. Adir, H. Liang, J. Shi, A. Schroeder, J. Conde, *Nat. Rev. Methods Primers* **2**/1, 24 (2022). https://doi.org/10.1038/s43586-022-00104-y
22 M. Hasanpoor, M. Aliofkhazraei, and H. J. P. M. S. Delavari, *Proc. Mater. Sci.* **11**, 320–325 (2015). https://doi.org/10.1016/j.mspro.2015.11.101
23 P. Ayyub, R. Chandra, P. Taneja, A. K. Sharma, and R. Pinto, *Appl. Phys. A* **73**/1, 67–73 (2001). https://doi.org/10.1007/s003390100833
24 J. M. De Teresa, *Nanofabrication: Nanolithography Techniques and their Applications* (IOP Publishing, Bristol, UK, 2020, p. 1). https://doi.org/10.1088/978-0-7503-2608-7ch1
25 D. Bokov, A. Turki Jalil, S. Chupradit, W. Suksatan, M. Javed Ansari, I. H. Shewael, G. H. Valiev, and E. Kianfar, *Adv. Mater. Sci. Eng.* **2021**, 1–21 (2021). https://doi.org/10.1155/2021/5102014
26 K. Hachem, M. J. Ansari, R. O. Saleh, H. H. Kzar, M. E. Al-Gazally, U. S. Altimari, S. A. Hussein, H. T. Mohammed, A. T. Hammid, and E. Kianfar, *BioNanoScience* **12**/3, 1032–1057 (2022). https://doi.org/10.1007/s12668-022-00996-w
27 S. Palit, and C. M. Hussain, *Nanomaterials in Diagnostic Tools and Devices* (Elsevier, 2020, pp. 395–415). https://doi.org/10.1016/b978-0-12-817923-9.00014-6
28 M. Zain, H. Yasmeen, S. S. Yadav, S. Amir, M. Bilal, A. Shahid, and M. Khurshid, *Nanotechnology for Hematology, Blood Transfusion, and Artificial blood* (Elsevier, 2022, pp. 215–235). https://doi.org/10.1016/b978-0-12-823971-1.00019-2
29 M. K. Yadav, R. Kumar, R. K. Ratnesh, J. Singh, R. Chandra, A. Kumar, V. Vishnoi, G. Singh, and A. K. Singh, *Mat. Sci. Eng. B* **303**, 117293 (2024). https://doi.org/10.1016/j.mseb.2024.117293
30 C. Y. Zhao, R. Cheng, Z. Yang, and Z. M. Tian, *Molecules* **23**/4, 826 (2018). https://doi.org/10.3390/molecules23040826
31 Z. Yang, Y. Ma, H. Zhao, Y. Yuan, and B. Y. Kim, *Wiley Interdiscip. Rev. Nanomed. Nanobiotechnol.* **12**/2, e1590 (2020). https://doi.org/10.1002/wnan.1590
32 Y. Jiang, M. Fan, Z. Yang, X. Liu, Z. Xu, S. Liu, G. Feng, S. Tang, Z. Li, Y. Zhang, S. Chen, C. Yang, W. C. Law, B. Dong, G. Xu, and K. T. Yong, *Biomater. Sci.* **10**/24, 6862–6892 (2022). https://doi.org/10.1039/d2bm01001a
33 C. Pang, K. S. Fan, L. Wei, and M. K. Kolar, *Wound Repair Regen.* **29**/2, 225–239 (2021). https://doi.org/10.1111/wrr.12881
34 R. Amin, S. Hwang, and S. H. Park, *Nano* **06**/2, 101–111 (2011). https://doi.org/10.1142/S1793292011002548
35 A. Haleem, M. Javaid, R. P. Singh, S. Rab, and R. Suman, *Glob. Health J.* **7**/2, 70–77 (2023). https://doi.org/10.1016/j.glohj.2023.02.008
36 J. Chao, D. Zhu, Y. Zhang, L. Wang, and C. Fan, *Biosens. Bioelectron.* **76**, 68–79 (2016). https://doi.org/10.1016/j.bios.2015.07.007
37 S. Pandit, D. Dasgupta, N. Dewan, and A. Prince, *J. Pharm. Innov.*, **5**/**6**A, 18 (2016).

38 A. L. Yarin, *Curr. Opin. Chem. Eng.* **29**, 17–25 (2020). https://doi.org/10.1016/j.coche.2020.02.004
39 C. Yu, L. Li, P. Hu, Y. Yang, W. Wei, X. Deng, L. Wang, F. R. Tay, J. Ma, *Adv. Sci. (Weinh)* **8**/14, 2100540 (2021). https://doi.org/10.1002/advs.202100540
40 Y. Jiang, M. Fan, Z. Yang, X. Liu, Z. Xu, S. Liu, G. Feng, S. Tang, Z. Li, Y. Zhang, S. Chen, C. Yang, W. C. Law, B. Dong, G. Xu, K. and T. Yong, *Biomater. Sci.* **10**/24, 6862–6892 (2022). https://doi.org/10.1039/d2bm01001a
41 H. Idrees, S. Z. J. Zaidi, A. Sabir, R. U. Khan, X. Zhang, and S. U. Hassan, *Nanomaterials (Basel)* **10**/10 (2020). https://doi.org/10.3390/nano10101970
42 J. W. Salameh, L. Zhou, S. M. Ward, C. F. Santa Chalarca, T. Emrick, and M. L. Figueiredo, *Wiley Interdiscip. Rev. Nanomed. Nanobiotechnol.* **12**/2, e1598 (2020). https://doi.org/10.1002/wnan.1598
43 A. Unnisa, N. H. Greig, and M. A. Kamal, *Neural Regen. Res.* **18**/10, 2127–2133 (2023). https://doi.org/10.4103/1673-5374.369096
44 R. K. Ratnesh, Hot injection blended tunable CdS quantum dots for production of blue LED and a selective detection of Cu^{2+} ions in aqueous medium, *J. Opt. Laser Technol.* **116**, 103–111 (2019). https://doi.org/10.1016/j.optlastec.2019.03.001
45 D. Mao, Y. Li, C. Lu, X. Liu, A. Sheng, K. Han, and X. Zhu, *ACS Appl. Bio Mater.* **3**/5, 2940–2947 (2020). https://doi.org/10.1021/acsabm.9b01219
46 X. Xu, L. Wang, K. Li, Q. Huang, and W. Jiang, *Anal. Chem.* **90**/5, 3521–3530 (2018). https://doi.org/10.1021/acs.analchem.7b05373
47 L. Song, Y. Zhuge, X. Zuo, M. Li, and F. Wang, *Adv. Sci. (Weinh)* **9**/18, e2200327 (2022). https://doi.org/10.1002/advs.202200327
48 S. Nummelin, B. Shen, P. Piskunen, Q. Liu, M. A. Kostiainen, and V. Linko, *ACS Synth. Biol.* **9**/8, 1923–1940 (2020). https://doi.org/10.1021/acssynbio.0c00235
49 H. Xu, M. Gao, X. Tang, W. Zhang, D. Luo, and M. Chen, *Small Methods* **4**/4, 1900506 (2020). https://doi.org/10.1002/smtd.201900506
50 M. Qasim, D. J. Lim, H. Park, and D. Na, *J. Nanosci. Nanotechnol.* **14**/10, 7374–7387 (2014). https://doi.org/10.1166/jnn.2014.9578
51 P. Nisar, N. Ali, L. Rahman, M. Ali, and Z. K. Shinwari, *J. Biol. Inorg. Chem.* **24**/7, 929–941 (2019). https://doi.org/10.1007/s00775-019-01717-7
52 D. Mundekkad, and W. C. Cho, *Int. J. Mol. Sci.* **23**/3, 1685 (2022). https://doi.org/10.3390/ijms23031685
53 U. Kim, C. Y. Kim, J. M. Lee, H. Oh, B. Ryu, J. Kim, and J. H. Park, *Pathol. Oncol. Res.* **26**/2, 977–984 (2020). https://doi.org/10.1007/s12253-019-00643-y
54 T. Druck, D. G. Cheung, D. Park, F. Trapasso, F. Pichiorri, M. Gaspari, T. Palumbo, R. I. Aqeilan, E. Gaudio, H. Okumura, R. Iuliano, C. Raso, K. Green, K. Huebner, and C. M. Croce, *Cell Death Dis.* **10**/3, 147 (2019). https://doi.org/10.1038/s41419-019-1414-7
55 S. Raj, S. Khurana, R. Choudhari, K. K. Kesari, M. A. Kamal, N. Garg, J. Ruokolainen, B. C. Das, D. Kumar, *Semin. Cancer Biol.* **69** (2021). https://doi.org/10.1016/j.semcancer.2019.11.002
56 M. J. López Grueso, R. M. Tarradas Valero, B. Carmona-Hidalgo, D. J. Lagal Ruiz, J. Peinado, B. McDonagh, R. Requejo Aguilar, J. A. Bárcena Ruiz, and C. A. Padilla Peña, *Antioxidants (Basel)* **8**/11, 505 (2019). https://doi.org/10.3390/antiox8110505
57 S. Khan, A. A. Ansari, A. A. Khan, M. Abdulla, O. Al-Obaid, and R. Ahmad, *Colloids Surf. B Biointerfaces* **153**, 320–326 (2017). https://doi.org/10.1016/j.colsurfb.2017.03.005
58 D. Mundekkad, and W. C. Cho, Nanoparticles in clinical translation for cancer therapy. *Int. Jo. Mol. Sci.* **23**/3, 1685 (2022). https://doi.org/10.3390/ijms23031685
59 I. Sondi, and B. Salopek-Sondi, *J. Colloid Interface Sci.* **275**/1, 177–182 (2004). https://doi.org/10.1016/j.jcis.2004.02.012
60 J. Kim, J. Lee, S. Kwon, and S. Jeong, *J. Nanosci. Nanotechnol.* **9**/2, 1098–1102 (2009). https://doi.org/10.1166/jnn.2009.c096

61 K. Bhardwaj, D. S. Dhanjal, A. Sharma, E. Nepovimova, A. Kalia, S. Thakur, S. Bhardwaj, C. Chopra, R. Singh, R. Verma, D. Kumar, P. Bhardwaj, and K. Kuča, *Int. J. Mol. Sci.* **21**/23, 9028 (2020). https://doi.org/10.3390/ijms21239028

62 M. A. Sharpe, R. Ollosson, V. C. Stewart, and J. B. Clark, *Biochem. J.* **366**/1, 97–107 (2002). https://doi.org/10.1042/BJ20020154

63 K. Kator, *Tanpakushitsu Kakusan Koso (Protein Nucleic Acid Enzyme)* **33**/16, 3127–3135 (1988).

64 H. Sies, and D. P. Jones, *Nat. Rev. Mol. Cell Biol.* **21**/7, 363–383 (2020). https://doi.org/10.1038/s41580-020-0230-3

65 Y. Xue, Q. Luan, D. Yang, X. Yao, and K. Zhou, *J. Phys. Chem. C* **115**/11, 4433–4438 (2011). https://doi.org/10.1021/jp109819u

66 M. G. D. B. Lanza and A. R. and D. Dos Reis, *Plant Physiol. Biochem.* **164**, 27–43 (2021). https://doi.org/10.1016/j.plaphy.2021.04.026

67 J. W. Alexander, *Surg. Infect.* **10**/3, 289–292 (2009). https://doi.org/10.1089/sur.2008.9941

68 J. J. Castellano, S. M. Shafii, F. Ko, G. Donate, T. E. Wright, R. J. Mannari, W. G. Payne, D. J. Smith, and M. C. Robson, *Int. Wound J.* **4**/2, 114–122 (2007). https://doi.org/10.1111/j.1742-481X.2007.00316.x

69 A. X. Zhai, X. H. Cai, and B. Du, *Trans. Nonferrous Met. Soc. China* **24**/5, 1452–1457 (2014). https://doi.org/10.1016/S1003-6326(14)63212-X

70 B. J. Wiley, Y. Chen, J. M. McLellan, Y. Xiong, Z. Y. Li, D. Ginger, and Y. Xia, *Nano Lett.* **7**/4, 1032–1036 (2007). https://doi.org/10.1021/nl070214f

71 C. J. Murphy, and N. R. Jana, *Adv. Mater.* **14**/1, 80–82 (2002). https://doi.org/10.1002/1521-4095(20020104)14:1<80::AID-ADMA80>3.0.CO;2-#

72 B. Khodashenas, and H. R. Ghorbani, *Arab. J. Chem.* **12**/8, 1823–1838 (2019). https://doi.org/10.1016/j.arabjc.2014.12.014

73 A. Pourjavadi, and R. Soleyman, *Mater. Res. Bull.* **46**/11, 1860–1865 (2011). https://doi.org/10.1016/j.materresbull.2011.07.040

74 C. M. Cobley, S. E. Skrabalak, D. J. Campbell, and Y. Xia, *Plasmonics* **4**/2, 171–179 (2009). https://doi.org/10.1007/s11468-009-9088-0

75 J. Natsuki, T. Natsuki, and Y. Hashimoto, *Int. J. Mater. Sci. Appl.* **4**/5, 325–332 (2015). https://doi.org/10.11648/j.ijmsa.20150405.17

76 A. Bagherivand, S. Jafarirad, R. Norouzi, and A. Karimi, *Biomater. Adv.* **157**, 213723 (2024). https://doi.org/10.1016/j.bioadv.2023.213723

77 A. Sajjad, S. H. Bhatti, and M. Zia, *Part. Sci. Technol.* **41**/5, 600–610 (2023). https://doi.org/10.1080/02726351.2022.2126340

78 A. Karuppaiah, K. Siram, D. Selvaraj, M. Ramasamy, D. Babu, and V. Sankar, *Mater. Today Commun.* **23**, 100884 (2020). https://doi.org/10.1016/j.mtcomm.2019.100884

79 M. Rozalen, M. Sánchez-Polo, M. Fernández-Perales, T. J. Widmann, and J. Rivera-Utrilla, *RSC Adv.* **10**/18, 10646–10660 (2020). https://doi.org/10.1039/c9ra08657a

80 H. Muthukumar, S. K. Palanirajan, M. K. Shanmugam, and S. N. Gummadi, *Biotechnol. Rep. (Amst)* **26**, e00469 (2020). https://doi.org/10.1016/j.btre.2020.e00469

81 P. G. Jamkhande, N. W. Ghule, A. H. Bamer, and M. G. Kalaskar, *J. Drug Deliv. Sci. Technol.* **53**, 101174 (2019). https://doi.org/10.1016/j.jddst.2019.101174

82 B. Zewde, A. Ambaye, J. Stubbs iii, and D. Raghavan, *Nanomed* **4**/1043, 1–14 (2016).

83 M. Sharifi-Rad, H. S. Elshafie, and P. Pohl, *J. Photochem. Photobiol. A* **448**, 115318 (2024). https://doi.org/10.1016/j.jphotochem.2023.115318

84 M. Sharifi-Rad, P. Pohl, F. Epifano, and J. M. Álvarez-Suarez, *Nanomaterials (Basel)* **10**/12, 2383 (2020). https://doi.org/10.3390/nano10122383

85 I. Macovei, S. V. Luca, K. Skalicka-Woźniak, C. E. Horhogea, C. M. Rimbu, L. Sacarescu, G. Vochita, D. Gherghel, B. L. Ivanescu, A. D. Panainte, C. Nechita, A. Corciova, and A. Miron, *Antioxidants (Basel)* **12**/4, 797 (2023). https://doi.org/10.3390/antiox12040797

86 A. A. Khashan, Y. Dawood, and Y. H. Khalaf, *Results Chem.* **5**, 100913 (2023). https://doi.org/10.1016/j.rechem.2023.100913
87 A. H. Labulo, O. A. David, and A. D. Terna, *Chem. Zvesti* **76**/12, 7313–7325 (2022). https://doi.org/10.1007/s11696-022-02392-w
88 S. Pathak, N. Kishore, G. Upadhyay, R. K. Ratnesh and R. Mishra, A compact size planar microstrip-fed patch antenna with hexagonal DGS slot for WLAN application, in *Recent Trends in Electronics and Communication, LNEE* (Springer, 2021, vol. 777, pp. 263–271). https://doi.org/10.1007/978-981-16-2761-3_25
89 K. I. Alsamhary, *Saudi J. Biol. Sci.* **27**/8, 2185–2191 (2020). https://doi.org/10.1016/j.sjbs.2020.04.026
90 M. Sharifi-Rad, and P. Pohl, *Nanomaterials (Basel)* **10**/4, 638 (2020). https://doi.org/10.3390/nano10040638
91 M. Asimuddin, M. R. Shaik, S. F. Adil, M. R. H. Siddiqui, A. Alwarthan, K. Jamil, and M. Khan, *J. King Saud Univ. Sci.* **32**/1, 648–656 (2020). https://doi.org/10.1016/j.jksus.2018.09.014
92 A. G. Femi-Adepoju, A. O. Dada, K. O. Otun, A. O. Adepoju, and O. P. Fatoba, *Heliyon* **5**/4, e01543 (2019). https://doi.org/10.1016/j.heliyon.2019.e01543
93 S. Behboodi, F. Baghbani-Arani, S. Abdalan, and S. A. Sadat Shandiz, *Biol. Trace Elem. Res.* **187**/2, 392–402 (2019). https://doi.org/10.1007/s12011-018-1392-0
94 A. D. Shirley, A. Dayanand, B. Sreedhar, and S. G. Dastager, *Dig. J. Nanomater. Biostruct.* **5**/2, 447–451 (2010). https://doi.org/10.15251/dig.j.nanomater.bios
95 R. Amatya, D. Lee, M. Sultana, K. A. Min, and M. C. Shin, *Heliyon* **9**/7, e17732 (2023). https://doi.org/10.1016/j.heliyon.2023.e17732
96 S. Ilbasmis-Tamer, M. Turk, Ş. Evran, I. H. Boyaci, H. Ciftci, and U. Tamer, *J. Drug Deliv. Sci. Technol.* **79**, 104077 (2023). https://doi.org/10.1016/j.jddst.2022.104077
97 R. Eshaghi Malekshah, B. Fahimirad, and A. Khaleghian, *Int. J. Nanomed.* **15**, 2583–2603 (2020). https://doi.org/10.2147/IJN.S231062
98 E. Dutková, Z. L. Bujňáková, O. Sphotyuk, J. Jakubíková, D. Cholujová, V. Šišková, N. Daneu, M. Baláž, J. Kováč, J. Kováč, J. Briančin, and P. Demchenko, *Nanomaterials (Basel)* **11**/1, 69 (2020). https://doi.org/10.3390/nano11010069
99 V. Vasilopoulos, M. Pitou, I. Fekas, R. Papi, A. Ouranidis, E. Pavlidou, P. Patsalas, and T. Choli-Papadopoulou, *ACS Omega* **5**/41, 26329–26334 (2020). https://doi.org/10.1021/acsomega.0c00834
100 B. K. Swamy, R. Bura, R. Godela, K. V. Ratnam, and K. S. Babu, *Int. J. Curr. Pharm. Rev. Res.* **13**/14, 156–160 (2021). https://doi.org/10.31782/IJCRR.2021.131411
101 K. Pratibha, K. Rajoriya, A. Singhal, R. Meena, and A. Kumari, *J. Herb. Med.* **44**, 100859 (2024). https://doi.org/10.1016/j.hermed.2024.100859
102 A. N. Labaran, Z. U. Zango, G. Tailor, A. Alsadig, F. Usman, M. T. Mukhtar, A. M. Garba, R. Alhathlool, K. H. Ibnaouf, and O. A. Aldaghri, *Sci. Rep.* **14**/1, 5589 (2024). https://doi.org/10.1038/s41598-024-56052-y
103 L. Natrayan, S. Kaliappan, A. Saravanan, A. S. Vickram, P. Pravin, M. Abbas, C. Ahamed Saleel, M. Alwetaishi, and M. S. M. Saleem, *Green Process. Synth.* **12**/1, 20230030 (2023). https://doi.org/10.1515/gps-2023-0030
104 P. Santhosh, L. A. Mukhtar, M. Kamaraj, T. G. Nithya, M. R. Ganesh, K. A. Aswathy, and M. Saravanan, *Biomass Convers. Biorefin.* 1–15 (2023). https://doi.org/10.1007/s13399-023-04700-0
105 M. Kenneth, and C. R. Nwodo, *GSC Biol. Pharm. Sci.* **22**/1, 302-311 (2023). https://doi.org/10.30574/gscbps.2023.22.1.0032
106 S. E. Gaber, A. H. Hashem, G. S. El-Sayyad, and M. S. Attia, *Biomass Convers. Biorefin.* 1–15 (2023). https://doi.org/10.1007/s13399-023-04550-w
107 S. Dhir, R. Dutt, R. P. Singh, M. Chauhan, T. Virmani, G. Kumar, A. Alhalmi, M. S. Aleissa, H. A. Rudayni, and M. Al-Zahrani, *Processes* **11**/9, 2698 (2023). https://doi.org/10.3390/pr11092698

108 A. A. Hussein, M. B. Abd El-Latif, M. I. Saad El-Din, N. S. El-Shenawy, O. Hammam, and A. M. Ibrahim, *Biol. Trace Elem. Res.* 1–11 (2023). https://doi.org/10.1007/s12011-023-03823-9

109 A. Jabeen, A. Khan, P. Ahmad, A. Khalid, Z. Majeed, Z. Anjum, Y. Modafer, O. A. Jefri, A. M. Alanazi, A. M. Saeedi, A. H. Alsehli, M. M. Alsowayigh, M. U. Khandaker, I. Boukhris, *J. Drug Deliv. Sci. Technol.* **87**, 104795 (2023). https://doi.org/10.1016/j.jddst.2023.104795.

110 D. Kirubakaran, K. Selvam, P. Prakash, M. S. Shivakumar, and M. Rajkumar, *OpenNano* **14**, 100188 (2023). https://doi.org/10.1016/j.onano.2023.100188

111 V. P. Sankara Narayanan, S. G. Kathirason, P. Elango, R. Subramanian, S. R, A. Gurusamy, *RSC Adv.* **13**/24, 16724–16740 (2023). https://doi.org/10.1039/d3ra00454f

112 N. Asif, R. Ahmad, S. Fatima, S. Shehzadi, T. Siddiqui, A. Zaki, and T. Fatma, *Sci. Rep.* **13**/1, 6246 (2023). https://doi.org/10.1038/s41598-023-33360-3

113 A. Stephen, S. R. Sai, N. Pal, and S. Arjun, *J. Pure Appl. Microbiol.* **17**/2, 1056–1064 (2023). https://doi.org/10.22207/JPAM.17.2.35

114 R. K. Ratnesh, M. K. Singh, V. Kumar, S. Singh, R. Chandra, M. Singh, and J. Singh, Mango leaves (Mangifera Indica) derived highly florescent green graphene quantum dot nanoprobes for enhanced on-off dual detection of cholesterol and Fe^{2+} ions based on molecular logic operation, *Biomat. Sci. Eng.* **7**, 4417–4426 (2024). https://doi.org/10.1021/acsabm.4c00292

115 Z. Wang, Z. Hou, P. Wang, F. Chen, and X. Luo, *Regen. Biomater.* **9**, rbac026 (2022). https://doi.org/10.1093/rb/rbac026

116 F. A. Alahdal, M. T. Qashqoosh, Y. K. Manea, R. K. Mohammed, and S. Naqvi, *Sustain. Mater. Technol.* **35**, e00540 (2023). https://doi.org/10.1016/j.susmat.2022.e00540

117 K. M. Doman, M. M. Gharieb, A. M. Abd El-Monem, and H. H. Morsi, *Int. J. Environ. Health Res.* **34**/2, 661–673 (2024). https://doi.org/10.1080/09603123.2022.2163987

118 E. Isiksel, A. Attar, O. Mutlu, and M. Altikatoglu Yapaoz, *Environ. Sci. Pollut. Res. Int.* **30**/1, 161–171 (2023). https://doi.org/10.1007/s11356-022-22149-y

119 R. Ghuglot, W. Titus, A. S. Agnihotri, V. Krishnakumar, G. Krishnamoorthy, and N. Marimuthu, *Biocatal. Agric. Biotechnol.* **32**, 101932 (2021). https://doi.org/10.1016/j.bcab.2021.101932

120 M. Hasanin, M. A. Al Abboud, M. M. Alawlaqi, T. M. Abdelghany, and A. H. Hashem, *Biol. Trace Elem. Res.* **200**/5, 2099–2112 (2022). https://doi.org/10.1007/s12011-021-02812-0

121 S. Noor, Z. Shah, A. Javed, A. Ali, S. B. Hussain, S. Zafar, H. Ali, and S. A. Muhammad, *J. Microbiol. Methods* **174**, 105966 (2020). https://doi.org/10.1016/j.mimet.2020.105966

122 D. Hemalatha, and S. Saraswathi, *Nanoscale Rep.* **3**/3, 22–27 (2020). https://doi.org/10.26524/nr.3.21

123 S. Tavakoli, M. Kharaziha, and S. Ahmadi, *J. Nanostruct.* **9**/1, 163–171 (2019). https://doi.org/10.22052/JNS.2019.01.018

124 S. Hemmati, S. Ahany Kamangar, A. Ahmeda, M. M. Zangeneh, and A. Zangeneh, *Appl. Organomet. Chem.* **34**/4, e5465 (2020). https://doi.org/10.1002/aoc.5465

125 M. N. Sarwar, H. G. Ali, S. Ullah, K. Yamashita, A. Shahbaz, U. Nisar, M. Hashmi, and I. S. Kim, *Polymers* **14**/7, 1361 (2022). https://doi.org/10.3390/polym14071361

126 G. F. Stiufiuc, and R. I. Stiufiuc, *Appl. Sci.* **14**/4, 1623 (2024). https://doi.org/10.3390/app14041623.

127 K. McNamara and S. A. Tofail, *Adv. Phys.* **2**/1, 54–88 (2017). https://doi.org/10.1080/23746149.2016.1254570

128 H. Li, H. Jin, R. Li, J. Hua, Z. Zhang, and R. Li, *Sci. Rep.* **14**/1, 1217 (2024). https://doi.org/10.1038/s41598-023-50368-x

129 Y. Gao, W. Cao, K. Wang, H. Shi, S. Wang, Q. Meng, K. Du, C. Wang, and J. Lin, *J. Alloys Compd.* **987**, 174144 (2024). https://doi.org/10.1016/j.jallcom.2024.174144

130 M. Y. Wu, R. L. Wu, C. Q. Zang, C. Y. Yu, and Y. F. Liu, *J. Nano Res.* **75**, 41–58 (2022). https://doi.org/10.4028/p-309506
131 A. Ghazzy, H. Nsairat, R. Said, O. A. Sibai, A. AbuRuman, A. S. Shraim, and A. Al Hunaiti, *Nanoscale Adv.* **6**/6, 1611–1642 (2024). https://doi.org/10.1039/d3na00903c
132 K. Selvam, C. Sudhakar, B. Senthilkumar, V. Sakthivel, A. Ragu Prasath, and V. Sangameshwaran, *Waste Biomass Valor.* 1–15 (2024). https://doi.org/10.1007/s12649-024-02445-z
133 P. Garg, T. Chandra, R. Ahlawat, N. Mittal, R. K. Ratnesh, and S. K. Tripathi, *Star Galaxy Image Classification Via Convolutional Neural Network, ICOSEC-2022* (IEEE, 2022, pp. 1156–1168). https://doi.org/10.1109/ICOSEC54921.2022.9952065
134 M. A. Ali and M. H. Alkhafaji, *Egypt. J. Bot.* **63**/3, 1201–1214 (2022). https://doi.org/10.21608/ejbo.2022.157735.2111
135 N. D. S. Zambri, N. I. Taib, F. Abdul Latif, and Z. Mohamed, *Molecules* **24**/20, 3803 (2019). https://doi.org/10.3390/molecules24203803
136 A. S. Y. Ting and J. E. Chin, *Water Air Soil Pollut.* **231**/6, 278 (2020). https://doi.org/10.1007/s11270-020-04658-z
137 K. Gupta, K. Saini, and J. Mathur, *ChemistrySelect* **8**/13, e202204421 (2023). https://doi.org/10.1002/slct.202204421
138 M. Sivakami, K. Renuka Devi, and R. Renuka, *Appl. Phys. A* **128**/4, 272 (2022). https://doi.org/10.1007/s00339-022-05437-9
139 N. A. Wani, W. I. Khanday, and S. Tirumale, *Matrix Sci. Pharma* **4**/2, 23–30 (2020). https://doi.org/10.4103/MTSP.MTSP_6_20
140 H. N. K. Al-Salman, Q. A. Qasim, B. A. Abbas, A. J. Hussein, F. H. Shari, M. S. Jabir, S. Albukhaty, G. M. Sulaiman, M. A. A.Garadi, and A. A. Swelum, *Biosynthesis of* Fe_3O_4*Nanoparticles Mediated Allium cepa Extract: Chemical Characterization and Antimicrobial Applications* (2023). https://doi.org/10.21203/rs.3.rs-3170152/v1
141 A. Zia, S. Tahir, S. Saba, A. Hussain, I. Arooj, and L. Aslam, *LGUJLS* **8**/1, 97–116 (2024). https://doi.org/10.54692/lgujls.2024.0801325
142 S. Singh, T. Goel, A. Singh, H. Chugh, N. Chakraborty, I. Roy, M. Tiwari, and R. Chandra, *Artif. Cells Nanomed. Biotechnol.* **52**/1, 46–58 (2024). https://doi.org/10.1080/21691401.2023.2295534
143 D. Kumar, N. Saini, N. Jain, R. Sareen, and V. Pandit, *Expert Opin. Drug Deliv.* **10**/3, 397–409 (2013). https://doi.org/10.1517/17425247.2013.749854
144 E. J. Park, U. Jeong, C. Yoon, and Y. Kim, *Environ. Toxicol.* **32**/4, 1363–1374 (2017). https://doi.org/10.1002/tox.22330
145 A. Jindal, Mainuddin, A. Kumar, R. K. Ratnesh, J. Singh, *Indian J. Microbiol.* (2024). https://doi.org/10.1007/s12088-024-01330-6
146 M. Aslam, A. Z. Abdullah, and M. Rafatullah, *J. Ind. Eng. Chem.* **98**, 1–16 (2021). https://doi.org/10.1016/j.jiec.2021.04.010
147 M. Safdar, M. Ozaslan, R. A. Khailany, S. Latif, Y. Junejo, M. Saeed, M. S. Al-Attar, and B. O. Kanabe, *J. Inorg. Organomet. Polym. Mater.* **30**/7, 2430–2439 (2020). https://doi.org/10.1007/s10904-019-01387-7
148 K. Divya, and M. S. Jisha, *Environ. Chem. Lett.* **16**/1, 101–112 (2018). https://doi.org/10.1007/s10311-017-0670-y
149 R. Solanki, H. Rostamabadi, S. Patel, and S. M. Jafari, *Int. J. Biol. Macromol.* **193**/A, 528–540 (2021). https://doi.org/10.1016/j.ijbiomac.2021.10.040
150 K. Langer, S. Balthasar, V. Vogel, N. Dinauer, H. Von Briesen, and D. Schubert, *Int. J. Pharm.* **257**/1–2, 169–180 (2003). https://doi.org/10.1016/s0378-5173(03)00134-0
151 B. García-Pinel, C. Porras-Alcalá, A. Ortega-Rodríguez, F. Sarabia, J. Prados, C. Melguizo, and J. M. López-Romero, *Nanomaterials (Basel)* **9**/4, 638 (2019). https://doi.org/10.3390/nano9040638

152 A. J. Sivaram, A. Wardiana, S. Alcantara, S. E. Sonderegger, N. L. Fletcher, Z. H. Houston, C. B. Howard, S. M. Mahler, C. Alexander, S. J. Kent, C. A. Bell, and K. J. Thurecht, *ACS Nano* **14**/10, 13739–13753 (2020). https://doi.org/10.1021/acsnano.0c06033
153 A. Plucinski, Z. Lyu, and B. V. Schmidt, *J. Mater. Chem. B* **9**/35, 7030–7062 (2021). https://doi.org/10.1039/d1tb00628b
154 Y. Li, H. J. Schluesener, and S. Xu, *Gold Bull.* **43**/1, 29–41 (2010). https://doi.org/10.1007/BF03214964
155 W. Cai, T. Gao, H. Hong, and J. Sun, *Nanotechnol. Sci. Appl.* 1, 17–32 (2008). https://doi.org/10.2147/nsa.s3788
156 N. Elahi, M. Kamali, and M. H. Baghersad, *Talanta* **184**, 537–556 (2018). https://doi.org/10.1016/j.talanta.2018.02.088
157 S. R. Kothiyal, R. K. Ratnesh, A. K. Shukla, *Enhancing Sensing with Surface Plasmons Resonance: Theoretical Insights and Design Strategies in Photonic Crystal Fiber, International Conference Artificial Intelligence and Information Technologies, (ICAIIT)-2024* (Taylor & Francis, 2023, pp. 279–285). https://doi.org/10.1201/9781003510833-46
158 X. Li, Y. Zhang, G. Liu, Z. Luo, L. Zhou, Y. Xue, and M. Liu, *RSC Adv.* **12**/13, 7635–7651 (2022). https://doi.org/10.1039/d2ra00566b
159 P. P. Mahamuni-Badiger, P. M. Patil, M. V. Badiger, P. R. Patel, B. S. Thorat-Gadgil, A. Pandit, and R. A. Bohara, *Mater. Sci. Eng. C Mater. Biol. Appl.* **108**, 110319 (2020). https://doi.org/10.1016/j.msec.2019.110319
160 P. Horky, S. Skalickova, L. Urbankova, D. Baholet, S. Kociova, Z. Bytesnikova, E. Kabourkova, Z. Lackova, N. Cernei, M. Gagic, V. Milosavljevic, V. Smolikova, E. Vaclavkova, P. Nevrkla, P. Knot, O. Krystofova, D. Hynek, P. Kopel, J. Skladanka, V. Adam, and K. Smerkova, *J. Anim. Sci. Biotechnol.* **10**, 17 (2019). https://doi.org/10.1186/s40104-019-0319-8
161 P. Uikey, and K. Vishwakarma, *Int. J. Emerg. Technol. Comput. Sci. Electron.* **21**/2, 239–242 (2016).
162 A. Kumar, V. D. Ashwlayan, R. K. Ratnesh, J. Singh, and M. Verma, *South Afr. J. Bot.* **169**, 197–209 (2024). https://doi.org/10.1016/j.sajb.2024.04.021
163 K. Nagpal, S. K. Singh, and D. N. Mishra, *Chem. Pharm. Bull. (Tokyo)* **58**/11, 1423–1430 (2010). https://doi.org/10.1248/cpb.58.1423
164 Y. Xu, and Y. Du, *Int. J. Pharm.* **250**/1, 215–226 (2003). https://doi.org/10.1016/s0378-5173(02)00548-3
165 D. Zhao, X. Zhao, Y. Zu, J. Li, Y. Zhang, R. Jiang, and Z. Zhang, *Int. J. Nanomed.* **5**, 669–677 (2010). https://doi.org/10.2147/ijn.s12918
166 Z. Yu, M. Yu, Z. Zhang, G. Hong, and Q. Xiong, *Nanoscale Res. Lett.* **9**/1, 343 (2014). https://doi.org/10.1186/1556-276X-9-343
167 S. Samimi, N. Maghsoudnia, R. B. Eftekhari, and F. Dorkoosh, *Char. Biol. Nanomater. Drug Deliv.* 47–76 (2019). https://doi.org/10.1016/b978-0-12-814031-4.00003-9
168 W. Li, and F. C. Szoka, *Pharm. Res.* **24**/3, 438–449 (2007). https://doi.org/10.1007/s11095-006-9180-5
169 Z. L. Tyrrell, Y. Shen, and M. Radosz, *Prog. Polym. Sci.* **35**/9, 1128–1143 (2010). https://doi.org/10.1016/j.progpolymsci.2010.06.003
170 R. W. Lee, D. B. Shenoy, and R. Sheel, *Handbook of Non-Invasive Drug Delivery Systems* (William Andrew Publishing, 2010, pp. 37–58). https://doi.org/10.1016/b978-0-8155-2025-2.10002-2
171 V. L. Sirisha, and J. S. D'Souza, *Mar. Omics* 663–702 (2016). https://doi.org/10.1201/9781315372303-52
172 Y. Qin, L. Xiong, M. Li, J. Liu, H. Wu, H. Qiu, H. Mu, X. Xu, and Q. Sun, *J. Agric. Food Chem.* **66**/17, 4373–4383 (2018). https://doi.org/10.1021/acs.jafc.8b00388
173 C. Lemarchand, R. Gref, and P. Couvreur, *Eur. J. Pharm. Biopharm.* **58**/2, 327–341 (2004). https://doi.org/10.1016/j.ejpb.2004.02.016

6 An Overview of Advanced Computing with Industry 4.0

Shubhanshu Rao, Gaurav Kumar, and Raj Kumar Singh

6.1 INTRODUCTION

To comprehend the significance of Industry 4.0, we must first contextualize it within the broader history of industrial revolutions. During the ancient era, most manufacturing was done by hand by experienced workers, which led to expensive and luxurious finished goods because it sometimes required many craftsmen and a lot of time to make a single piece. Near the end of the 18th century, Britain experienced the first industrial revolution, which quickly extended to other nations around the globe. It resulted in long-lasting changes to society's institutions as well as to business and economics. During this period, mass manufacturing became heavily dependent on machinery rather than manual labour. The term 'urbanization' originated with the growth of surrounding villages as industries expanded and people moved to areas near the industries to earn a living, as shown in Figure 6.1. A few of the prominent discoveries of this era were the steam engine, founded by James Watt, and textile machinery.

The first industrial revolution was soon followed by the second one, which lasted between the 19th century and early 20th century. It etched its name in history as the technical revolution; it witnessed a rise in electricity and electrical technologies, which, in turn, altered manufacturing and communication technologies. These improvements enabled businesses to achieve unprecedented economic growth, and the production of steel for use in manufacturing and building increased.

This was soon succeeded by the third industrial revolution, regarded as the digital revolution, marked by the rapid development of digital technologies. This was mostly fuelled by new innovations like the development of the internet and computers, which changed the industrial landscape, and these developments led to significant advancements in industrial automation. Automation software was developed to streamline the workflow of various industries. It gave birth to the knowledge economy, and soon intellectual capital became a vital aspect; access to education and possessing digital skills became essential for people to participate in the modern world.

Today, the world we live in is witnessing the 4th industrial revolution, which we refer to as Industry 4.0. It is said to have originated in Germany in 2011, but it was not until 2016 that it gained popularity.

 DOI: 10.1201/9781003634553-6

FIGURE 6.1 Earlier settlements around the industries [1].

6.1.1 What Is Industry 4.0?

In the year 2016, Klaus Schwab, the executive chairman of the World Economic Forum (WEF), made people familiar with the term 'Industry 4.0' at the Annual Meeting of the World Economic Forum that same year in Davos-Klosters, Switzerland, which was based on the theme of 'Mastering the Fourth Industrial Revolution'. Industry 4.0 represents a vision of smart, interconnected factories capable of autonomous decision-making and real-time optimization.

The core objective of Industry 4.0 revolves around leveraging an array of cutting-edge technologies, including Internet of Things (IoT), additive manufacturing, Big Data analytics, and artificial intelligence, to revolutionize the manufacturing landscape, as seen in Figure 6.2. This entails creating a dynamic ecosystem characterized by heightened responsiveness, productivity, and adaptability to meet the evolving demands of the modern market.

By seamlessly integrating these advanced technologies into their existing operational infrastructure, industries position themselves to unlock a myriad of substantial benefits.

This integration not only helps enhance the efficiency of their systems but also provides them with a significant competitive advantage in the marketplace. Through the use of advanced artificial intelligence capabilities, industries can optimize their decision-making processes, allowing for more informed and timely actions.

This has resulted in industries across various sectors actively pursuing the adoption and incorporation of Industry 4.0 technologies, as they have recognized the potential of these advancements to drive innovation, improve operational performance,

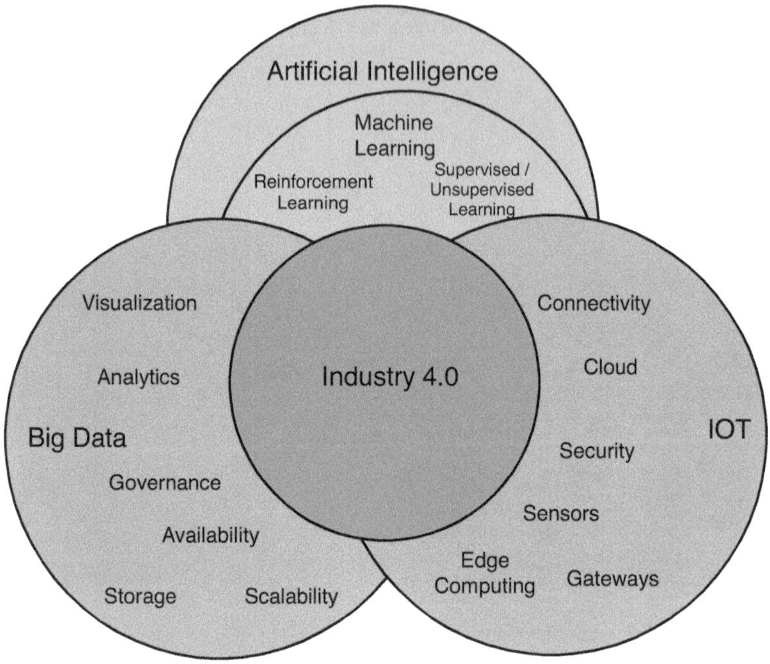

FIGURE 6.2 Industry 4.0 and its key features [2].

and secure a strategic position in today's increasingly competitive landscape. Thus, Industry 4.0 represents a shift in manufacturing paradigms. As the industry and technology come together, it promises the start of a new era of unprecedented efficiency, agility, and growth. Data stands as the cornerstone of these cutting-edge technologies. Just as the human body relies on blood to function properly, these technologies depend extensively on data to work effectively. It serves as the foundation upon which advanced analytics and machine learning algorithms thrive, enabling them to deliver accurate predictions and valuable insights. Industries have recognized the importance and the need for data, and they are continuously trying to use its potential to their advantage. Industries are implementing various methods to collect, analyse, and interpret the data they have captured in order to gain vital insights and a better understanding of the various aspects of their interests. In today's data-driven world, data is the spark that drives innovation and progress. Several prominent concepts have gained recognition, such as predictive maintenance (PdM), which leverages the power of data and machine learning techniques and has changed maintenance practices. Digital twin is another concept that has gained widespread attention; it creates virtual replicas of physical assets with the help of CAD modelling and uses machine learning to provide deeper insights to stakeholders about their assets. These concepts offer revolutionary potential for improving operational efficiency and optimizing maintenance practices through data-driven insights and virtual representations of the assets.

6.1.2 IoT and Its Vital Role in Industry 4.0

Before we delve into PdM, it is important to understand the role of IoT in Industry 4.0 and its significance. In London, around the year 1999, the term IoT was first introduced during a presentation given by Kevin Ashton, who happened to be a computer scientist [3]. Since then, IoT has rapidly gained momentum within the contemporary wireless telecommunications landscape, characterized by the widespread integration of sensors, RFID tags, cell phones, and other interconnected devices [4]. The core strength of the IoT concept lies in its capacity to profoundly influence numerous dimensions of everyday life and shape the behaviours of potential users [5].

IoT technologies encompass various components for the acquisition of data, which include sensors, data storage, a transmission network, processing of information, and operational applications [3]. Within this realm, the Industrial Internet of Things, which is a specific subset of the Internet of Things designed with industry applications in mind, leverages IoT methodologies to enhance machine-to-machine connectivity, streamline data handling, and optimize productivity [3].

In industrial contexts, the data generated often takes the form of time series data, characterized by recordings made at regular intervals over time. Each data point within this dataset is linked to a specific timestamp, indicating when the observation was taken and the corresponding value associated with the parameter being monitored by the sensor. Time series data collected from industrial machinery comprises diverse measurements such as temperature, pressure, vibration, humidity, power consumption, and production output. These data points provide critical insights into the operational condition and overall health of the equipment. They play a crucial role in performance monitoring, facilitating fault detection, and optimizing maintenance practices within industrial environments.

6.1.3 PdM

As we discussed data and how it can be collected using IoT devices, data on its own has no meaning until it is used to tell a story. For the data collected by industries to be useful, techniques like PdM come in handy, which is why they are being incorporated into industries these days. Figure 6.3 shows the rise in the market of Industry 4.0 and the predicted increase as well. PdM can help industries reduce maintenance costs by up to 25%–30% and can also reduce breakdowns by 70%–57%. Moreover, it is highly reliable and leads to better product quality and less waste of raw materials [6].

Maintenance strategies have also gradually evolved over time as seen in Figure 6.4, and currently, it is a continuous process [7]. Some of the previous maintenance practices include calendar-based maintenance, which is scheduled maintenance performed x number of times over a y duration, and preventive maintenance, which focuses on preventing failures and decreasing the number and duration of unscheduled shutdowns. Condition-Based Maintenance (CBM): In this method, rather than following set timetables, maintenance is carried out in accordance with the actual state of the equipment. Reactive maintenance, usually referred to as breakdown

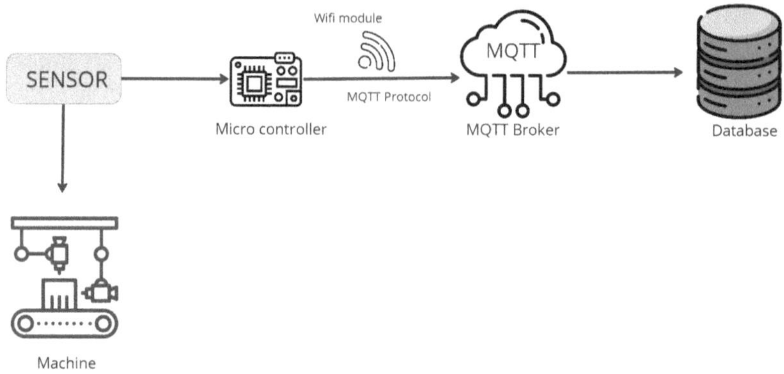

FIGURE 6.3 Connection from machine to sensor to the database.

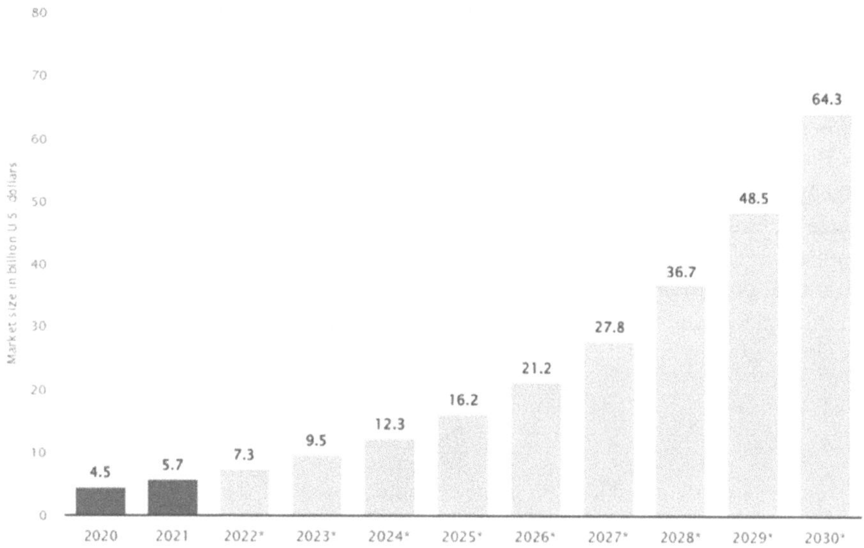

FIGURE 6.4 Estimated market size for maintenance 4.0 starting in 2020 [11].

maintenance, is maintenance performed after a breakdown or failure occurs. PdM is a type of proactive maintenance strategy that utilizes data to analyse equipment behaviour and leverages machine learning to anticipate possible equipment malfunctions before they happen.

This allows companies to schedule their repairs and maintenance in advance, reducing unplanned downtime and avoiding costly repairs. It significantly aids industries in slashing their maintenance expenses by 50%–65% compared to traditional preventive maintenance or corrective maintenance approaches. While the initial setup cost for PdM may be higher than that of its counterparts, this investment gradually decreases over time, leading to substantial cost savings in the long run [8–10].

It also includes methods like remaining useful life (RUL), prognostics and health management, and CBM [11].

RUL provides the duration of time a piece of equipment can operate without being replaced. Service life prediction plays a critical role in prognostication and systems health management by offering early warnings of potential failures, thereby enhancing overall operational efficiency [12].

Prognostics and health management was introduced in 1990 after the US Army's JSF project was launched, initially designed to cater exclusively to the aviation industry. However, due to its effectiveness and demonstrated benefits, it quickly gained significant popularity and achieved widespread adoption across various fields [13]. It has since become a cornerstone technique for enhancing and maintaining the availability of engineered systems.

A maintenance technique known as CBM recommends maintenance actions based on information obtained from condition monitoring. By only performing maintenance in reaction to physical assets' aberrant behaviours, CBM seeks to reduce unnecessary maintenance activities. It can be divided into two parts: diagnostics, which involve identifying and isolating a fault when it occurs, and prognostics, which estimate how likely the fault is to occur [14].

6.1.4 What Is Digital Twin and Its Use Cases?

The history of digital twin dates back to the 1960s during the space exploration missions when NASA spearheaded the development of digital twin technology. This innovative approach involved creating precise replicas of spacecraft here on Earth since the two were identical in build. Because of this, they were dubbed 'Twins', which gave rise to the concept of a 'digital twin'. As a result, we can define a 'twin' as any prototype designed to mimic actual operating conditions in order to simulate behaviours in real time. These replicated versions were utilized by NASA personnel, particularly those serving on flight crews, for comprehensive study and simulation purposes. This played a pivotal role in enhancing mission preparedness and advancing space exploration capabilities.

In addition to the iconic Apollo mission, another noteworthy example of an early digital twin is the 'Iron Bird'. Primarily utilized by the aviation industry, the Iron Bird served as an engineering tool aimed at validating aircraft systems. This system replicated the conditions that real aircraft would encounter, with factors such as Mach number, airspeed, and air temperature meticulously generated by computer simulations. Essentially, the Iron Bird provided a realistic environment for testing and validating aircraft systems, contributing significantly to the advancement of aviation technology [15].

Digital twin has evolved over the following years, as shown in Figure 6.5, from the concept proposed by Grieves initially to the new 5D models.

A digital twin is an electronic copy of a real thing that has been through its whole lifecycle. The object's sensors provide real-time data that the digital twin uses to mimic its behaviour and monitor operations. This technology can simulate a wide range of real-world objects, from tiny pieces of machinery in a factory to large-scale

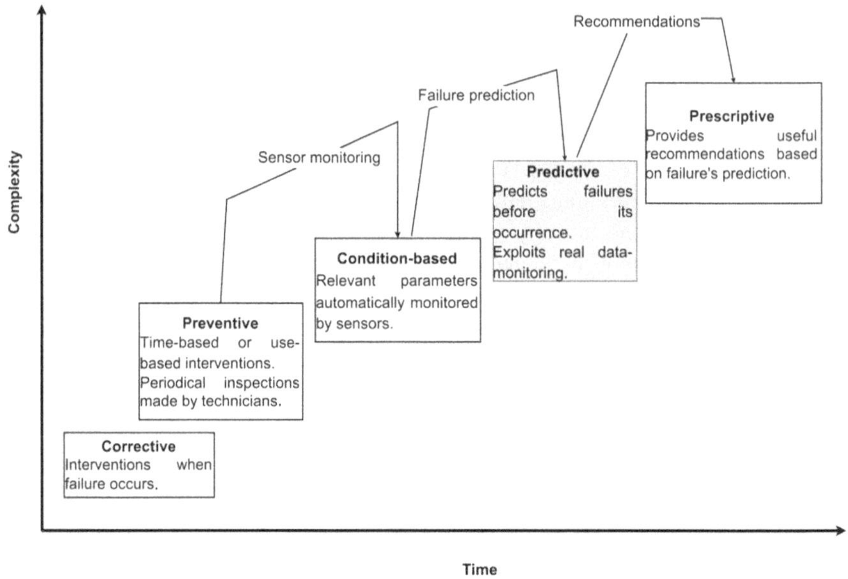

FIGURE 6.5 Different maintenance strategies before and after PdM [16].

structures such as wind turbines and even whole cities. With the use of digital twin technology, users can monitor asset performance, identify any issues, and make informed decisions about maintenance and lifecycle management. Digital twins enhance performance as they make it possible to comprehend asset behaviour, making it easier to adjust and improve performance metrics. They have great predictive capabilities, as they can anticipate possible problems or failures by evaluating previous patterns and real-time data, thus reducing downtime. Figure 6.6 shows the various sectors, such as manufacturing, healthcare, aerospace, and others, in which digital twin technology is being used.

With digital twin technology, it has become possible to remotely monitor assets and receive real-time updates on their performance. There are various types of digital twins, such as:

1. **Product twins:** These replicas emulate individual products or parts, furnishing a detailed digital copy. Commonly applied in manufacturing and engineering, they refine product design and functionality.
2. **Process twins:** Simulating entire systems or workflows, like manufacturing lines or supply chains, these twins offer a comprehensive overview. They empower organizations to enhance efficiency and detect potential challenges.
3. **System twins:** Mimicking intricate systems such as aircraft or industrial machinery, these twins replicate the behaviour of the entire system, including subsystems and components. They offer a holistic perspective on system performance.

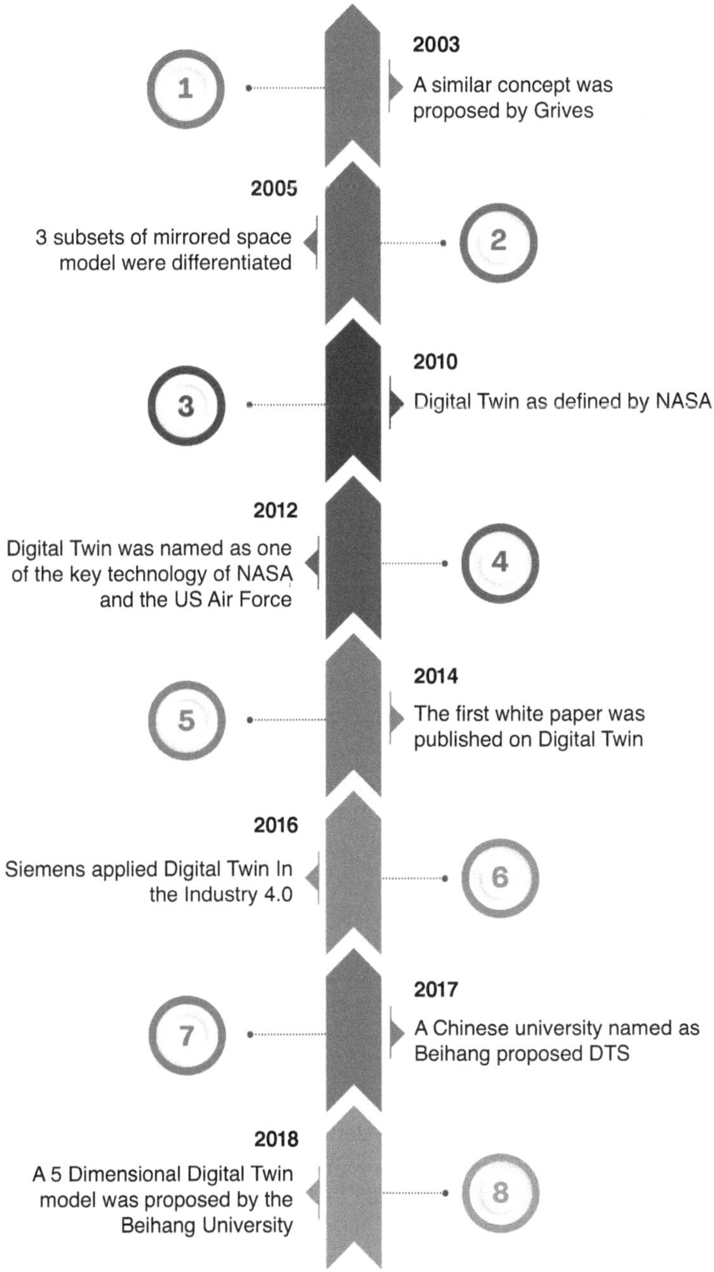

FIGURE 6.6 How digital twin evolved over the years [17].

4. **Performance twins:** These twins specialize in real-time data monitoring and analysis, identifying trends and anomalies in system or asset behaviour. Their use aids in performance optimization and pre-emptive issue prediction.
5. **Physics-based twins:** Utilizing advanced modelling and simulation, these twins create precise digital models of physical systems. Widely employed in engineering and manufacturing, they refine product design and functionality.

6.2 METHODOLOGY

In this section, we will go through some of the methods to implement the techniques that we have discussed in the above section. We will also explore some of the machine learning algorithms in depth.

6.2.1 IoT

Monitoring various parameters of machines or equipment is a crucial aspect of Industry 4.0. To measure these parameters, sensors are deployed to capture relevant data. The sensors should support standard industry protocols like SECS/GEM, OPC, and TCP/IP, enabling all the equipment to be connected to a key manufacturing execution system [3]. However, the placement of these sensors is a critical consideration, as they need to be strategically positioned to capture accurate readings.

As seen in Figure 6.7, once the data is collected by sensors, it is transmitted to a microcontroller for interpretation. Subsequently, this data is sent to the internet via a Wi-Fi module, utilizing the MQTT protocol for secure IoT data transmission. It is a lightweight messaging protocol designed to facilitate communication between IoT and machine-to-machine applications. It operates using the publisher and subscriber model, often known as the pub-sub model, in which some devices publish messages to particular topics, and other devices act as subscribers, subscribing to these topics to receive relevant content.

The data is then sent to an MQTT broker, which acts as an intermediary between the publishers and subscribers. MQTT brokers manage the connection between clients, handle authentication, security, and quality of service levels, and route messages from publishers to the appropriate subscribers. Commonly used MQTT brokers include MQTT, RabbitMQ, HiveMQ, and EMQX, each offering different pricing and capabilities based on requirements.

After passing through the broker, the data is stored in a database for future use. Document-type databases, a subset of NoSQL databases like Elasticsearch, are preferred for their ability to facilitate fast data retrieval when needed. This comprehensive IoT infrastructure enables efficient data collection, transmission, and storage, laying the foundation for PdM and other Industry 4.0 applications [18,19].

6.2.2 How Is PdM Done?

A PdM program fundamentally comprises three primary steps: data acquisition, data processing, and maintenance decision-making. The initial step of data acquisition captures relevant information about the equipment's performance and condition through

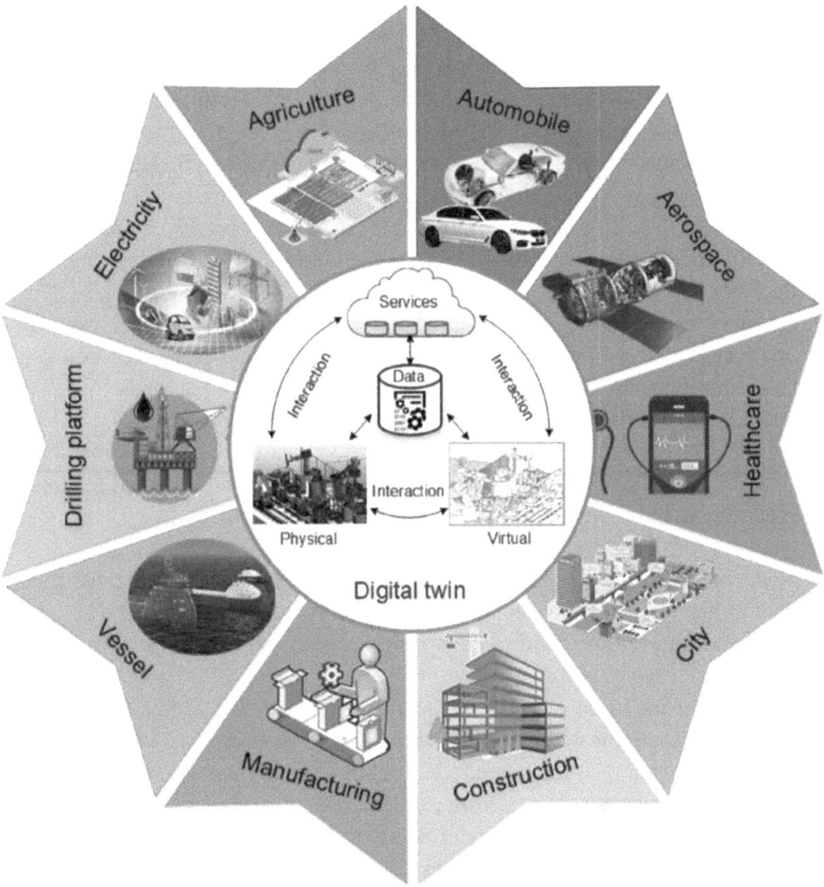

FIGURE 6.7 Digital twin technology's area of application [17].

various sensors and monitoring devices, as discussed in Section 6.3. Subsequently, in the data processing phase, the raw data uploaded to the document-oriented database is analysed and interpreted using advanced algorithms, analytical techniques, and machine learning algorithms to identify patterns, trends, and potential issues, and to forecast the future behaviour of the component. The ML algorithms can be further classified as:

1. Supervised learning, often referred to as the predictive technique, is a straightforward approach in pattern recognition. It revolves around predicting the outcome of a function by examining a predefined set of training data [20]. It is usually the most commonly used ML algorithm in PdM; for this, we require labelled data that indicates at what value of a parameter the breakdown occurred. Common supervised learning algorithms, like decision trees, logistic regression, commonly known as support vector machines, and random forests can be trained on the data to predict when equipment

is likely to fail. The calculation of RUL traditionally involves regression models; recent advancements have demonstrated the efficacy of alternative techniques. For instance, deep learning frameworks such as CNN and hierarchical Bayesian regression have emerged as viable approaches to predict the RUL of components, as shown by [21].

2. Semi-supervised learning is useful when the available labelled data is sparse and obtaining more labelled data is expensive. Utilizing both labelled and unlabelled data can enhance model performance, possibly by leveraging the unlabelled data to refine the representations of the underlying data distribution.
3. Reinforcement learning is a learning framework that interacts with the environment to improve a policy concerning a given goal. According to this paradigm, an agent observes the environment and responds to it. The agent modifies the state of the environment by acting to receive a reward signal [22]. By assigning integer or scalar values linked to specific states or state–action pairings, the reward function acts as the impartial feedback mechanism from the environment [23].

Finally, in the maintenance decision-making stage, based on the insights gleaned from data analysis, informed decisions are made regarding the timing and nature of maintenance activities, aiming to optimize asset performance and minimize downtime [6].

For PdM to work more effectively, we need a significant amount of data from various manufacturing or working processes [24,25]. In industries where machinery is consistently operational, data availability is not typically a concern, given the substantial data generated during machine runtime. Notably, recurrent neural networks (RNNs) and long short-term memory (LSTM) models stand out as popular choices for analysing and forecasting time series data. We will go through various methods that are in practise for analysing the time series data, as shown in Figure 6.8. These models offer sophisticated capabilities for uncovering patterns and trends within sequential data, facilitating accurate predictions and informed maintenance

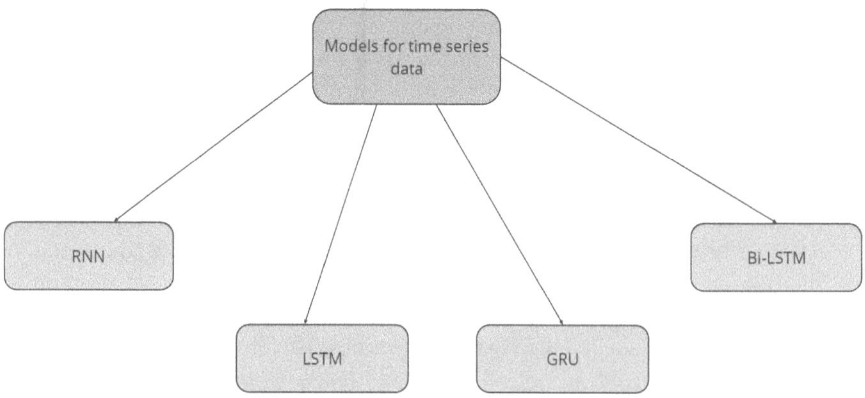

FIGURE 6.8 Different models used for working with time series data.

decisions. Based on the output of these network architectures, the future performance of the equipment is calculated, and as per the predicted values, it can be determined whether a piece of machinery requires maintenance. This allows the company to prevent breakdowns, thereby keeping the process running and saving costs associated with sudden stoppages.

6.2.2.1 RNN

RNN was originally developed in the 1980s and constitutes a distinct neural network architecture within the realm of deep learning. The structure of an RNN is shown in Figures 6.9, and Figure 6.10 shows the unfolded structure of the RNN.

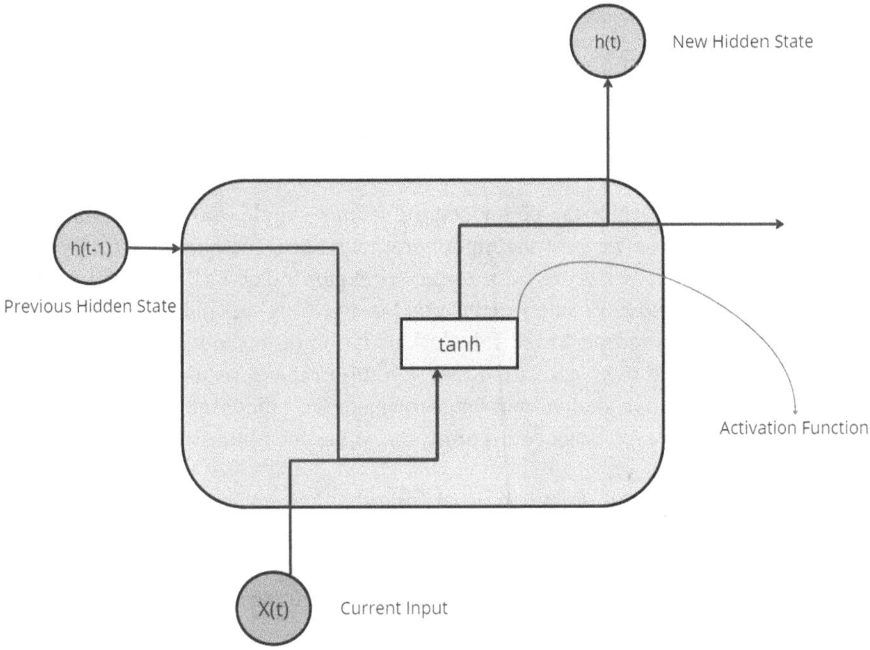

FIGURE 6.9 Single RNN unit.

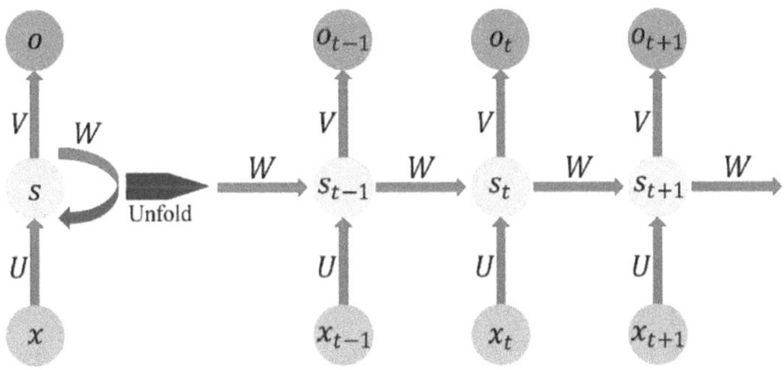

FIGURE 6.10 Structure of RNN memory passing [28].

In unfolding or unrolling, we simply make a copy of the network and direct the feedback loop to the copy; this is a crucial aspect of the RNN's functioning. Like other deep learning models, an RNN comprises interconnected neurons with associated weights, biases, layers, and activation functions. However, what sets RNNs apart is their internal memory mechanism, which enables them to effectively capture sequential data [26]. This internal memory is calculated by equation 6.1:

$$h_t = g\left(WX_t + U_f h_{t-1} + b\right) \tag{6.1}$$

Here, the internal memory is denoted by h_t, g is the activation function, b is the bias term, X is the input fed into the RNN, and U and W are the weights of the layer h [27].

This feature allows RNNs to retain information about previous inputs, enabling them to process sequences of data with temporal dependencies. Unlike feedforward neural networks, where information flows in one direction only and each input is regarded as independent, RNNs can maintain a memory of past states while processing current inputs. This capability makes RNNs a preferred choice for tasks that involve sequential data.

The working of an RNN can be understood from a simple example, as illustrated in Figure 6.11. Suppose we give the input parameters of some quantity from yesterday; we want to predict what the value of the same parameter will be tomorrow. The first RNN block, which has some weights and biases (Weight 1 and Bias 1), takes the input from yesterday and performs calculations based on these weights and biases. It then sends the result to an activation function. From this activation function, there are two connections: one shown by a red arrow, and the other yielding the predicted value for today. However, since we require the value for tomorrow, we ignore this output.

The path shown by the red arrow is the one that helps the next unit predict the desired output, which is the value for tomorrow. The value obtained from the first block is sent to a weight W2 and then fed to the second network to perform the necessary computations, thus predicting the value of the desired parameter for tomorrow.

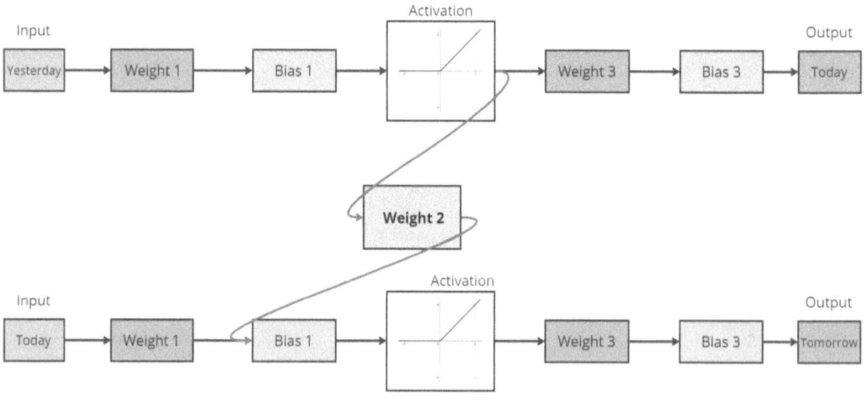

FIGURE 6.11 RNN working illustration.

However, RNNs have certain limitations; they have short-term memory, which makes them unable to retain information over long durations. Therefore, it is not common practice to use RNNs; instead, networks like LSTM, bidirectional LSTM, and gated recurrent unit (GRU) are more commonly used for their enhanced capabilities to store data over longer periods and produce better results. RNNs also suffer from the problem of vanishing gradients.

6.2.2.2 LSTM

LSTM is an enhanced RNN variant that is effective at storing long-term data necessary for various computations and analyses. It was first introduced in 1997 [29].

Figure 6.12 shows an LSTM model architecture where each input is sent to the LSTM model and then, in the end, it is passed through a dense layer having the number of neurons equal to the number of output units, which is one in our case, as the model needs to predict a single value. The activation function used in this final layer is a softmax function.

The structure of the simple LSTM is given in Figure 6.13. It can be seen that an LSTM has three gates:

1. The forget gate is the first gate in an LSTM, and it decides what information should be forgotten. The data is transmitted from the current input and the prior hidden state.
2. The input gate is the second gate in the structure. Its function is to decide what piece of information will enter the state of the LSTM. It first passes the

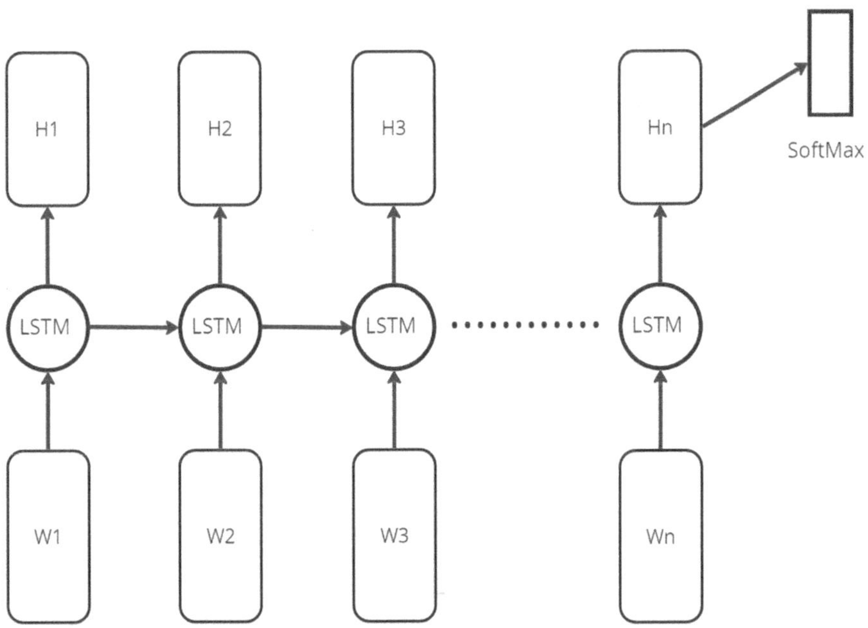

FIGURE 6.12 Structure of an LSTM network model.

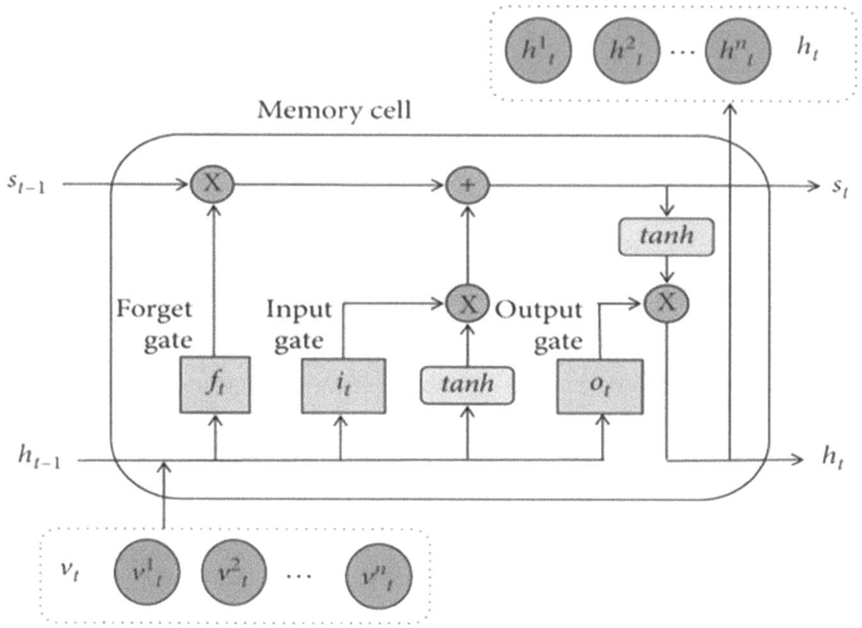

FIGURE 6.13 Structure of a single LSTM unit [30].

hidden state and current input to the sigmoid and tanh activation functions. The sigmoid has a range of 0–1, and tanh has a range of −1 to 1. The outputs from both activations are then multiplied and added to the previous state to get the new state.

3. The output gate is the last gate in the LSTM block, and it decides what the next hidden state should be.

The line S_t in Figure 6.13 denotes the long-term memory, and the line ht denotes short-term memory. LSTM has a unique blend of hidden units, elementwise products, and inter-unit summations. It employs gating mechanisms to regulate memory cells, which are engineered to preserve information over extended durations without alteration [31].

$$i_t = \sigma\left(W_i X_t + U_i h_{t-1} + b_i\right) \tag{6.2}$$

$$f_t = \sigma\left(W_f X_t + U_i h_{t-1} + b_f\right) \tag{6.3}$$

$$O_t = \sigma\left(W_o X_t + U_o h_{t-1} + b_o\right) \tag{6.4}$$

$$C_t' = \tanh\left(W_c X_t + U_c h_{t-1} + b_c\right) \tag{6.5}$$

$$C_t = f_t * C_{t-1} + i_t * C_t^{'} \tag{6.6}$$

$$h_t = \tanh(C_t) * O_t \tag{6.7}$$

$$\tanh x = \frac{e^x - e^{-x}}{e^x + e^{-x}} \tag{6.8}$$

$$\sigma = \frac{1}{1 + e^{-x}} \tag{6.9}$$

The range of the tanh function and σ in equations 6.8 and 6.9 is −1 to 1 and 0–1, respectively, and can be seen from Figures 6.14 and 6.15.

In the above equations 6.2–6.7, the variables i_t represent the input gate at time t, and similarly, f_t and O_t are the forget and output gates at some time t. W is the weight, and the subscript denotes the corresponding gate to which the weights are associated. Similarly, U is the weight matrix that maps the hidden layer output to the gates. b_o, b_f, b_i, and b_c are biases. h_t and C_t are the outputs of the cell and the layer, respectively.

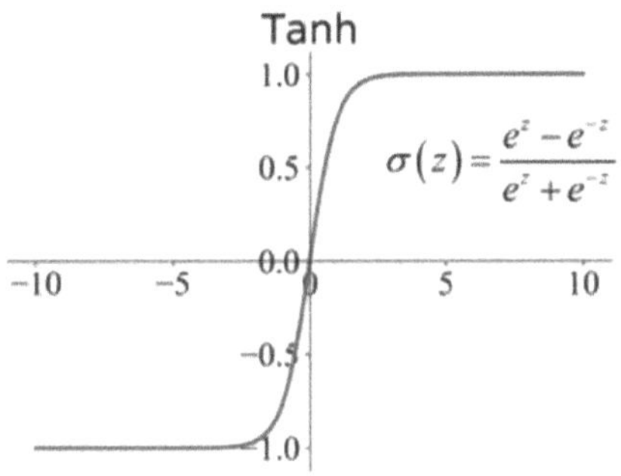

FIGURE 6.14 Tanh activation [32].

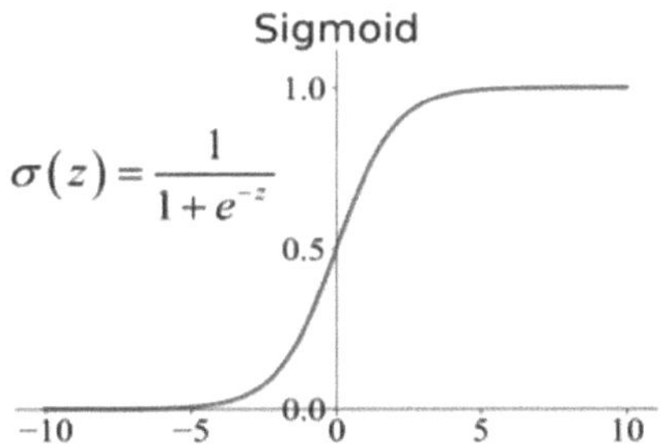

FIGURE 6.15 Sigmoid activation [28].

6.2.2.3 GRU

The GRU [33] was introduced in 2014. It represents another variant of RNNs, featuring built-in gating mechanisms. GRU addresses the issue of short-term memory encountered in traditional RNNs. Structurally, it bears resemblance to LSTM, as illustrated in Figure 6.16, yet it has a simpler architecture.

Unlike LSTM, GRU contains two gates: an Update gate and a Reset gate. In addition to these two gates, it lacks a discrete cell state and has current memory content. As seen in the following equation, the Update gate is essential in deciding how much of the past should be kept and combined with the present inputs at each particular time step.

Equation 6.10 shows the computation of the update gate:

$$Z_t = \sigma(W_z[h_{t-1}, X_t] + b_z) \tag{6.10}$$

Here, h_{t-1} is the preceding hidden state and X_t is the current input.

The reset gate, as the name suggests, somewhat resets the information. It is the deciding entity that decides what past data should be forgotten. It is calculated by equation 6.11:

$$r_t = \sigma\left(W_r[h_{t-1}, X_t] + b_r\right) \tag{6.11}$$

The contents of the current memory are calculated with equation 6.12:

$$H_t' = \tanh\left(W_h[r_t h_{t-1}, X_t]\right) \tag{6.12}$$

The last memory state h_t is calculated by equation 6.13:

$$H_t = (1 - Z_t)h_{t-1} + Z_t H_t' \tag{6.13}$$

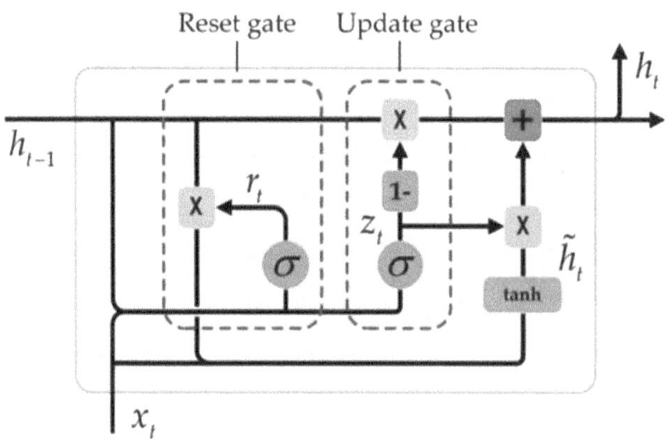

FIGURE 6.16 Structure of a single unit of GRU [34].

The two gates allow GRU to capture information over a period of time, and it has also been seen that GRU performs better than LSTM on smaller datasets [35]. GRU provides a streamlined alternative to LSTM, featuring a reduced number of tensor operations. This simplified architecture enables quicker training processes. Both LSTM and GRU have their own merits and demerits; selecting one purely depends on the use case and the problem that needs to be tackled [26,36].

6.2.2.4 Bilateral LSTM (Bi-LSTM)

Bi-LSTM is an upgraded version of traditional LSTM, which processes the data only in one particular direction – specifically, the forward direction – whereas Bi-LSTM has two layers of LSTM stacked one on top of the other, capturing features in not just one but two directions: both backward and forward. The structure of Bi-LSTM is given in Figure 6.17. The way it works is quite similar to the principle on which LSTM operates; one unit of the Bi-LSTM processes the data in one direction, let us assume forward, while the other processes the data in the opposite direction. This feature enables it to capture both past and future contexts at the same time; both the forward and backward processing are done in the same time interval.

The output of the forward layer is calculated in the time interval $T-n$ and $T-1$with inputs in positive order, whereas the output of the backward layer is computed in the same interval with the help of inverted inputs [38].

To prevent overfitting in the networks, dropout can be used, randomly dropping units from the network. By 'dropping', it means we are temporarily disregarding that unit from the network, including both its incoming and outgoing connections, as shown in Figure 6.18.

Bi-LSTM parameters can be calculated using the following equations 6.13–6.15:

$$f_t = \sigma(W_{\text{ef}}X_t + W_{\text{ef}}h_{t-1} + W_{\text{cf}}C_{t-1} + b_f) \tag{6.13}$$

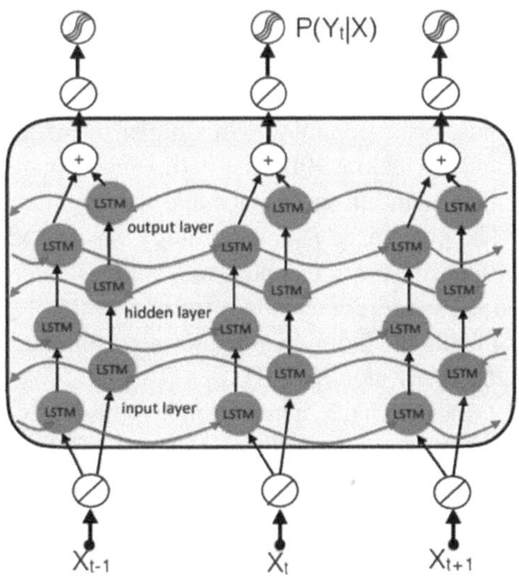

FIGURE 6.17 Structure of Bi-LSTM [37].

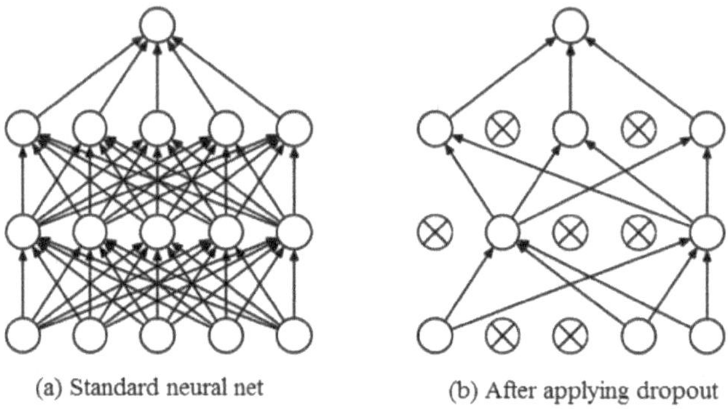

FIGURE 6.18 Dropout vs no dropout [39].

$$i_t = \sigma\left(W_{\mathrm{xi}}X_t + W_{\mathrm{hi}}h_{t-1} + W_{\mathrm{ci}}C_{t-1} + b_i\right) \tag{6.14}$$

$$o_t = \sigma\left(W_{\mathrm{xo}}X_t + W_{\mathrm{ho}}h_{t-1} + W_{\mathrm{co}}C_{t-1} + b_o\right) \tag{6.15}$$

In the above equations, the terms f_t, o_t, and i_t are the forget, output, and input gates, respectively, W is the weighted neural cells, b is the bias term, and h_{t-1} and C_{t-1} are the short and long memory vectors, respectively.

These algorithms have proven successful in predicting the future behaviour of equipment whose data is given as input, allowing for any necessary steps to mitigate the situation to be taken.

6.3 NANOTECHNOLOGY IN INDUSTRY 4.0

In the 20th century, computers used to take up a lot of space and resources; devices like ENIAC weighed tons and covered a vast area. But today, a computer can reside in the palm of our hands. Nanotechnology, in simpler terms, can be understood as the control of atomic matter below 100 nm. At this level, materials stop following the laws of physics; here, quantum mechanics and the combination of physics start playing a major role. The concept of nanotechnology was introduced in 1959 by the well-known scientist Richard Feynman, and it was not until later years that Kim Eric Drexler, an American engineer, coined the term nanotechnology. Nanotechnology covers a wide range of fields, including nanomaterials, nanomedicine, nanoelectronics, nanofabrication, and many more [40,41].

With the advent of Industry 4.0, nanotechnology industry can benefit greatly from it. Industry 4.0 can help increase the accuracy and efficiency of manufacturing nanodevices. Moreover, by leveraging the power of machines, the properties of nanomaterials can be predicted. This can help enhance research and development. The use of digital twin technology can create a virtual replica of nanoscale devices and materials, enabling virtual testing and efficient optimization of the device or material.

This makes nanotechnology a promising avenue for the advancement of Industry 4.0 technologies. It is not just the nanotechnology industry that can benefit from advances in Industry 4.0; the fourth industrial revolution itself can have a significant impact with nanotech. The core of Industry 4.0 is the sensors that capture data from various machines. The use of nanomaterials in sensors can enhance the overall efficiency of the sensor, with the added benefit of a smaller size of the material.

6.4 CONCLUSION

We have examined how important IoT is to Industry 4.0, highlighting its significance for productivity enhancement and data collection. We have also emphasized how digital twin technology is emerging as a game-changing instrument for industrial process validation and optimization, and how PdM solutions can improve operational efficiency, reducing downtime through various assessments, leading to tangible improvements in overall productivity. Furthermore, we delved into various machine learning networks, such as RNN, LSTM, GRU, and Bi-LSTM, which have given us the ability to process sequential data for PdM efficiently. In summary, this study highlights the significance of utilizing cutting-edge technology and machine learning approaches to stimulate creativity and productivity in manufacturing processes, thereby paving the way for a more environmentally friendly and fruitful future.

In the coming years, Industry 4.0 is expected to witness more developments and further advancements. With technologies like digital twin, the outlook of the factory floor is subject to complete change as it is a field currently under study and is expected to see further improvements and refinements. Companies like Nvidia have their own software that can be used to create digital replicas of the world.

REFERENCES

1. E. Kregczy, *European Industrial Landscape by Edmund Kregczy* (Uploaded by Mark Cartwright), 2023. https://www.Worldhistory.Org/Image/17318/Industrial-Landscape-by-Kregczy/
2. Z. Jan, F. Ahamed, W. Mayer, N. Patel, G. Grossmann, M. Stumptner, A. Kuusk, Artificial intelligence for Industry 4.0: systematic review of applications, challenges, and opportunities, *Expert Systems with Applications* 216 (2023). https://doi.org/10.1016/j.eswa.2022.119456.
3. A. Chehri, A. Zimmermann, R. Schmidt, Y. Masuda, Theory and practice of implementing a successful enterprise IoT strategy in the Industry 4.0 era, *Procedia Computer Science* (2021) 4609–4618. https://doi.org/10.1016/j.procs.2021.09.239
4. M. Okano, *IOT and Industry 4.0: The Industrial New Revolution, ICMIS-17, International Conference on Management and Information Systems*, Bangkok, Volume 17, 2017.
5. L. Atzori, A. Iera, G. Morabito, The internet of things: a survey, *Computer Networks* 54 (2010) 2787–2805. https://doi.org/10.1016/j.comnet.2010.05.010
6. S. Selcuk, Predictive maintenance, its implementation and latest trends, *Proceedings of the Institution of Mechanical Engineers, Part B: Journal of Engineering Manufacture* 231 (2017) 1670–1679. https://doi.org/10.1177/0954405415601640
7. J. Lee, J. Ni, J. Singh, B. Jiang, M. Azamfar, J. Feng, Intelligent maintenance systems and predictive manufacturing, *Journal of Manufacturing Science and Engineering* 142 (2020). https://doi.org/10.1115/1.4047856

8. K. Adu-Amankwa, A. K. A. Attia, M. N. Janardhanan, I. Patel, A predictive maintenance cost model for CNC SMEs in the era of Industry 4.0, *The International Journal of Advanced Manufacturing Technology* 104 (2019) 3567–3587. https://doi.org/10.1007/s00170-019-04094-2
9. M. K. Yadav, R. Kumar, R. K. Ratnesh, J. Singh, R. Chandra, A. Kumar, V. Vishnoi, G. Singh, A. K. Singh, Revolutionizing technology with spintronics: devices and their transformative applications, *Materials Science & Engineering B* 303 (2024), 117293. https://doi.org/10.1016/j.mseb.2024.117293
10. S. Pathak, N. Kishore, G. Upadhyay, R. K. Ratnesh, R. Mishra, A compact size planar microstrip-fed patch antenna with hexagonal dgs slot for wlan application, in: *Recent Trends in Electronics and Communication*, LNEE, Springer, 2021, 777, 263–271. https://doi.org/10.1007/978-981-16-2761-3_25
11. M. Achouch, M. Dimitrova, K. Ziane, S. Sattarpanah Karganroudi, R. Dhouib, H. Ibrahim, M. Adda, On predictive maintenance in Industry 4.0: overview, models, and challenges, *Applied Sciences (Switzerland)* 12 (2022). https://doi.org/10.3390/app12168081
12. J. Levitt, *Complete Guide to Preventive and Predictive Maintenance*, 2nd edition, Industrial Press, UK, 2011.
13. H. Meng, Y.-F. Li, A review on prognostics and health management (PHM) methods of lithium-ion batteries, *Renewable and Sustainable Energy Reviews* 116 (2019) 109405. https://doi.org/10.1016/j.rser.2019.109405
14. A. K. S. Jardine, D. Lin, D. Banjevic, A review on machinery diagnostics and prognostics implementing condition-based maintenance, *Mechanical Systems and Signal Processing* 20 (2006) 1483–1510. https://doi.org/10.1016/j.ymssp.2005.09.012
15. S. Boschert, R. Rosen, Digital twin-the simulation aspect, in: *Mechatronic Futures: Challenges and Solutions for Mechatronic Systems and Their Designers*, Springer International Publishing, 2016, 59–74. https://doi.org/10.1007/978-3-319-32156-1_5
16. P. Nunes, J. Santos, E. Rocha, Challenges in predictive maintenance – a review, *CIRP Journal of Manufacturing Science and Technology* 40 (2023) 53–67. https://doi.org/10.1016/j.cirpj.2022.11.004
17. Q. Qi, F. Tao, T. Hu, N. Anwer, A. Liu, Y. Wei, L. Wang, A. Y. C. Nee, Enabling technologies and tools for digital twin, *Journal of Manufacturing Systems* 58 (2021) 3–21. https://doi.org/10.1016/j.jmsy.2019.10.001
18. P. Garg, T. Chandra, R. Ahlawat, N. Mittal, R. K. Ratnesh, S. K. Tripathi, *Star Galaxy Image Classification Via Convolutional Neural Network, ICOSEC-2022*, IEEE, 2022, 1156–1168. https://doi.org/10.1109/ICOSEC54921.2022.9952065
19. S. R. Kothiyal, R. K. Ratnesh, A. K. Shukla, Enhancing sensing with surface plasmons resonance: theoretical insights and design strategies in photonic crystal fiber, in: *International Conference Artificial Intelligence and Information Technologies, (ICAIIT)-2024*, Taylor & Francis, 2023, 279–285. https://doi.org/10.1201/9781003510833-46
20. G. Taranto-Vera, P. Galindo-Villardón, J. Merchán-Sánchez-Jara, J. Salazar-Pozo, A. Moreno-Salazar, V. Salazar-Villalva, Algorithms and software for data mining and machine learning: a critical comparative view from a systematic review of the literature, *Journal of Supercomputing* 77 (2021) 11481–11513. https://doi.org/10.1007/s11227-021-03708-5
21. M. Kitai, T. Kobayashi, H. Fujiwara, R. Tani, M. Numao, K.-I. Fukui, A framework for predicting remaining useful life curve of rolling bearings under defect progression based on neural network and Bayesian method, *IEEE Access* 9 (2021) 62642–62652. https://doi.org/10.1109/ACCESS.2021.3073945
22. Y. Matsuo, Y. LeCun, M. Sahani, D. Precup, D. Silver, M. Sugiyama, E. Uchibe, J. Morimoto, Deep learning, reinforcement learning, and world models, *Neural Networks* 152 (2022) 267–275. https://doi.org/10.1016/j.neunet.2022.03.037

23. R. S. Sutton, A. G. Barto, *Reinforcement Learning: An Introduction*, MIT Press, US, 1998.
24. T. Zonta, C. A. da Costa, R. da Rosa Righi, M. J. de Lima, E. S. da Trindade, G. P. Li, Predictive maintenance in the Industry 4.0: a systematic literature review, *Computers & Industrial Engineering* 150 (2020). https://doi.org/10.1016/j.cie.2020.106889
25. K. S. Kiangala, Z. Wang, Initiating predictive maintenance for a conveyor motor in a bottling plant using Industry 4.0 concepts, *The International Journal of Advanced Manufacturing Technology* 97 (2018) 3251–3271. https://doi.org/10.1007/s00170-018-2093-8
26. F. M. Shiri, T. Perumal, N. Mustapha, R. Mohamed, *A Comprehensive Overview and Comparative Analysis on Deep Learning Models: CNN, RNN, LSTM, GRU*, Cornell University, 2023.
27. H. Apaydin, H. Feizi, M. T. Sattari, M. S. Colak, S. Shamshirband, K.-W. Chau, Comparative analysis of recurrent neural network architectures for reservoir inflow forecasting, *Water (Basel)* 12 (2020) 1500. https://doi.org/10.3390/w12051500
28. W. Bao, J. Yue, Y. Rao, A deep learning framework for financial time series using stacked autoencoders and long-short term memory, *PLoS One* 12 (2017) e0180944. https://doi.org/10.1371/journal.pone.0180944
29. S. Hochreiter, J. Schmidhuber, Long short-term memory, *Neural Computation* 9 (1997) 1735–1780. https://doi.org/10.1162/neco.1997.9.8.1735
30. C. Yang, J. Zhai, G. Tao, Deep learning for price movement prediction using convolutional neural network and long short-term memory, *Mathematical Problems in Engineering* 2020 (2020) 1–13. https://doi.org/10.1155/2020/2746845
31. I. H. Witten, E Frank, *Data mining: Practical machine learning tools and techniques*, Elsevier, San Francisco, CA, 2002.
32. J. Feng, X. He, Q. Teng, C. Ren, H. Chen, Y. Li, Reconstruction of porous media from extremely limited information using conditional generative adversarial networks, *Physical Review E* 100 (2019) 033308. https://doi.org/10.1103/PhysRevE.100.033308
33. K. Cho, B. van Merrienboer, D. Bahdanau, Y. Bengio, On the properties of neural machine translation: encoder–decoder Approaches, in: *Proceedings of SSST-8, Eighth Workshop on Syntax, Semantics and Structure in Statistical Translation, Association for Computational Linguistics*, Stroudsburg, PA, USA, 2014, 103–111. https://doi.org/10.3115/v1/W14-4012
34. P. Li, A. Luo, J. Liu, Y. Wang, J. Zhu, Y. Deng, J. Zhang, Bidirectional gated recurrent unit neural network for Chinese address element segmentation, *ISPRS International Journal of Geo-Information* 9 (2020) 635. https://doi.org/10.3390/ijgi9110635
35. N. Gruber, A. Jockisch, Are GRU cells more specific and LSTM cells more sensitive in motive classification of text?, *Frontiers in Artificial Intelligence* 3 (2020). https://doi.org/10.3389/frai.2020.00040
36. S. Abbaspour, F. Fotouhi, A. Sedaghatbaf, H. Fotouhi, M. Vahabi, M. Linden, A comparative analysis of hybrid deep learning models for human activity recognition, *Sensors* 20 (2020) 5707. https://doi.org/10.3390/s20195707
37. D. Liciotti, M. Bernardini, L. Romeo, E. Frontoni, A sequential deep learning application for recognising human activities in smart homes, *Neurocomputing* 396 (2020) 501–513. https://doi.org/10.1016/j.neucom.2018.10.104
38. H. Apaydin, H. Feizi, M. T. Sattari, M. S. Colak, S. Shamshirband, K. W. Chau, Comparative analysis of recurrent neural network architectures for reservoir inflow forecasting, *Water (Switzerland)* 12 (2020). https://doi.org/10.3390/w12051500
39. N. Srivastava, G. Hinton, A. Krizhevsky, R. Salakhutdinov, *Dropout: A Simple Way to Prevent Neural Networks from Overfitting*, MIT Press, Microtome Publishing, JMLR, Inc., 2014.

40. R. K. Ratnesh, M. K. Singh, J. Singh, Enhancing ZnO/Si heterojunction photodetector performance for ultra high responsivity across wide spectral range, *Journal of Material Science: Materials in Electronics* 35 (2024) 756. https://doi.org/10.1007/s10854-024-12516-5
41. R. K. Ratnesh, M. K. Singh, V. Kumar, S. Singh, R. Chandra, M. Singh, J. Singh, Mango leaves (Mangifera Indica) derived highly florescent green graphene quantum dot nanoprobes for enhanced on-off dual detection of cholesterol and Fe^{2+} ions based on molecular logic operation, *Biomaterials Science & Engineering, American Chemical Society* 7 (2024) 4417–4426. https://doi.org/10.1021/acsabm.4c00292

7 Advancement of a Transmission System with Novel Methods and Tools

Abdullah Bin Queyam, Pawan Kumar Pathak, Ramesh Kumar, and Mohammad Aljaidi

7.1 INTRODUCTION

7.1.1 BACKGROUND

Solar energy, as an extremely abundant source of renewable energy, provides sustainable energy security without damaging the environment. The two main technologies for solar power are photovoltaic (PV) power systems and concentrated power systems (CPSs) [1]. In a PV system, solar energy is converted into electricity through the photoelectric effect, allowing energy to pass through the semiconductor material and generate electricity when exposed to light. The device responsible for this conversion is called a PV cell or solar cell. In CPSs, sunlight is concentrated using lenses and mirrors to produce heat, which is then used to generate steam. This steam drives turbines that generate electricity, making CPSs a thermal-to-work conversion system similar to thermal power systems [2].

Two types of systems are employed to tap the electricity generated in the solar cells: off-grid (stand-alone) and grid–tie systems. Off-grid systems operate independently from the main power grid, relying on batteries to store surplus energy for use when sunlight is not available. In contrast, grid–tie systems are connected to the main power grid, allowing surplus electricity produced by the solar cells to be fed back into the grid for economic compensation [3].

Power conversion efficiency is a crucial measure for assessing a solar cell's performance, indicating the percentage of sunlight converted into electricity. A higher efficiency signifies that more sunlight is effectively transformed into usable electrical power. Various factors influence a solar cell's efficiency, including its material, design, and environmental conditions. Monocrystalline silicon solar cells, widely used, typically have efficiencies ranging from 15% to over 25% [4]. Gallium arsenide (GaAs) solar cells, typically utilized in space applications, can attain efficiencies exceeding 30%. Amorphous silicon solar cells, recognized for their flexibility and lightweight nature, have efficiencies starting from 5% to 12% [5]. Silicon heterojunction solar cells, which combine amorphous and crystalline silicon, have proven efficiencies of over 26%. Other types of solar cells include cadmium telluride (CdTe)

DOI: 10.1201/9781003634553-7

solar cells, which have efficiencies of around 9%–22%, and cadmium indium gallium selenide (CIGS) solar cells, which have efficiencies ranging from 10% to 20% [6].

Basic solar cells, which use natural molecules to absorb sunlight and generate power, have efficiencies starting from 5% to 15%. Perovskite solar cells (PSCs), a promising emerging technology, have shown overwhelming development in recent years, reaching efficiencies surpassing 25% or even achieving 29%. PSCs offer potential benefits over conventional silicon cells, including lower production costs and the ability to be manufactured using simpler techniques [7]. Continued research and development are focused on enhancing the efficiency of these types of solar cell technology to make solar energy more affordable within the electricity market.

Perovskite-sensitized solid-state solar cells in their optimized form can deliver efficiencies as high as 20% [8]. Previously, perovskite was used as a sensitizer in a dye-sensitized cell, and it was discovered early on that the operation of perovskite was quite similar to that of other cells. In some research, perovskite PVs were treated as a $p-i-n$ diode [9]. An equivalent circuit model with a three-diode lumped parameter for organic solar cells capable of electronically explaining the S-shaped curve was also proposed [10]. Researchers have proposed a closed-form solution to Mazhari's model and predicted that higher annealing temperatures are helpful in suppressing the S-shaped *I–V* characteristics of organic solar cells [11]. Power conversion efficiency has significantly increased as a result of research on PSCs, ranging from 3.8% in 2009 [12] to 22.7% recently [13,14].

7.1.2 Structural Comparison

The structure of a monocrystalline silicon solar cell is illustrated in Figure 7.1. It utilizes an aluminum back surface field on a P-type silicon wafer, which is 200 μm thick and has a resistivity of 1–3 Ω. This back surface field is created by establishing

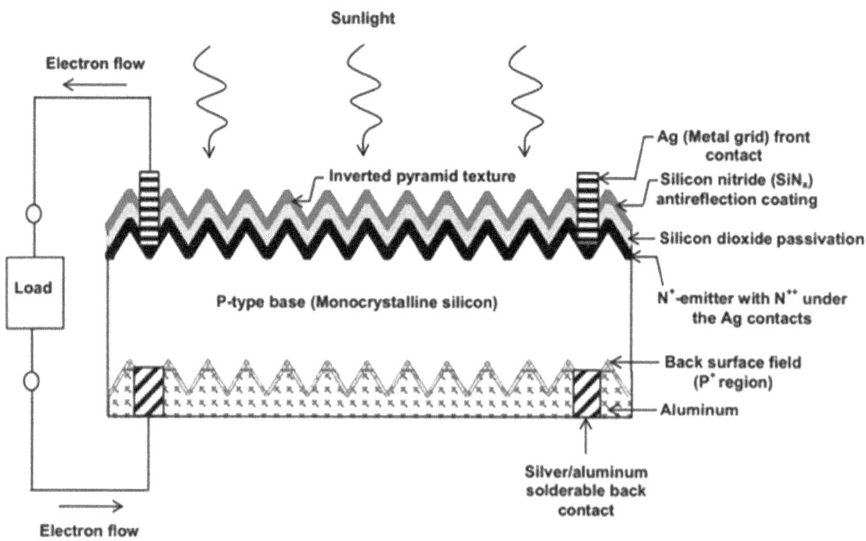

FIGURE 7.1 Monocrystalline silicon solar cell [15].

a heavily doped (P+) region at the rear surface, resulting in a high-low (P+/P–) junction that acts as a barrier to the flow of minority carriers. The manufacturing process involves texturizing the wafer surface, diffusing phosphorus to create an N+ emitter, removing phosphosilicate glass, growing SiO_2, depositing a SiN_x antireflecting coating, screen-printing Ag front contacts, screen-printing and drying Ag/Al back contacts, and sintering an Al film. The solar cell is irradiated from the side of the Ag grid, and the electrical output is collected between the Ag front and the Ag/Al back contacts [15].

The material perovskite has the same crystal structure as the naturally occurring mineral calcium titanate ($CaTiO_3$). Perovskites are compounds with the chemical formula ABX_3, where A and B are cations, with A being larger than B, and X is an anion, as illustrated in Figure 7.2a. Halide perovskites, a type of perovskite like oxide perovskites, have gained attention for their use in solar cells. One commonly used organic–inorganic perovskite is methylammonium lead trihalide, which contains the methylammonium cation and a halogen anion such as iodine, bromine, or chlorine ($CH_3NH_3PbI_3$). Perovskite materials offer several advantages, including low cost, strong optical absorption, bandgap tuning, long carrier diffusion lengths, and ease of device fabrication.

The structure of the PSC described in Figure 7.2b is manufactured on a glass substrate coated with indium-doped tin oxide or fluorine-doped tin oxide [15]. A layer of N-type semiconductor is deposited on this substrate as an electron transport layer (ETL), with compact TiO_2 (c-TiO_2) commonly used for this purpose, although other materials such as SnO_2, C_{60}, and its derivatives can also be used. The perovskite absorber layer (typically 300 ~ 500 nm) is then formed over the ETL, followed by the application of a P-type semiconductor as a hole transport layer (HTL). The final layer is a metal layer over the HTL. The perovskite layer, ETL, and HTL are applied using spin-coating for small-area solar cells and screen-printing or spray-coating for larger-size cells.

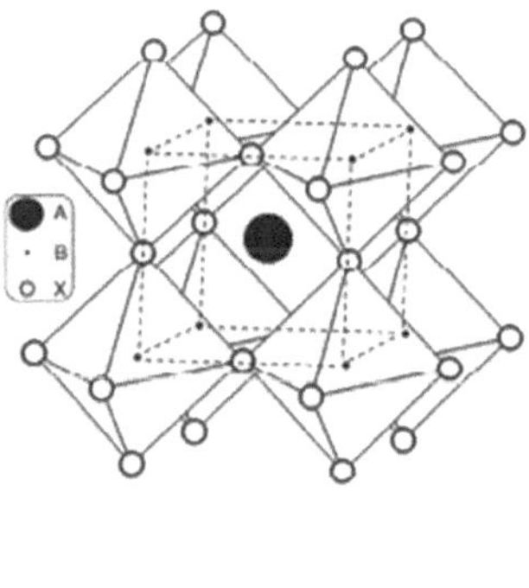

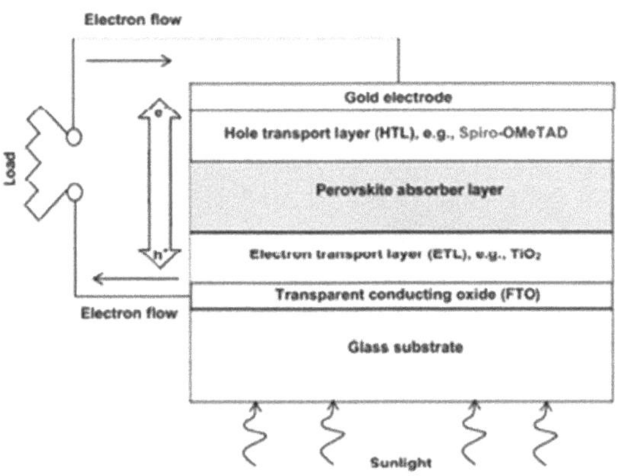

FIGURE 7.2 (a) Crystal structure of a perovskite solar cell (PSC) with the chemical formula ABX_3, (b) Structure of a perovskite solar cell (PSC) [15].

7.1.3 Equivalent Circuit Model of a PSC

The current–voltage (*I*–*V*) characteristic of a solar cell is limited in providing a comprehensive understanding of the mechanisms related to charge transport, recombination, and storage. Therefore, it is crucial to evaluate the cell's response to small perturbations of real inputs such as voltage, temperature, or light intensity. These mechanisms are typically represented by an electrical equivalent circuit, which describes the physical processes using passive electrical elements, voltage sources, and current sources.

The perovskite cell has been represented as a straightforward $p-i-n$ diode [9]. As shown in Figure 7.3a and b, numerous publications have employed the one-diode and two-diode models [16]. In addition to these regularly used models, PSCs have been modeled using various organic cell models. PSCs also have an asymmetrical photovoltage rise and decay curve, which differs from prior forms of PV cells, such as dye-sensitized solar cells. This unconventional phenomenon was explained by an electrical equivalent circuit model for the perovskite cell, including electronic components and ionic components [17], as illustrated in Figure 7.3c.

In this study, our focus will be on using the single diode model because the equivalent circuit for the perovskite-based solar cell simplifies the illustration of the cell's behavior, allowing us to investigate its overall performance.

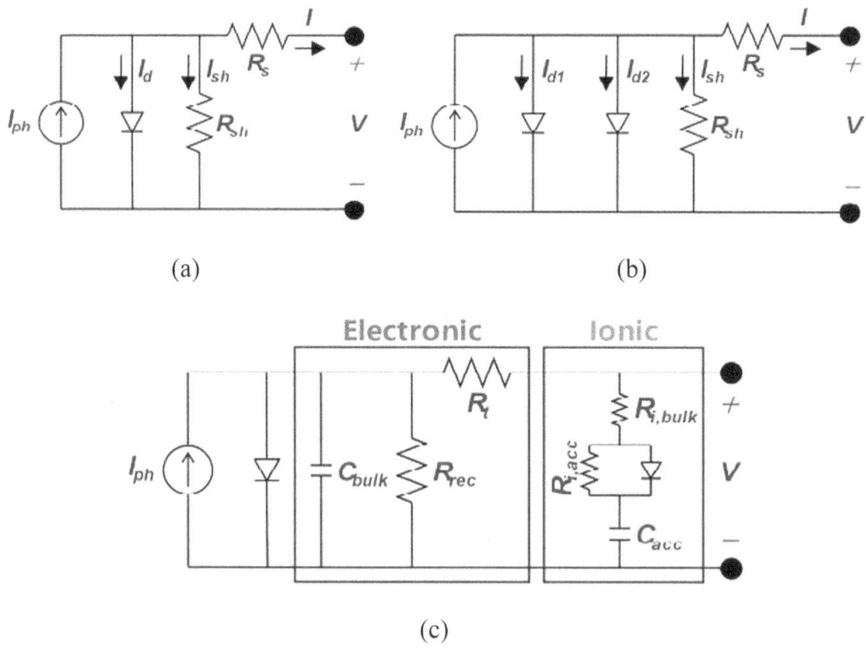

FIGURE 7.3 A perovskite solar cell circuit model (equivalent) (a) One-diode model, (b) Two-diode model [16], (c) Equivalent circuit for perovskite solar cell (photo-voltage rise and falling-off) [17].

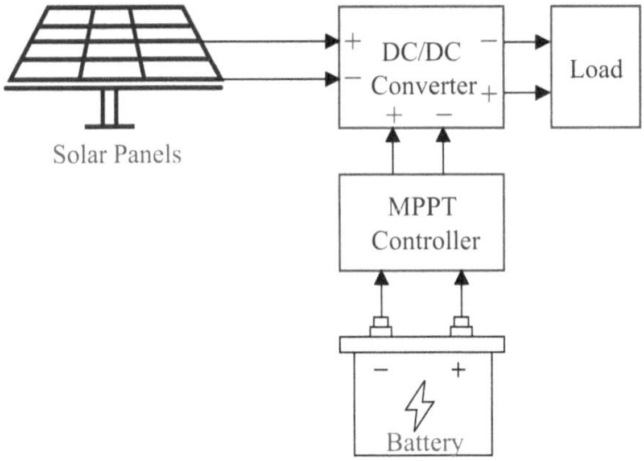

FIGURE 7.4 Block diagram of a PV system.

7.2 MATERIALS AND METHODS

7.2.1 Maximum Power Point Tracking (MPPT) System

MPPT controllers are an important part of PV systems, which improve the efficiency of solar energy cells by incorporating a sophisticated algorithmic approach that continuously screens and regulates the operating point of the PV modules to extract excess available energy from the PV array, as shown in Figure 7.4 of the block diagram of the MPPT-based PV system.

The MPPT controller tracks the point of maximum power in the *I–V* curve, irrespective of environmental factors, including varying light conditions and temperature, as shown in Figure 7.5.

In this study, we have considered a single-diode simplified electrical equivalent model for the maximum power output of a PV module to estimate its performance [18–,19,20]. Equation 7.1 given below is used in the evaluation and design of solar cells to determine their maximum power output under specific operating conditions.

$$P_{max} = V_{OC} \times I_{SC} \times FF \tag{7.1}$$

where P_{max} is the maximum output power, V_{OC} is the open-circuit voltage at the initial conditions, I_{SC} is the short-circuit current when the voltage across the solar cell is zero, and FF is the fill factor, a parameter that reflects losses in the solar cell due to unfavorable behaviors such as recombination and resistance. It is the ratio between the nominal and maximum power standards [21].

7.2.2 Particle Swarm Optimization (PSO)

PSO is a nature-inspired optimization algorithm used to discover approximate answers to optimization problems by simulating the behavior of a population of particles moving through a search space to find the most effective solution. The set of

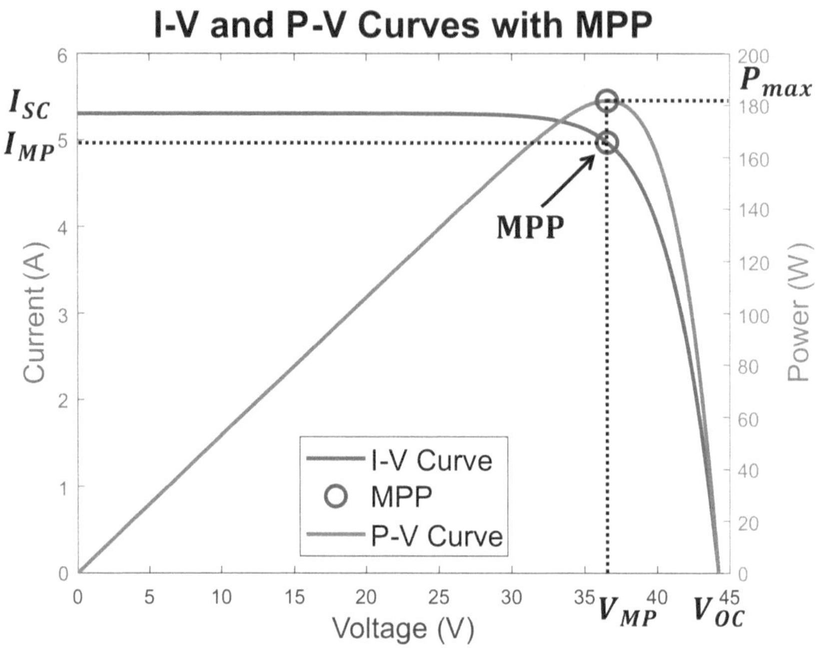

FIGURE 7.5 I–V and P–V curves of a photovoltaic module and MPP.

rules is primarily based on the social behavior of birds and fish, wherein particles in a swarm collaborate to find the best solution.

The position of each PSO particle is represented as a vector $x_i = (x_{i1}, x_{i2}, ..., x_{id})$, where i is the particle index is and d is the dimensionality of the problem. The velocity of each particle is represented as $v_i = (v_{i1}, v_{i2}, \ldots, v_{id})$. At each iteration, the velocity and position of each particle are updated using the following equations:

$$v_{id}(t+1) = w \cdot v_{id}(t) + c_1 \cdot r_1 \cdot (p_{id} - x_{id}(t)) + c_2 \cdot r_2 \cdot (p_{gd} - x_{id}(t)) \tag{7.2}$$

$$x_{id}(t+1) = x_{id}(t) + v_{id}(t+1) \tag{7.3}$$

where w is the inertia weight, controlling the impact of the previous velocity on the current velocity.

The inertia weight w is typically decreased linearly over iterations to balance exploration and exploitation.

Several studies have been conducted to enhance the performance of the MPPT control strategies using PSO [22], other hybrid metaheuristic techniques [23], PSO dynamic swarm size reduction [24] and fuzzy logic [25].

7.2.3 Gravitational Search Algorithm (GSA)

Rashedi, Nezamabadi-Pour, and Saryazdi introduced the GSA, a metaheuristic optimization algorithm, in 2009 [26]. The approach was created to resolve optimization problems by simulating the gravitational forces among masses. It draws inspiration from the regulation of gravity and the behavior of celestial entities within the universe. The fundamental idea behind GSA is to represent potential solutions to optimization problems as masses in a multidimensional space. The optimization process is then modeled as the motion of these masses under the influence of gravitational forces. The key components of the GSA include

- **Mass representation:** Each potential solution is represented as a mass in the search space.
- **Gravitational force:** The gravitational force between two masses is calculated based on their masses and the distance between them. Higher masses experience stronger attractive forces, while greater distances result in weaker forces.

$$F_{ij} = G \cdot \frac{m_i \cdot m_j}{r_{ij}^2} \tag{7.4}$$

- **Acceleration:** The gravitational force is used to calculate the acceleration of each mass, which, in turn, determines its movement in the search space.

$$a_i = \sum_{j=1, j \neq i}^{N} \frac{F_{ij}}{m_i} \tag{7.5}$$

- **Updating positions:** Masses update their positions based on the calculated accelerations, and the process is iteratively repeated to improve the solution.

$$x_i(t+1) = x_i(t) + v_i(t+1) \tag{7.6}$$

where $x_i(t)$ is the position of mass i at time t and $v_i(t+1)$ is the velocity update based on acceleration.

Like other metaheuristic algorithms, the effectiveness of GSA depends on various factors such as parameter settings, problem characteristics, and the nature of the optimization landscape. Researchers continue to explore and enhance the performance of the GSA and its variants for different types of optimization problems.

7.2.4 PV Model

At the heart of the MPPT system is the PV Model, an electrical equivalent representation of a PSC that exhibits electrical characteristics similar to real-world PV cells, as shown in Figure 7.6. This model is particularly sensitive to two key environmental inputs: irradiance and temperature.

The equations governing a solar cell's output current are fundamental to understanding its behavior. The output current I_{PV} is determined by the difference between the PV current I_{ph} and the sum of the diode current I_d and the parallel resistance current I_{sh}.

$$I_{PV} = I_{ph} - I_d - I_{sh} \tag{7.7}$$

The diode current I_d is influenced by the diode factor n, the reverse saturation current I_o, the electron charge q, the Boltzmann constant k, and the p–n junction temperature T.

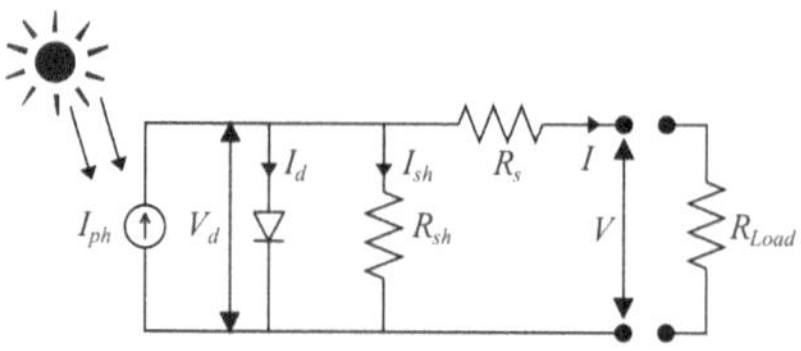

FIGURE 7.6 Equivalent circuit of a solar cell.

$$I_d = I_o\left(e^{\frac{q(V_{PV}+I_{PV}R_s)}{nkT}} - 1\right) \tag{7.8}$$

By substituting the diode current equation into the output current equation, we get

$$I_{PV} = I_{ph} - I_o\left[e^{\frac{q(V_{PV}+I_{PV}R_S)}{nkT}} - 1\right] - \frac{V_{PV}+I_{PV}R_s}{R_{sh}} \tag{7.9}$$

Considering that R_s is negligible and R_{sh} is significant, we simplify the equation to

$$I_{PV} = I_{ph} - I_o\left[e^{AV_{PV}} - 1\right] \approx I_{ph} - I_o e^{AV_{PV}} \tag{7.10}$$

where $= \frac{q}{nkT}$, this equation 7.10 helps determine the cell's output current. Moreover, by substituting zero PV voltage and current into the equation, we can calculate the short-circuit current I_{SC} and the open-circuit voltage V_{OC}:

$$I_{SC} = I_{ph} \tag{7.11}$$

$$I_o = I_{SC}\exp(-AV_{OC}) \tag{7.12}$$

Additionally, the maximum power point (MPP; I_{MP}, V_{MP}) can be expressed using these equations as follows:

$$I_{MP} = I_{SC}\left[1 - \exp\left(A\left(V_{MP} - V_{OC}\right)\right)\right] \tag{7.13}$$

$$A = \frac{1}{\left(V_{MP} - V_{OC}\right)}\ln\left(1 - \frac{I_{MP}}{I_{SC}}\right) \tag{7.14}$$

Lastly, the output voltage V_{PV} can be determined using the circuit parameters V_{OC}, I_{SC}, I_{MP}, and V_{MP}.

$$V_{PV} = V_{OC}\left[1 + \frac{1}{A}\ln\left(1 - \frac{I}{I_{SC}}\right)\right] \tag{7.15}$$

This PV model is an assembly of several sub-components, including a source that simulates solar irradiance, a thermal model to account for temperature variations,

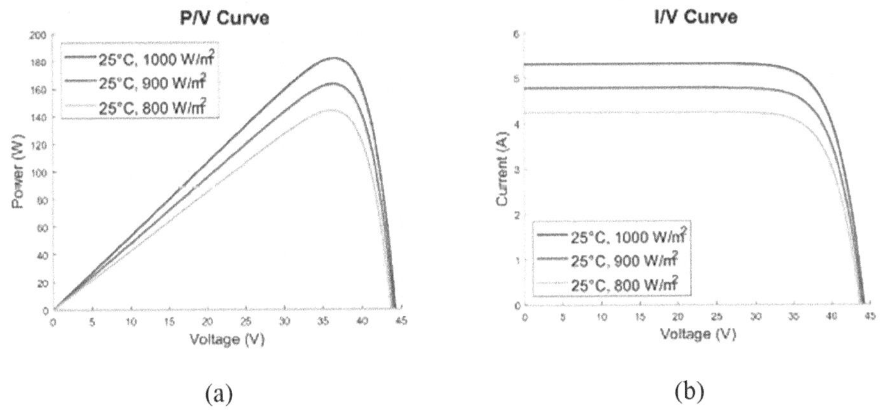

FIGURE 7.7 (a) P–V curve and (b) I–V curve on different solar irradiance.

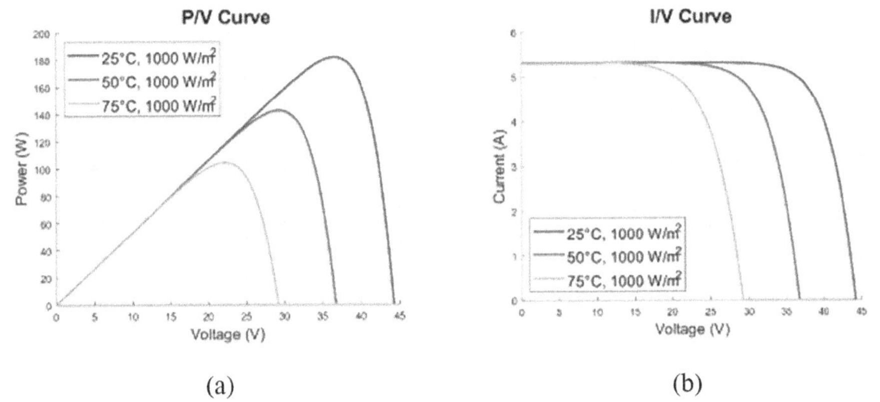

FIGURE 7.8 (a) P–V curve and (b) I–V curve on different temperatures.

and an equivalent circuit comprising diodes, resistors, and capacitors. These components collectively mimic the power–voltage (*P–V*) and *I–V* characteristics of a solar panel. By altering the irradiance and temperature inputs, one can observe the corresponding changes in the PV cell's output as illustrated in Figure 7.7 and 7.8, thus enabling a dynamic analysis of the system's behavior.

7.2.5 Commonly Used MPPT Techniques

The primary challenge faced by commercial solar PV systems is their lower conversion efficiency. To improve the system's effectiveness, MPPT algorithms are used. The PV system is expected to achieve maximum efficiency when operating at its MPP [27].

Perturb and observe (PO): The PO method is a simpler MPPT tracking technique that adjusts the operating voltage of the solar panels by perturbing the voltage and observing the change in power output. The algorithm then reverses the direction of the perturbation to find the MPP. However, it suffers from oscillations around the MPP, especially under rapidly changing irradiance conditions.

Incremental conductance (IC): The principle behind the IC algorithm is based on the fact that at the MPP, the IC (dI/dV) of the PV array is equal to the negative ratio of the array current (I) to voltage (V). In other words, the algorithm seeks to make the (dI/dV) match the ($-I/V$). At the MPP, the rate of change of power with respect to voltage (dP/dV) is zero, which implies that dI/dV should be equal to $-I/V$.

In this study, we have implemented a modified IC algorithm for MPPT in a PV system. The modified IC function first calculates the duty cycle D based on the current voltage V and current I readings [28]. It uses a continuous variable to store the previous voltage, power, current, and duty cycle values to track changes. The algorithm adjusts the duty cycle incrementally based on changes in power and voltage, aiming to track the MPP of the PV system [29,30]. It also consists of safeguards to prevent the duty cycle from exceeding predefined limits (D_{max} and D_{min}) and to hold balance under positive conditions in versatile applications [31–36].

7.3 PROPOSED METHODOLOGY

7.3.1 MPPT Algorithms Integration in MATLAB®/Simulink

The MPPT algorithms are carried out as distinct functional blocks within the Simulink model. Every block is designed to interface seamlessly with the PV model, as illustrated in Figure 7.9. The schematics depict the use of the PO, IC, PSO, and GSA techniques with a rotary switch to select the preferred MPPT algorithm. These algorithms are represented in a way that abstracts the complex optimization processes into manageable blocks, focusing on the interaction between the PV system and control logic.

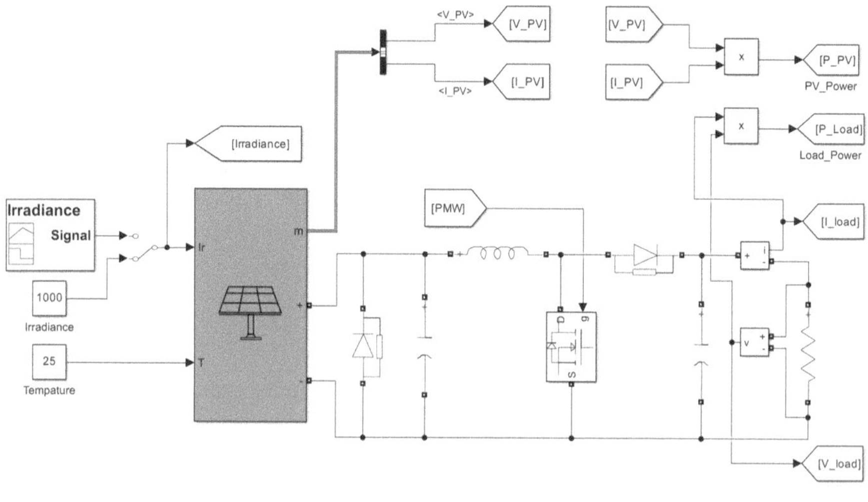

FIGURE 7.9 Implementation of PV system in MATLAB/Simulink.

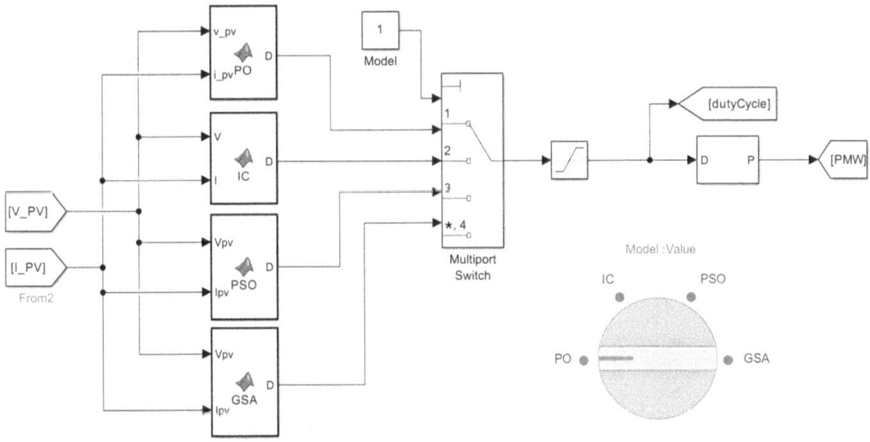

FIGURE 7.10 MPPT algorithm integration.

The algorithms receive inputs directly from the PV model, such as current and voltage, and apply their respective optimization techniques to iteratively adjust the duty cycle of a DC–DC converter.

Solar PV is designed to run at MPP for maximum power generation. To track the MPP of PV systems, the power converter operates at the appropriate duty cycle. To track MPP, the duty cycle must adjust in accordance with changes in solar insolation. Since boost converters are simpler and more reliable than other DC–DC converter configurations that have been investigated thus far, they are frequently selected and considered in the work being presented.

This process is inherently aimed at finding the optimal operating point that maximizes the power output from the PV module, irrespective of the static nature of the system's environmental inputs. The rotary switch allows the user to change which algorithms will be implemented in the system, as can be seen in Figure 7.10.

The visualization of simulation outcomes in a Simulink model of an MPPT system is pivotal to understanding the efficiency and effectiveness of the MPPT algorithms being used. The schematics provided suggest a systematic method of collecting and displaying data that highlights the dynamic performance of the system, particularly the power outputs from the PO, IC, and PSO algorithms.

In the Simulink environment (Figure 7.11), 'Display' blocks are utilized to provide immediate visual feedback of key performance indicators such as voltage, current, power (from the PV system as well as from the boost converter), and the duty cycle. These blocks are connected directly to the outputs of the MPPT algorithms and the load, ensuring that any changes in the system's behavior are quickly reflected in the displayed values. The use of a 'Workspace' block serves as a bridge between the Simulink model and the MATLAB workspace. This block is particularly useful for exporting simulation data for further analysis within the MATLAB environment. By feeding data from the MPPT algorithms into the Workspace block, it becomes possible to store the entire simulation dataset in a MATLAB variable.

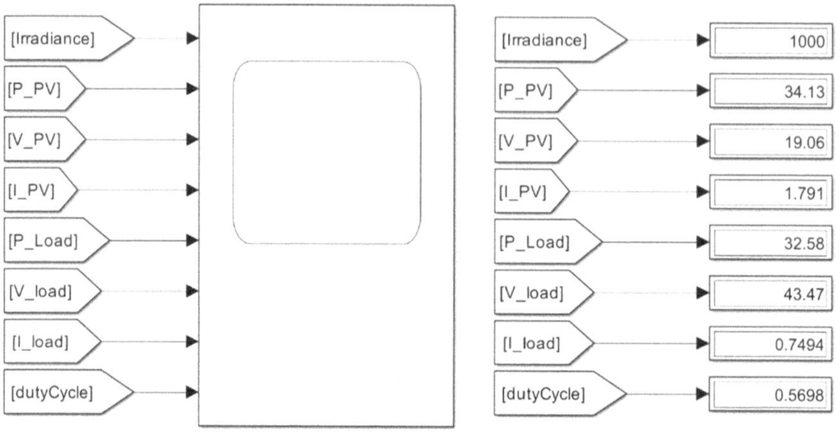

FIGURE 7.11 Block design for visualization.

7.3.2 PSO Implementation

The implementation of the PSO algorithm follows a well-defined process, as illustrated in the flowchart provided in Figure 7.12 and discussed further.

7.3.2.1 Initialization

Firstly, a set of persistent variables is initialized if they are not already set. These variables store the state of the PSO algorithm between function calls:

- **localbest or personalbest:** Stores the best position found by each particle.
- **globalbest:** Stores the best position found by any particle.
- **k:** A counter to limit the number of iterations the algorithm runs.
- **p:** Array to store the best power output found by each particle.
- **dc:** Array to store the duty cycle for each particle.
- **Pbest:** The best power output found by the algorithm so far.
- **Pprev:** The power output from the previous iteration.
- **dcurrent:** The current duty cycle being used.
- **u:** An index to cycle through each particle.
- **v:** Array to store the velocity of each particle.
- **temp:** A variable used as a conditional flag in the algorithm.

The duty cycles (**dc**) are initialized with random values within specific ranges to ensure a diverse starting point for each particle's exploration. The variable **dcurrent** is set to an initial guess of 0.5, and **globalbest** is assigned this value as the initial best-known position.

7.3.2.2 Iterative Loop

In this stage, the algorithm starts by setting **D** to the current duty cycle (**dcurrent**). It then performs a series of checks and updates. If **temp** is negative, it increments **temp** and exits early. This mechanism could serve as a delay or a precondition check before

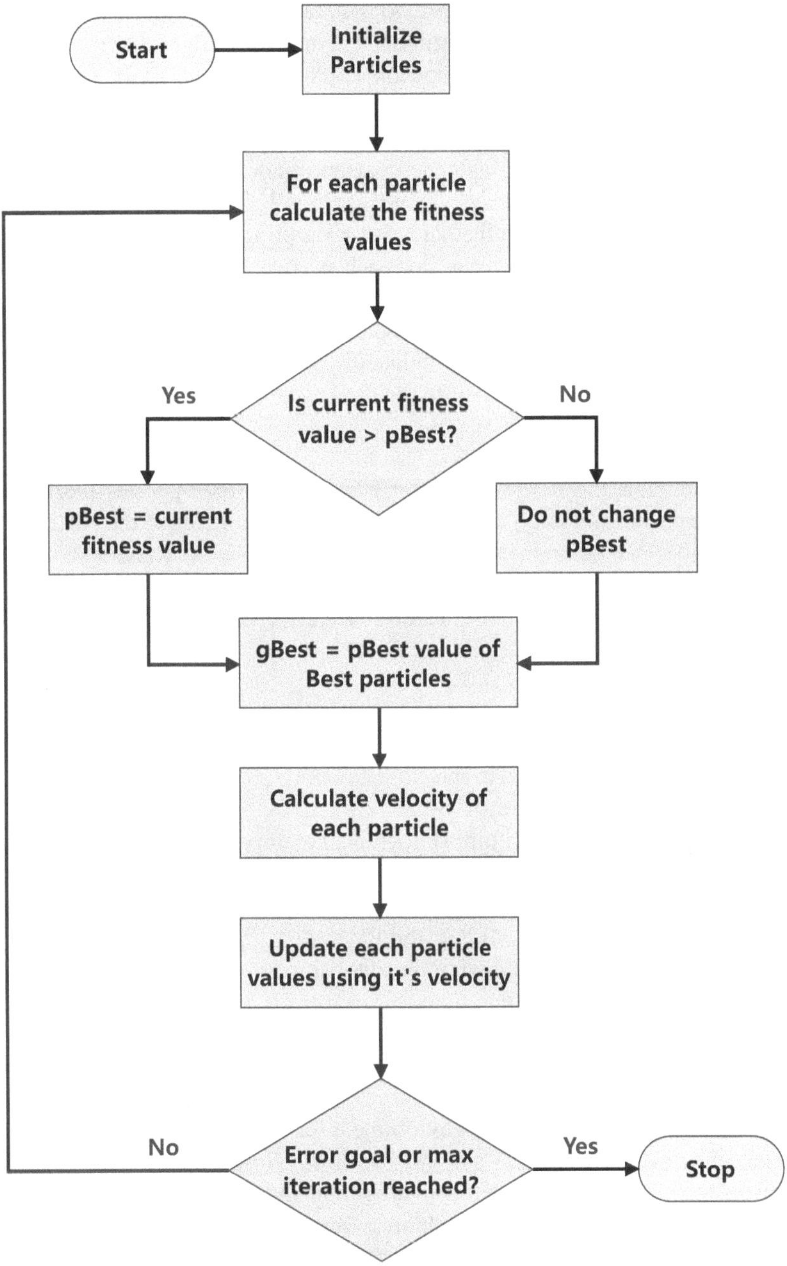

FIGURE 7.12 Flow chart of PSO algorithm.

the algorithm proceeds. If the current power P is greater than **Pbest**, it updates **Pbest**. This step is equivalent to evaluating a particle's fitness and updating its personal best. If the iteration counter **k** is less than 3000, it increments **k** and exits early, indicating that the algorithm should continue for more iterations.

If the change in power from the previous iteration is negligible, it suggests the possibility of being stuck at a local optimum. Therefore, the algorithm re-randomizes the duty cycles (**dc**) to potentially escape and explore new areas of the search space.

7.3.2.3 Particle Updates

Next, the algorithm iterates through each particle (indexed by **u**) to update its personal best (**p**) and its best-known position (**localbest**). If **u** equals 4, it signifies the end of a complete cycle through the particles, prompting an update of **globalbest** to the best position found in this cycle. For each particle, the function calculates the new velocity using the update velocity function, which is influenced by the particle's current velocity, the distance from its personal best, and the distance from the global best. The new velocity is then used to update the particle's position (duty cycle). This position update is constrained to ensure the duty cycle remains between 0 and 1.

7.3.2.4 Velocity Update

This stage computes the new velocity for a particle. It incorporates inertia (via a fraction of the current velocity), cognitive component (attraction to the particle's personal best), and social component (attraction to the global best). Random coefficients **c1** and **c2** introduce 20 stochastic elements to the cognitive and social components, allowing the particles to explore the search space effectively.

7.3.2.5 Duty Update

In this stage, the duty cycle of a particle is updated based on its velocity. It ensures that the new duty cycle is within the valid range (0–1). If the updated value goes beyond this range, it is corrected to the closest bound (set to 1 if it's above 1, or taken as the absolute value if it's below 0).

By the end of the PSO algorithm, **D** contains the duty cycle corresponding to the best solution found by the swarm (when u equals 4) or the next value to try (when u is less than 4). This value will be used by the MPPT system to adjust the converter's operation, aiming to reach the MPP of the PV system. The loop continues with the next call to the PSO function, allowing the algorithm to refine its search for the optimal duty cycle over successive iterations.

7.3.3 GSA Implementation

The MATLAB system provided the environment for implementing the GSA for optimizing the duty cycle in an MPPT system. The GSA is inspired by the laws of physics and utilizes the metaphor of gravitational attraction and mass interactions to navigate the search space toward the optimal solution, which in this case is the MPP of a PV system. In MATLAB the GSA algorithm is implemented, as described in Figure 7.13.

7.3.3.1 Initialization

The commencement of the GSA function is marked by the inspection of the counter variable. The absence of its instantiation, determined by an empty state, signals the inaugural execution of the algorithm. This phase is characterized by the initialization of an array, which encapsulates the positional data of each agent within the search

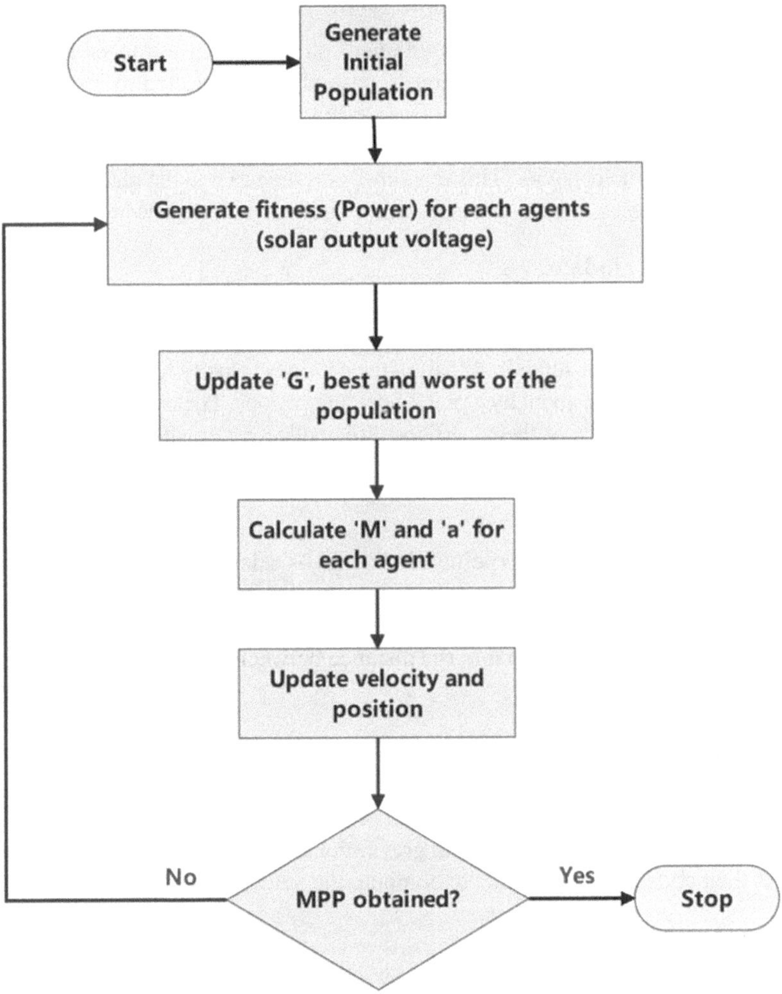

FIGURE 7.13 Flow chart of GSA algorithm.

space. These initial positions, derived through a randomized allocation, span a vast spectrum of potential solutions, thereby facilitating comprehensive exploration.

7.3.3.2 Iterative Loop

Post-initialization, the agents embark on their iterative journey through the search space, propelled through the GSA algorithm. Each iteration represents an incremental time step, advancing the agents' search for the optimal solution. During every iteration, the algorithm engages in a thorough evaluation of fitness, denoted by strength output of each agent. This method involves the assessment of the current output compared with the previously established personal records.

7.3.3.3 Force Calculation

With the agents' masses therefore ascertained, the algorithm advances to the force computation stage. Here, the gravitational constant G is subjected to an exponential decay over time, embodying the gradual convergence of the algorithm. The forces acting among pairs of agents are calculated, incorporating a random detail to saturate the device with stochastic behavior. This randomness is essential to the algorithm capability to navigate complex search spaces, averting untimely convergence to local optima.

7.3.3.4 Position Update

The calculated forces precipitate an update in the agents' velocities and finally, their positions – conceptualized as duty cycles inside the MPPT. The update mechanism is infused with randomness, retaining a part of the previous velocity while assimilating the new acceleration brought about by gravitational interactions. The resultant movement of the agents is a confluence of their inertia and the collective gravitational pull, guiding them toward regions of the search space characterized by superior power outputs.

7.3.3.5 Distance Function

Euclidean distance between two points, **d1** and **d2**, is calculated, which constitute the positions (duty cycles) of two agents inside the search area. In the context of GSA, this distance plays a role in the force calculation between agents, as the gravitational pressure is inversely proportional to the distance between the agents.

7.3.3.6 Velocity Update

The velocity of an agent is updated based on its current velocity and the acceleration due to the gravitational force. A stochastic element is added by multiplying the current velocity with a random number, which provides variability to the agent's movement. The acceleration, a result of the gravitational pressure divided by the agent's mass, is then added to this product to compute the new velocity.

7.3.3.7 Duty Update

A new position of an agent is calculated by adding the velocity to the current position (duty cycle). The resulting duty cycle is then checked to ensure that it is within the valid range for the system. If the resulting responsibility cycle exceeds 1, it is set to 1, and if it is less than 0.1, it is set to 0.1. These bounds ensure that the duty cycle remains within a physically realizable range for the PV system.

By harnessing the ideas of gravitational interaction and mass relationships, the GSA presents an effective and powerful method of tracking the maximum power peak of the PV system. It reflects a fine-tuned process that adapts to the unique characteristics of the search space, methodically progressing toward the peak power output, which is the ultimate goal of the MPPT system.

7.4 RESULTS

In this study, we delve into the algorithmic efficacy of four predominant strategies in the domain of PV system optimization: PO, IC, PSO, and GSA. The data are visually represented across several key performance indicators: power, voltage, current (for PV module and load), and duty cycle.

Firstly, we have compared the convergence of PSO and GSA to an optimal power level at constant solar irradiance of 1000 W/m^2 and at 25°C temperature. The graph in Figure 7.14 indicates that GSA quickly stabilizes to an optimal power level, exhibiting minor fluctuations, suggestive of a robust and effective tracking approach. In contrast, PSO's power graph is marked by a more varied performance, with power levels taking a longer time to stabilize, reflecting its expansive exploration of the search space.

Examining the duty cycle changes from Figure 7.15 at a constant solar irradiance of 1000 W/m^2 at 25°C, GSA demonstrates a swift approach to identifying an optimal duty cycle, as evidenced by fewer and smaller adjustments. On the other hand, PSO shows larger and more frequent changes in duty cycle, implying a more extensive search behavior.

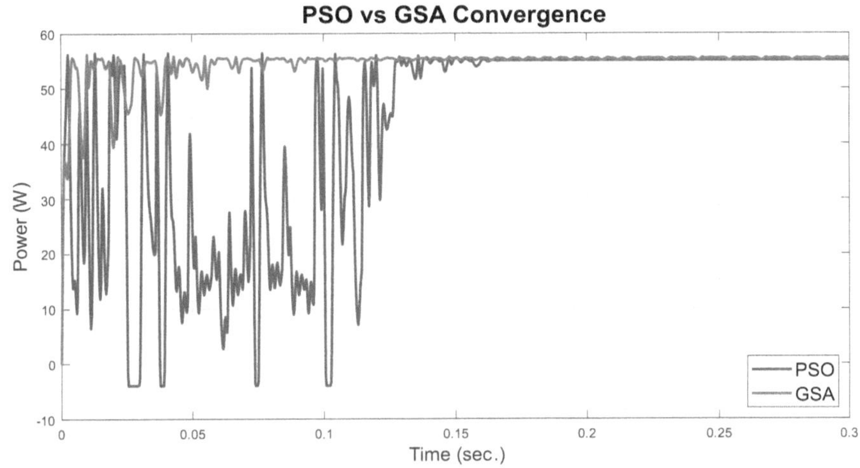

FIGURE 7.14 Maximal power convergent pattern between PSO and GSA.

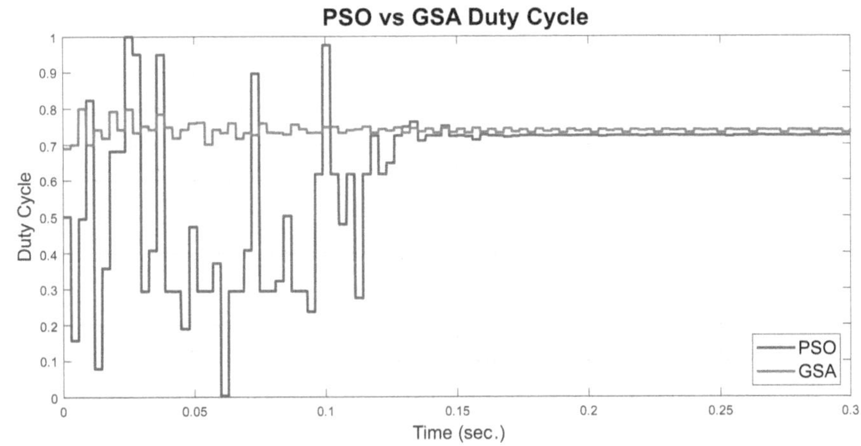

FIGURE 7.15 Duty patterns of PSO and GSA.

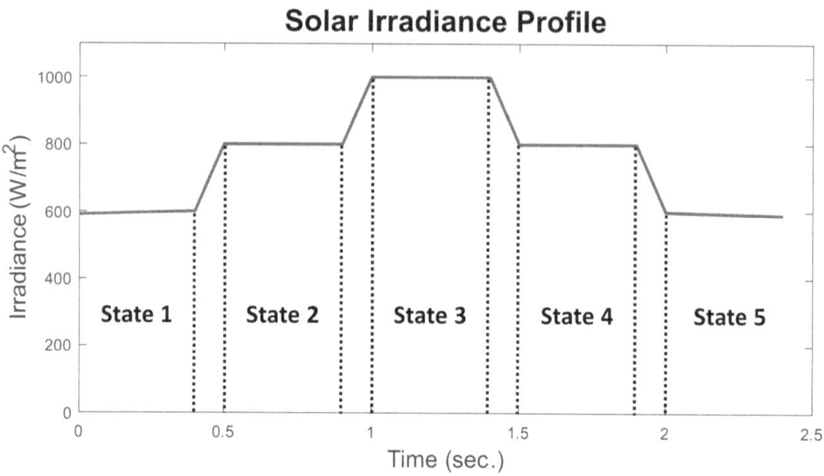

FIGURE 7.16 Solar irradiance profile.

Next, we have evaluated and compared the performance of the four MPPT strategies at different solar irradiance levels but at a constant temperature of 25°C as depicted by the solar profile given in Figure 7.16. The profile of solar radiation varied between 590 and 1000 W/m^2. To perform a complete investigation, the variation in the solar irradiance profile was kept trapezoidal in character, with five states: 590 W/m^2 at 25°C, 800 W/m^2 at 25°C, 1000 W/m^2 at 25°C, 800 W/m^2 at 25°C, and 590 W/m^2 at 25°C.

Figure 7.17 shows the maximal power convergence pattern between PO, IC, PSO, and GSA and indicates that GSA quickly stabilizes to an optimal power level with the highest amplitude, while PSO also shows minor fluctuations but has an average power amplitude level. IC also lags behind in optimal power level attainment. PO, on the other hand, has maximum fluctuations, which is an inherent property of this algorithm. This comparison concludes that out of the four MPPT strategies, GSA shows very robust and optimal performance in tracking PV power.

Finally, we have also compared the output from the boost converter, showing all parameters in Figure 7.18.

The results show that the solar power system we designed can effectively track the MPP even when the sunlight intensity changes rapidly (as seen in Figure 7.18 and Table 7.1). Figure 7.18 also shows that under stable conditions, the PO technique causes more fluctuation in power output compared to the IC, PSO, and GSA algorithms. This fluctuation reduces the average power output of the PO technique, making the system less efficient

7.5 DISCUSSION

The comparative finding of GSA and PSO within the realm of MPPT unveils distinct operational profiles for each algorithm. GSA's proficiency is evident in the data, which highlights its rapid convergence to the MPP, suggesting an innate

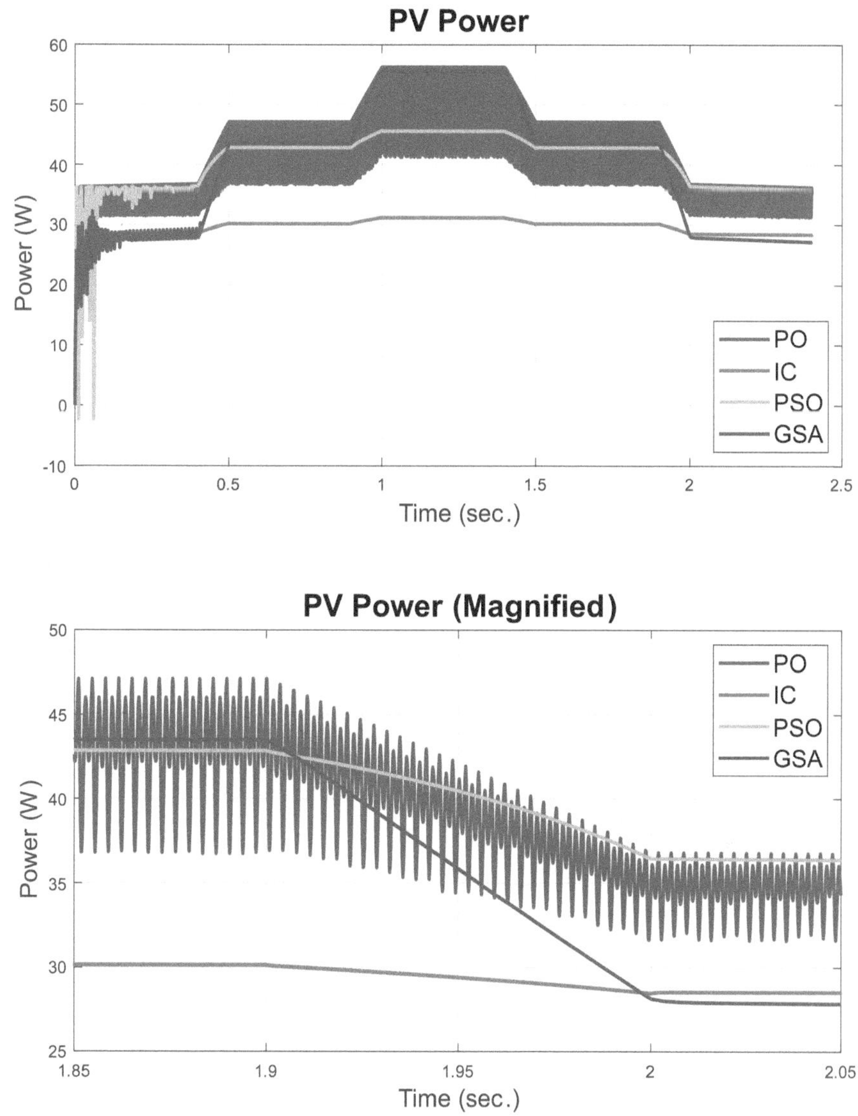

FIGURE 7.17 Maximum power at PV using PO, IC, PSO, and GSA.

capability to efficiently hone in on the optimal solution. The consistent maintenance of a steady state following convergence further exemplifies GSA's stability, marking it as an effective tool for MPPT where rapid and reliable tracking is indispensable. Such performance is advantageous in dynamic environments where swift adaptation to changing solar irradiance and temperature conditions is essential. In evaluation, PSO's performance is defined by its exploratory nature and it also shows an oscillatory convergence pattern, as observed in the experimental data. However, the GSA algorithm can become computationally demanding because the number of agents

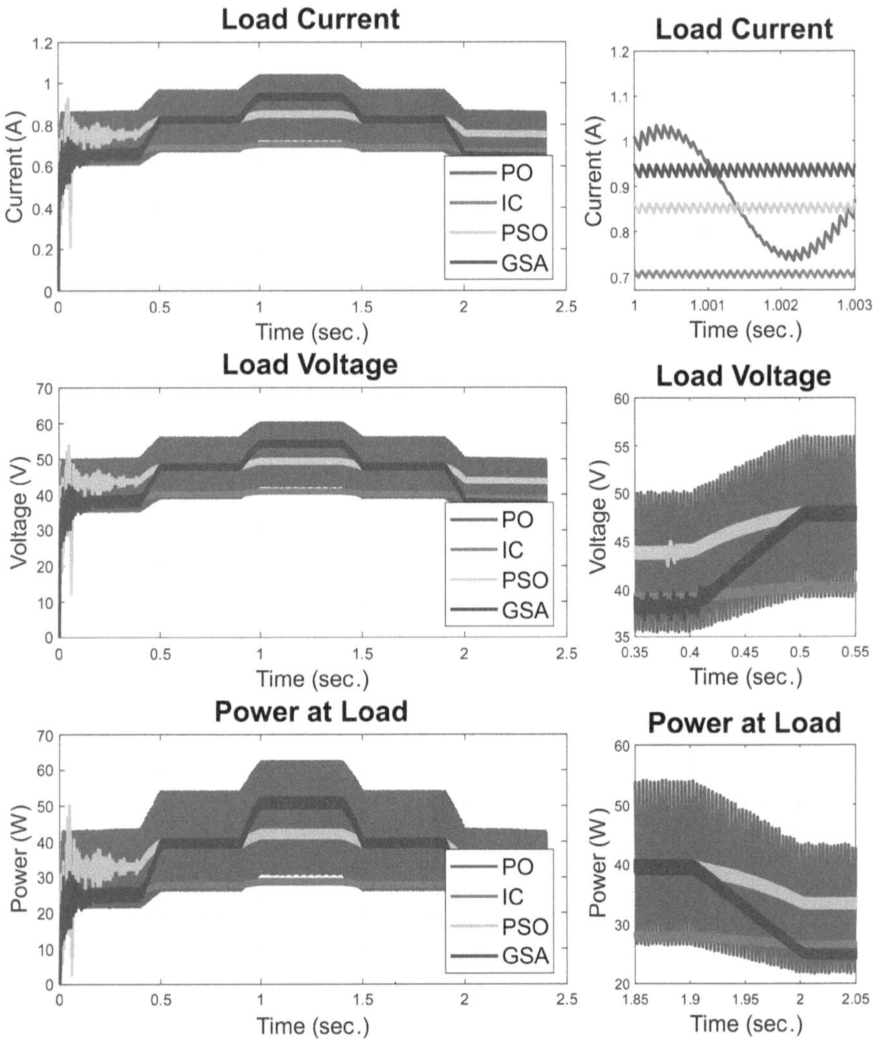

FIGURE 7.18 Output parameters of boost converter, i.e., output current, output voltage, and maximum power at load using PO, IC, PSO, and GSA.

increases due to the need to calculate gravitational forces between all pairs of particles. In GSA, there is additionally a risk of premature convergence that may lead to suboptimal solutions if the gravitational constant is not appropriately controlled. Moreover, initially, GSA's performance may be substantially low because of the initial distribution of agents, highlighting its sensitivity to initial conditions. While PSO is praised for its simplicity and ease of implementation, characterized by the need for fewer parameters to adjust, this simplicity makes PSO handy for a variety of applications. Nevertheless, PSO faces challenges, such as the risk of getting stuck in local optima in complex landscapes, which necessitates maintaining swarm

TABLE 7.1
Comparative Performance Analysis Using PO, IC, PSO, and GSA MPPT Techniques

Parameter	States	PO	IC	PSO	GSA
Efficiency	State 1	92.2%	91.9%	93.2%	91.6%
	State 2/4	95.6%	90.6%	96.6%	96.6%
	State 3	96.7%	91.7%	96.9%	**98.5%**
	State 5	92.7%	92.8%	94.7%	92.7%
Settling time	Power	Not settled	0.03	0.16	0.1
	Voltage	Not settled	0.03	0.16	0.1
	Current	Not settled	0.03	0.16	0.1
Ripple in voltage and current (output of a boost converter)	State 1	High	Minimum	Very high	High
	State 2/4	High	Minimum	Minimum	Minimum
	State 3	High	Minimum	Minimum	Minimum
	State 5	High	Minimum	Minimum	Minimum
Range of voltage variation (V)		42–60	40–41	48–50	**53–55**
Range of current variation (A)		0.73–1.03	0.69–0.71	0.84–0.86	**0.92–0.94**
Control strategy		Sampling	Sampling	Intelligent Control	Intelligent Control

variety. The algorithm's overall performance is also highly sensitive to parameter settings, including inertia weight and acceleration coefficients, which can notably impact outcomes.

7.6 CONCLUSION

In conclusion, the utility of PSO and GSA in MPPT systems has provided a profound and insightful evaluation where each algorithm demonstrates its unique strengths and limitations, highlighting that the importance of selecting the right algorithm is primarily based on the specific requirements of the application. The implementation of GSA for MPPT systems, in particular, has shown substantial capability for future applications. Its robust global search functionality, derived from the concepts of gravity and mass interactions, makes it incredibly powerful for exploring the search space associated with MPPT optimization problems concerning nanotechnology. This study can serve as a valuable resource for research in the area of MPPT optimization. As the demand for green renewable energy systems continues to grow, future research can build upon these findings by exploring hybrid procedures, refining algorithm-based parameters, and investigating the scalability and adaptability of these algorithms to exceptional MPPT system configurations with nanotechnology, thereby enhancing the efficiency and effectiveness of renewable energy harvesting.

REFERENCES

1. Y. Xu, J. Pei, J. Yuan, and G. Zhao, "Concentrated solar power: technology, economy analysis, and policy implications in China," *Environmental Science and Pollution Research*, vol. 29, no. 1, pp. 1324–1337, 2021, https://doi.org/10.1007/s11356-021-15779-1
2. A. B. Awan, "Comparative analysis of 100 MW concentrated solar power plant and photovoltaic plant," In *5th International Conference on Energy, Environment and Sustainable Development (EESD-2018)*, 2019, https://doi.org/10.1063/1.5115363
3. J. M. Tabora, U. C.Paixão Júnior, C. E. M. Rodrigues, et al., "Hybrid system assessment in on-grid and off-grid conditions: a technical and economical approach," *Energies*, vol. 14, no. 17, p. 5284, 2021, https://doi.org/10.3390/en14175284
4. T. Saga, "Advances in crystalline silicon solar cell technology for industrial mass production," *NPG Asia Materials*, vol. 2, no. 3, pp. 96–102, 2010, https://doi.org/10.1038/asiamat.2010.82
5. S. Moon, K. Kim, Y. Kim, J. Heo, and J. Lee, "Highly efficient single-junction GaAs thin-film solar cell on flexible substrate," *Scientific Reports*, vol. 6, no. 1, 2016, https://doi.org/10.1038/srep30107
6. M. A. Scarpulla, B. McCandless, A. B. Phillips, et al., "CdTe-based thin film photovoltaics: Recent advances, current challenges and future prospects," *Solar Energy Materials and Solar Cells*, vol. 255, p. 112289, 2023, https://doi.org/10.1016/j.solmat.2023.112289
7. M. A. Green, A. Ho-Baillie, and H. J. Snaith, "The emergence of perovskite solar cells," *Nature Photonics*, vol. 8, no. 7, pp. 506–514, 2014, https://doi.org/10.1038/nphoton.2014.134
8. N. G. Park, "Organometal perovskite light absorbers toward a 20% efficiency low-cost solid-state mesoscopic solar cell," *The Journal of Physical Chemistry Letters*, vol. 4, no. 15, pp. 2423–2429, 2013, https://doi.org/10.1021/jz400892a
9. K. Miyano, N. Tripathi, M. Yanagida, and Y. Shirai, "Lead halide perovskite photovoltaic as a model *p–i–n* diode," *Accounts of Chemical Research*, vol. 49, no. 2, pp. 303–310, 2016, https://doi.org/10.1021/acs.accounts.5b00436
10. B. Mazhari, "An improved solar cell circuit model for organic solar cells," *Solar Energy Materials and Solar Cells*, vol. 90, no. 7–8, pp. 1021–1033, 2006, https://doi.org/10.1016/j.solmat.2005.05.017
11. G. Huang, F. Yu, and C. Xu, "An analytical solution to lumped parameter equivalent circuit model of organic solar cells," *Crystals*, vol. 8, no. 5, p. 224, 2018, https://doi.org/10.3390/cryst8050224
12. A. Kojima, K. Teshima, Y. Shirai, and T. Miyasaka, "Organometal halide perovskites as visible-light sensitizers for photovoltaic cells," *Journal of the American Chemical Society*, vol. 131, no. 17, pp. 6050–6051, 2009, https://doi.org/10.1021/ja809598r
13. National Renewable Energy Laboratory (NREL) Efficiency Chart. https://www.nrel.gov/pv/assets/images/efficiency-chart.png (accessed: April, 2024).
14. J. Luo, J. H. Im, M. T. Mayer, et al., "Water photolysis at 12.3% efficiency via perovskite photovoltaics and Earth-abundant catalysts," *Science*, vol. 345, no. 6204, pp. 1593–1596, 2014, https://doi.org/10.1126/science.1258307
15. V. K. Khanna, *Nano-Structured Photovoltaics: Solar Cells in the Nanotechnology Era*. 2022. https://doi.org/10.1201/9781003215158
16. M. A. Cappelletti, G. A. Casas, A. P. Cédola, E. L. Peltzer, and B. M. Soucase, "Study of the reverse saturation current and series resistance of p-p-n perovskite solar cells using the single and double-diode models," *Superlattices and Microstructures*, vol. 123, pp. 338–348, 2018, https://doi.org/10.1016/j.spmi.2018.09.023
17. F. Ebadi, N. Taghavinia, R. Mohammadpour, A. Hagfeldt, and W. Tress, "Origin of apparent light-enhanced and negative capacitance in perovskite solar cells," *Nature Communications*, vol. 10, no. 1, 2019, https://doi.org/10.1038/s41467-019-09079-z

18. P. K. Pathak and A. K. Yadav, "Design of battery charging circuit through intelligent MPPT using SPV system," *Solar Energy*, vol. 178, pp. 79–89, 2019, https://doi.org/10.1016/j.solener.2018.12.018
19. P. K. Pathak, A. K. Yadav, and P. A. Alvi, "Advanced solar MPPT techniques under uniform and non-uniform irradiance: a comprehensive review," *Journal of Solar Energy Engineering*, vol. 142, no. 4, 2020, https://doi.org/10.1115/1.4046090
20. P. K. Pathak, S. Padmanaban, A. K. Yadav, P. A. Alvi, and B. Khan, "Modified incremental conductance MPPT algorithm for SPV-based grid-tied and stand-alone systems," *IET Generation, Transmission & Distribution*, vol. 16, no. 4, pp. 776–791, Oct. 2021, https://doi.org/10.1049/gtd2.12328
21. V. Prakash, K. C. Sharma, R. Bhakar, and H. Tiwari, "Modified interval-based generator scheduling for PFR adequacy under uncertain PV generation," IET *Generation, Transmission & Distribution*, vol. 13, no. 16, pp. 3725–3733, 2019, https://doi.org/10.1049/iet-gtd.2018.6391
22. M. Abdulkadir, A. Halim, and S. T. Yusuf, "An improved PSO-based MPPT control strategy for photovoltaic systems," *International Journal of Photoenergy*, vol. 2014, pp. 1–11, 2014, https://doi.org/10.1155/2014/818232
23. S. Chtita, S. Motahhir, A. El Hammoumi, et al., "A novel hybrid GWO–PSO-based maximum power point tracking for photovoltaic systems operating under partial shading conditions," *Scientific Reports*, vol. 12, no. 1, 2022, https://doi.org/10.1038/s41598-022-14733-6
24. A. O. Baatiah, A. M. Eltamaly, and M. A. Alotaibi, "Improving photovoltaic MPPT performance through PSO dynamic swarm size reduction," *Energies*, vol. 16, no. 18, pp. 6433–6433, 2023, https://doi.org/10.3390/en16186433
25. O. B. Belghith, L. Sbita, and F. Bettaher, "MPPT design using PSO technique for photovoltaic system control comparing to fuzzy logic and P&O controllers," *Energy and Power Engineering*, vol. 8, no. 11, pp. 349–366, 2016, https://doi.org/10.4236/epe.2016.811031
26. E. Rashedi, H. Nezamabadi-pour, and S. Saryazdi, "GSA: a gravitational search algorithm," *Information Sciences*, vol. 179, no. 13, pp. 2232–2248, 2009, https://doi.org/10.1016/j.ins.2009.03.004
27. M. L. Katche, A. B. Makokha, S. O. Zachary, and M. S. Adaramola, "A comprehensive review of maximum power point tracking (MPPT) techniques used in solar PV systems," *Energies*, vol. 16, no. 5, p. 2206, 2023, https://doi.org/10.3390/en16052206
28. I. O. Nyarko, M. A. Elgenedy, I. Abdelsalam, and K. H. Ahmed, "Modified variable step-size incremental conductance MPPT technique for photovoltaic systems," *Electronics*, vol. 10, no. 19, p. 2331, 2021, https://doi.org/10.3390/electronics10192331
29. P. Singh, N. Shukla, and P. Gaur, "Modified variable step incremental-conductance MPPT technique for photovoltaic system," *International Journal of Information Technology*, 2020, https://doi.org/10.1007/s41870-020-00450-8
30. N. Hadjidj, M. Benbrahim, D. Ounnas, and L. H. Mouss, "Analysis and design of modified incremental conductance-based MPPT algorithm for photovoltaic system," *Journal of Renewable Energies*, vol. 25, no. 2, pp. 187–198, Dec. 2022, https://doi.org/10.54966/jreen.v25i2.1081
31. Yadav, M. K., Kumar, R., Ratnesh, R. K., Singh, J., Chandra, R., Kumar, A., and Singh, A. K., "Revolutionizing technology with spintronics: devices and their transformative applications," *Materials Science and Engineering: B*, vol. 303, 117293, 2024, https://doi.org/10.1016/j.mseb.2024.117293
32. Kumar, R., and R. Mahadeva, "An experimental measurement and control of human body stomach using electrical impedance tomography," *Journal of Circuits, Systems and Computers*, vol. 30, no. 6, p. 2150103, 2021, https://doi.org/10.1142/S0218126621501036

33. Singla, M. K., Gupta, J., Nijhawan, P., Singh, P., Giri, N. C., Hendawi, E., and Abu El-Sebah, M. I., "Parameter estimation techniques for photovoltaic system modelling," *Energies*, vol. 16, p. 6280, 2023, https://doi.org/10.3390/en16176280
34. Gupta, J., Singla, M. K., Mohit, B., Anand, V., and Sinha, C., Hybrid algorithm for optimizing parameter in the double diode model. In *2023 IEEE 3rd International Conference on Applied Electromagnetics, Signal Processing, & Communication (AESPC),* 2023, November, pp. 1–6. IEEE. https://doi.org/10.1080/15325008.2023.2257701
35. Quttoum, A. N., Alsarhan, A., Moh'd, A., Alshareet, O., Nawaf, S., Khasawneh, F., Aljaidi, M., Alshammari, M., and Awasthi, A., "ABLA: application-based load-balanced approach for adaptive mapping of datacenter networks," *Electronics*, vol. 12, no. 17, p. 3689, 2023, https://doi.org/10.3390/electronics12173689
36. M. Aljaidi, N. Aslam, G. Samara, S. Almatarneh, K. AL- Qawasmi, and A. Alqammaz, "EV charging station placement and sizing techniques: survey, challenges and directions for future work," In *2022 International Arab Conference on Information Technology (ACIT)*, Abu Dhabi, United Arab Emirates, 2022, pp. 1–6, 10.1109/ACIT57182.2022.9994128

8 Nanotechnology Revolutionizes Communication Engineering

Mohsin Afroz, Mohd Maaz, Muhammad Adeeb Ansari, Ratneshwar Kumar Ratnesh, Subodh Kumar Tripathi, and Neha Mittal

8.1 INTRODUCTION

Nanotechnology is a multidisciplinary field that involves the manipulation, measurement, modelling, and manufacturing of structures, devices, and systems within the scale of 1–100 nm [1]. Materials at this scale often possess distinctive and superior properties in comparison with bulk materials, leading to new possibilities in different fields. It includes nanowire fabrication techniques, lithography, molecular self-assembly, and atomic layer deposition [2]. This technology, therefore, allows for the development of new materials, nanomachines, and nano-devices, which are primarily used in electronics, medicine, diagnostics, military applications, and the food industry [3].

Nanotechnology has an extensive history beginning from the Big Bang, from which nanoparticles (NPs) and nanostructured materials are thought to have originated, and these now play critical roles in various industries today [4]. Research on nanomaterials dates back to ancient civilizations, where humans unknowingly utilized nanomaterials such as pigments in pottery. In his famous lecture entitled "There's Plenty of Room at the Bottom," delivered in 1959, physicist Richard Feynman put forward the concept of nanotechnology, discussing how it would be possible to manipulate atoms one by one to create new materials and devise machines. Already in 1974, Norio Taniguchi proposed the name "nanotechnology" to refer to the manipulation of materials at the nanoscale [5]. Nevertheless, it was only in the 1980s that it started appearing as a separate scientific discipline [6]. In the course of the 1980s and 1990s, significant strides were made in nanotechnology in terms of studying the properties of nanosized materials and creating new ways of manipulating matter at this level [7].

DOI: 10.1201/9781003634553-8

In 1981, Gerd Binnig and Heinrich Rohrer developed the scanning tunneling microscope at IBM's Zurich Research Laboratory, which has been recognized as one of the key achievements in this area. This instrument allowed for the imaging and manipulation of individual atoms and molecules for the first time, opening up new opportunities for nano-science research [8].

Carbon nanotubes (CNTs), cylindrical structures made up of carbon molecules with distinct mechanical properties and electrical conductivity, were discovered by Sumio Iijima have become and some of the most investigated nanomaterials in this field since then [9].

With advances in nanofabrication techniques and the discovery of new types of nanomaterials, there was an explosion of research on nanotechnology during the early 2000s. The US National Nanotechnology Initiative (NNI) was established in 2000 to bring together funding and coordination efforts regarding nano research across various federal agencies. This move has substantially contributed to the development of the nanotechnology industry globally and the manufacture of many products as well as applications based on NPs [10]. Figure 8.1 shows the evolution of nanomaterial discovery in the modern nanotechnology era.

Nowadays, nanotechnology is a thriving field with applications in different industries such as medicine, electronics, materials, and energy science. Nanotechnology researchers have continued to explore the potential of the technology to create new materials, systems, and devices with unprecedented properties and functionalities. It

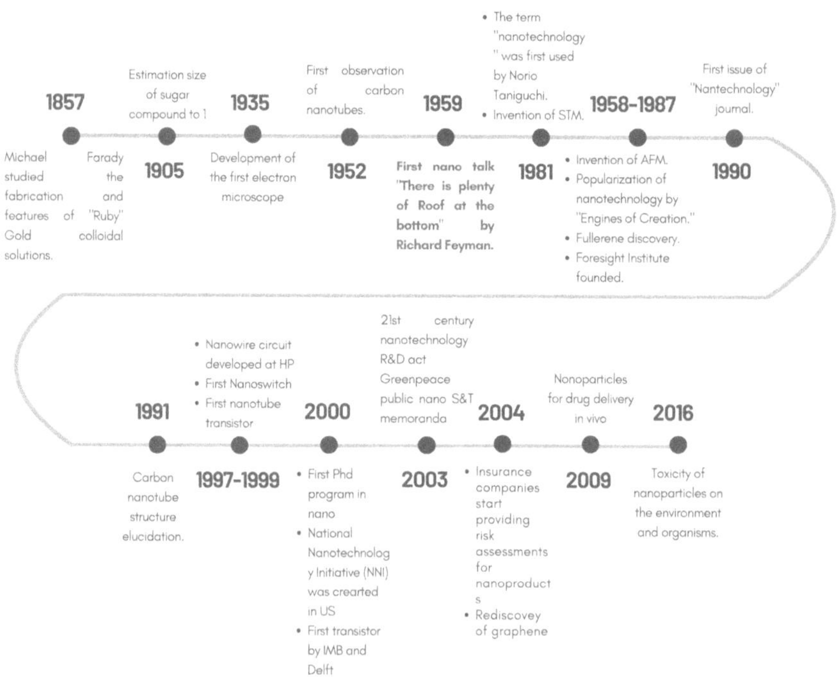

FIGURE 8.1 Evolution of nanomaterial discovery in the modern nanotechnology era.

aids in the manufacture of smaller, faster, and more efficient devices used in communication engineering [11]. This has resulted in tiny, power-efficient, and low-cost integrated circuits and sensors that have transformed communication systems for good [12]. Nanotechnology makes it possible to produce materials with exceptional properties at the nanoscale, thereby improving the performance of electronic components [13]. For this reason, it is used in the healthcare system to create targeted drug delivery systems or disease monitoring robots called nanorobots [14]. It integrates various fields, including materials science, physics, biology, and computer engineering, where new possibilities for computing systems can be found along with human enhancement technologies.

Nanotechnology in electronics and communication engineering has seen significant advancements and holds promising future prospects. Current research focuses on utilizing low-dimensional nanomaterials for photodetectors [15], enhancing EMI shielding with polymer composites [16], and developing nanobioelectronic devices for medical applications. Nanoelectronics offers breakthroughs in integrated circuits, cryptographic hardware, and security primitives [17]. The field of nanotechnology is expected to revolutionize communications and computing, with applications ranging from medicine to industrial sectors [18]. However, challenges like device performance sensitivity, large-scale manufacturing, and fabrication process robustness need to be addressed for industrial implementation. Future directions include optimizing polymer composites for better EMI shielding, advancing nanobioelectronic devices for clinical use, and exploring nanotechnology's potential in transforming electronic devices for diverse applications.

8.2 SOURCES OF NANOMATERIALS

An overview of the diverse sources of nanomaterials is presented in this section. Nanomaterials are derived from a variety of sources, ranging from natural phenomena to artificial synthesis in laboratories, as shown in Figure 8.2 [19,20]. Understanding

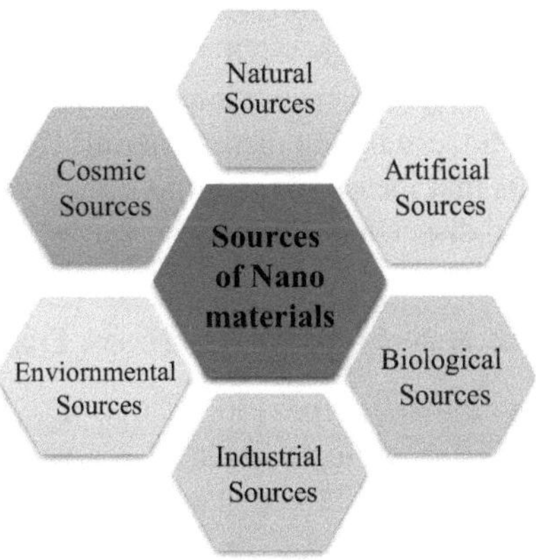

FIGURE 8.2 Different sources of nanomaterials.

these sources is essential for elucidating the wide-ranging applications and implications of nanomaterials in various fields.

1. **Natural sources:** Nanomaterials are generated in phenomena like supernovae in the skies, biosphere, hydrosphere, lithosphere, and atmosphere. These natural synthesis mechanisms involve various reactions such as biological, photochemical, chemical, thermal, volcanic eruptions, winds, mechanical, geothermal, and biogeochemical processes [21].
2. **Artificial sources:** Nanomaterials are also created in laboratories to meet market demand. There are numerous synthetic techniques available for the fabrication of nanomaterials [22].
3. **Biological sources:** Biological systems such as plants, animals, and microorganisms can also serve as sources of nanomaterials [23]. For example, certain plants can accumulate NPs in their tissues, and microorganisms can produce NPs through their metabolic processes.
4. **Industrial sources:** Nanomaterials can be generated as byproducts of industrial processes, such as combustion, chemical reactions, and grinding. These processes can unintentionally produce NPs that are released into the environment [24].
5. **Environmental sources:** Natural phenomena like weathering, erosion, and volcanic activity can release NPs into the environment. These NPs can then be transported through air, water, or soil [25].
6. **Cosmic sources:** Cosmic sources, like dust and meteorites can also contribute to the presence of NPs in Earth's atmosphere and on its surface [26].

8.3 CLASSIFICATION OF NANOMATERIALS

Nanomaterials are categorized according to factors such as size, composition, origin, porosity, dispersion, and phases. Size-based classification divides nanomaterials into zero-, one-, two-, and three-dimensional types denoted as OD, 1D, 2D, and 3D, respectively [27]. Composition wise classification includes organic-, inorganic-,or composite-based nanomaterials with carbon-based ones like fullerene, graphene, nanodiamonds, and CNTs forming a category [28]. Nanomaterials can also be classified based on their origin, porosity levels, dispersion characteristics, and phases, offering an in-depth framework for understanding these materials [29]. This section provides a taxonomy of nanomaterials that focuses on their porosity levels, phases, dimensionality, origin, composition, and dispersion. The categorization is illustrated in Figure 8.3.

8.3.1 Classification of Nanomaterials Based on Dimensionality

Nanomaterials are grouped into four types [30,31]: 0D, 1D, 2D, and 3D depending on their dimensional characteristics. Refer to Table 8.1 for a comparison of nanomaterials based on dimensionality.

1. **0D:** Nanomaterials have all three dimensions at the nanometer scale, such as graphene quantum dots, carbon quantum dots, and fullerenes.

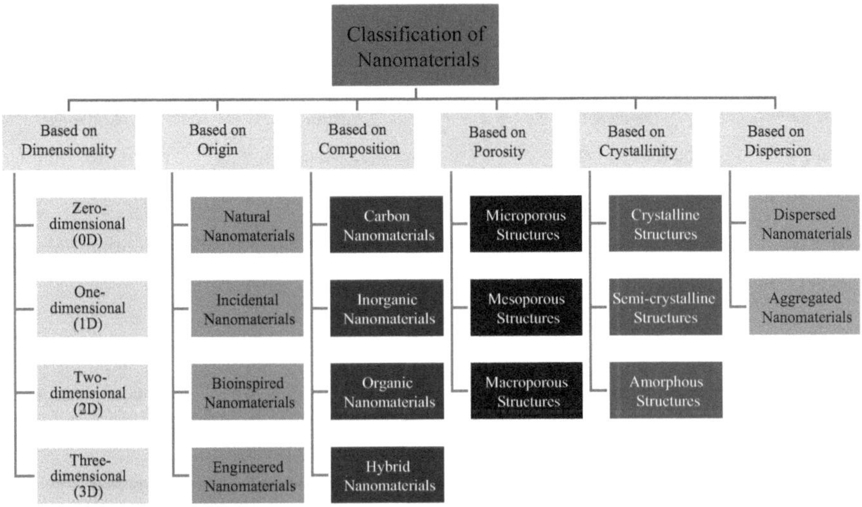

FIGURE 8.3 Different classifications of nanomaterials [29].

TABLE 8.1
Comparison of Different Types of Nanomaterials Based on Dimensionality [30,31]

Dimensionality	Description	Examples	Applications
0D	All three dimensions in nanometer scale (<100 nm)	Fullerenes, graphene quantum dots, carbon quantum dots	Optoelectronic and biomedical applications
1D	One dimension >100 nm	Nanotubes, nanowires, nanofibers, polymer nanofiber mats	Electron sources, decontamination, catalysis, filtration, tissue engineering, wound dressings
2D	Two dimensions >100 nm	Graphene, graphene oxide, silicate clays, transition metal dichalcogenides	Various physical, chemical, optical, and biological applications
3D	Three dimensions >100 nm	Box-shaped graphene nanostructures, bundles of nanowires, nanotubes	Electrochemical energy conversion, storage, water treatment, solar cells

2. **1D:** Nanomaterials have two dimensions at the nanometer scale, like nanowires and nanotubes.
3. **2D:** Nanomaterials have one dimension at the nanometer scale, such as graphene and boron nitride nanosheets.
4. **3D:** Nanomaterials have no dimensions at the nanometer scale, but they possess nanoscale features within their structure, like NPs and nanoclusters.

TABLE 8.2
Comparison of Different Types of Nanomaterials Based on Origin [32–35]

Origin	Description	Examples	Sources
Natural	Produced by natural processes without human intervention	Foraminifera structures, lotus leaf wax crystals, spider silk	Forest fires, volcanic ash, weathering of rocks
Incidental	Unintentionally produced by human activities or anthropogenic processes	Welding gases, vehicle exhaust gases, and solid fuel heating emissions	Industrial processes, combustion during cooking
Bioinspired	Engineered to mimic the properties of natural nanomaterials or living matter	Color-changing nanomaterials inspired by chameleons, mechanochromic elastomers	Advanced nanofabrication technologies
Engineered	Specifically designed and produced for particular applications	Aerosol-produced fumed silica, engineered carbon nanostructures	Nanomedicine, electronics, environmental remediation
Anthropogenic	Refers to both incidental and engineered nanomaterials resulting from human activities	Incidental and engineered nanomaterials	Increasing human activities

8.3.2 Classification of Nanomaterials Based on Their Origin

There are four categories of nanomaterials, which are categorized according to their source or origin: incidental, bioinspired, and engineered nanomaterials [32,33]. Table 8.2 provides a comparison of nanomaterials based on their origin.

1. **Natural nanomaterials:** These types of nanomaterials are not related to human activity; rather, they are produced by natural processes. Examples include foraminifera structures, virus capsids, wax crystals on leaves, spider silk, butterfly wing scales, and natural colloids.
2. **Incidental nanomaterials:** Human activities such as welding, vehicle exhaust, and combustion during cooking inadvertently produce these types of nanomaterials. Forest fires also generate a wide range of incidental nanomaterials.
3. **Bioinspired nanomaterials:** These nanomaterials are inspired by biological systems and mimic their structures and functions. Examples include synthetic spider silk, lotus leaf-inspired self-cleaning surfaces, and gecko-inspired adhesives.
4. **Engineered nanomaterials:** These nanomaterials are intentionally designed and fabricated for specific applications. They are synthesized

through various techniques and can have tailored properties. Examples include NPs used in drug delivery systems, nanocomposites for energy storage, and nanosensors for environmental monitoring.

5. **Anthropogenic nanomaterials:** This term refers to both incidental and engineered nanomaterials [34,35] resulting from human activities.

8.3.3 Classification of Nanomaterials Based on Their Chemical or Elemental Composition

Nanomaterials can be classified based on their chemical or elemental composition [36] into carbon, inorganic, organic, and hybrid nanomaterials. In Table 8.3, a comparison of different nanomaterials based on composition [37] is depicted.

1. **Carbon nanomaterials:** These nanomaterials are primarily composed of carbon atoms and include structures like graphene, CNTs, and fullerenes.
2. **Inorganic nanomaterials:** These nanomaterials are composed of elements other than carbon, such as metals, metal oxides, and metal sulfides. Examples include silver NPs, titanium dioxide NPs, and zinc oxide NPs.
3. **Organic nanomaterials:** These nanomaterials are composed of organic compounds, including polymers, lipids, and proteins. They can be synthesized or derived from natural sources. Examples include polymer NPs, liposomes, and protein-based NPs.
4. **Hybrid nanomaterials:** These nanomaterials are composed of a combination of different chemical components, such as inorganic–organic hybrids or organic–inorganic hybrids. They offer unique properties and functionalities. Examples include silica-coated NPs and polymer–metal hybrids.

8.3.4 Classification of Nanomaterials Based on Their Function of Porosity

Nanomaterials can be classified based on their porosity [39] into three main classes: microporous, mesoporous, and macroporous structures. In Table 8.4, a comparison of different nanomaterials based on porosity is depicted [40].

1. **Microporous materials:** These are nanoporous materials with pore diameters less than 2 nm. Examples include zeolites and metal–organic frameworks.
2. **Mesoporous materials:** These are nanoporous materials with pore diameters ranging from 2 nm to 50 nm in size. They find applications in separation processes, solid-phase organic synthesis, and catalysts.
3. **Macroporous materials:** These are nanoporous materials with pores diameter larger than 50 nm in size. They can have hierarchical structures with pores of different length scales. However, obtaining materials with hierarchical porosity is challenging.

TABLE 8.3
Comparison of Different Types of Nanomaterials Based on Chemical Composition [36–38]

Composition	Description	Examples	Properties	Applications
Carbon nanomaterials	Made of sp2-bonded carbon atoms, including nanodiamonds, fullerenes, graphene, carbon nanotubes, carbon nanofibers, and nano-graphite	Nanomedicine, electronics, reinforcement materials, sustainable energy	High strength, electrical conductivity, thermal conductivity	Nanomedicine, electronics, reinforcement materials, sustainable energy
Organic nanomaterials	Have functional properties due to chemical association of carbon with other elements, such as lipid and polymer nanoparticles	Drug delivery, imaging, medical diagnostics	Biodegradable, non-toxic, compatibility with biological systems	Drug delivery, imaging, medical diagnostics
Inorganic nanomaterials	Formed by non-carbon elements, including metals, metal oxides, and metal salts	Catalysis, electronics, wastewater treatment, antibacterial materials	High surface reactivity, varied crystallinity, magnetic/optical properties	Catalysis, electronics, wastewater treatment, antibacterial materials
Hybrid nanocomposites	Multiphase materials that have at least one nano-sized component (1–100 nm) or nanometric phase separation	Ceramic matrix composites, metal nanocomposites, polymer nanocomposites	Enhanced mechanical, thermal, electrical properties	Aerospace industry, tissue engineering, automotive industry

TABLE 8.4
Comparison of Nanomaterials Based on Porosity [39,40]

Porosity	Description	Examples	Properties	Applications
Microporous	Nanoporous materials with pore diameter less than 2 nm	Zeolites, metal–organic frameworks	High surface area, selective adsorption of molecules based on size and shape	Gas separation, adsorption, catalysis
Mesoporous	Nanoporous materials with pores diameter between 2 and 50 nm. Can have disordered or ordered mesostructure	MCM-41, MCM-48, MCM-50, SBA-15, SBA-16	Increased specific surface area, dual-porosity system (e.g., SBA-15 and SBA-16 for drug delivery)	Separation, solid phase synthesis, ion exchange, catalysis
Macroporous	Nanoporous materials with pores diameter larger than 50 nm in size	Hierarchical porous materials	Higher porosity allows for easy transport of molecules, can be challenging to create materials with hierarchical porosity	Filters, scaffolds for tissue engineering, catalyst supports, materials for controlled drug release and delivery

8.3.5 Classification of Nanomaterials Based on Their Function of Crystallinity

Nanomaterials can be classified based on their crystallinity [41] into crystalline, semi-crystalline (polycrystalline), and amorphous structures [42]. In Table 8.5, a comparison of different nanomaterials based on crystallinity is depicted.

1. **Crystalline nanomaterials:** These nanomaterials have atoms arranged in a periodic manner, forming a crystal lattice. They exhibit well-defined crystal structures and properties. Examples include single-crystal NPs.
2. **Semi-crystalline nanomaterials:** These nanomaterials contain both crystalline and amorphous regions. They have small regions of single-crystal material within a larger polycrystalline structure. Examples include polycrystalline NPs.
3. **Amorphous nanomaterials:** These nanomaterials do not have a periodic structure and lack long-range order. They are disordered in nature and exhibit unique properties. Examples include amorphous NPs.

8.3.6 Classification of Nanomaterials Based on Their Function of Dispersion

Nanomaterials can be classified into dispersed [43,44] and aggregated nanomaterials based on their dispersibility and solvent type. In Table 8.6, a comparison of different nanomaterials based on dimensionality is depicted.

1. **Dispersed nanomaterials:** These nanomaterials exhibit good dispersibility and are well-distributed in a solvent. They can be further classified as well-dispersed, having a polydispersity index (PDI) range of 0.1–0.3, and moderately dispersed, having a PDI range 0.3–0.5 based on their PDI.
2. **Aggregated nanomaterials:** These nanomaterials tend to form aggregates or clusters due to poor dispersibility. They have a higher PDI range of 0.5–1, indicating a broader particle size distribution.

The dispersion stability of NP dispersions [45] can be measured by calculating the zeta potential. Electrostatic repulsion will be high among adjacent NPs if the value for zeta potential is also high, resulting in better dispersion stability. NP dispersions [46] with high zeta potential (negative or positive) values are electrically stabilized, while those with low zeta potential values tend to coagulate.

8.4 FABRICATION AND ENGINEERING OF NANOMATERIALS

8.4.1 Classification of Synthesis Methods Based on Raw Materials

The specific features of nanomaterials are closely tied to the synthesis method used [47,48]. Currently, methods are broadly categorized into "bottom–up" and "top–down" approaches, regardless of the NP origin. Table 8.7 shows the important differences between the various approaches to the synthesis of nanomaterials.

TABLE 8.5
Comparison of Nanomaterials Based on Crystallinity [41,42]

Crystallinity	Description	Examples	Properties	Applications
Crystalline	Nanomaterials in which the atoms are arranged in a periodic manner, exhibiting a clear crystal structure	Metal nanoparticles	Well-defined properties based on crystal structure, uniformity in composition and properties	Electronics, catalysis, sensing
Semi-crystalline	Nanomaterials containing many small regions of single-crystal material, but not displaying a completely periodic structure throughout	Nanocomposites, core–shell nanoparticles	Combination of properties from different crystalline regions, can have enhanced mechanical, electrical, or optical properties	Structural materials, composites, coatings
Amorphous	Nanomaterials that do not display a periodic structure, with atoms arranged in a random or disordered manner	Amorphous silica nanoparticles	Lack of long-range order, isotropic properties, can have high surface area and be easily functionalized	Drug delivery, fillers in composites, coatings
Multiphase	Nanomaterials containing various crystals or semi-crystalline materials, with properties determined by the individual phases present	Core–shell nanoparticles, nanocomposites	Combination of properties from different phases, can tailor properties for specific applications	Biomedical imaging, catalysis, sensors

TABLE 8.6
Comparison of Nanomaterials Based on Dispersion [43,45]

Classification	Description	Examples
Well-dispersed	Nanomaterials with a low PDI (0.1–0.3), showing a narrow particle size distribution and high dispersibility	Monodisperse nanoparticles
Moderately dispersed	Nanomaterials with a moderate PDI (0.3–0.5), indicating a broader size distribution and moderate dispersibility	Polydisperse nanoparticles
Aggregated	Nanomaterials with a high PDI (0.5–1), showing significant aggregation and poor dispersibility	Aggregated nanoparticles
Isometric	Nanomaterials where all particles have the same size, leading to a uniform dispersion	Uniformly sized nanoparticles
Inhomogeneous	Nanomaterials with particles of different sizes, leading to a non-uniform dispersion	Heterogeneously sized nanoparticles

TABLE 8.7
Comparison of Different Synthesis Method of Nanomaterials Based on Raw Material [49–54]

Synthesis Method	Bottom–up Approaches	Top–down Approaches	Hybrid Approaches
Approach	Assembly of atoms or molecules	Segmentation of solid materials	Combination of top–down and bottom–up methods
Advantages	Synthesize nanostructures with fewer defects and high chemical homogeneity	Suitable for producing nanostructures such as thin films and nanoparticles with distinct properties compared to bulk counterparts	Potential for controlled synthesis with integration capabilities
Limitations	Requires compatible molecules and surfaces	Resolution constraints limit application in certain areas, especially at smaller scales	Equipment costs and use of toxic chemicals pose challenges
Example	Sol–gel, micro-emulsion, co-precipitation	Lithography, etching	Lithography with ion growth layering

1. **Bottom–up approaches:** These methods involve fabricating NPs by assembling individual atoms or molecules. Leveraging the self-assembly capacity of atoms, ions, or molecules, bottom–up approaches utilize physical and chemical interactions (e.g., hydrogen bonds and van der Waals forces) to construct macroscopic structures [49]. Wet chemical synthesis methods, such as sol–gel, micro-emulsion, and co-precipitation, are prime examples. These methods offer the advantage of synthesizing nanostructures with fewer defects and high chemical homogeneity, albeit with limitations such as the need for compatible molecules and surfaces [50].

2. **Top–down approaches:** In contrast, top–down approaches involve the segmentation of solid, large uniform materials into smaller fractions to form NPs [51]. Techniques like lithography and etching are suitable for producing nanostructures, especially thin films and NPs larger than 100 nm, which exhibit distinct properties compared to their bulk counterparts. While top–down methods are favored for electronic circuitry fabrication due to their integration capabilities, they are limited by resolution constraints [52].
3. **Hybrid approaches:** Hybrid methods, which combine elements of both top–down and bottom–up approaches, provide a unique synthesis route. For example, lithography serves as a top–down method, while ion growth layering represents a bottom–up approach. Despite their potential, hybrid methods face challenges such as equipment costs and the use of highly toxic chemicals [53,54].

8.4.2 Classification Based on the Nature of Driving Forces

The synthesis of nanomaterials [55] can also be classified based on the driving forces involved. This classification encompasses mechanical, physical, chemical, physicochemical, and biological methods [56]. Table 8.8 shows the important differences between the various approaches to the synthesis based on the nature of driving forces.

1. **Mechanical methods:** These include grinding and milling [57], which involve the high-energy grinding of macro- and micro-metric particles to obtain NPs with non-uniform dimensions. While mechanical methods offer the advantage of minimal waste emission, they lack control over NP size.

TABLE 8.8
Comparison of Different Synthesis Method of Nanomaterials Based on Nature of the Driving Forces [55–61]

Synthesis Method	Nature of Deriving Forces	Advantages	Limitations
Mechanical methods	Grinding, milling	Absence of waste emission	Lack of control over nanoparticle size
Physical methods	Exposure to energy (e.g., mechanical and thermal)	Production of pure nanoparticles with desired size	Expensive equipment, hazardous chemicals required
Chemical methods	Chemical reactions	Control over nanoparticle size, morphology, and stability	Use of hazardous chemicals
Physicochemical methods	Combination of physical and chemical processes	Modulation of nanoparticle properties, faster reactions	Expensive equipment, hazardous chemicals, technical limitations
Biological methods	Activity of microbes, plants	Environmentally friendly, consume less energy, produce benign products	Lack of stability compared to physical and chemical methods

2. **Physical methods:** In physical methods, bulk starting materials are converted into smaller materials through exposure to energy sources [58] like mechanical, chemical, thermal, or laser irradiation. Techniques such as physical vapor deposition, laser ablation, and spinning enable the production of highly pure NPs with precise control over size and morphology.
3. **Chemical methods:** Chemical techniques [59] involve the nucleation and growth of precursor species to fabricate nanomaterials. Methods like chemical vapor deposition and sol–gel processing allow for the controlled synthesis of NPs with specific size, morphology, and stability properties.
4. **Physicochemical methods:** Combining physical and chemical processes [60], physicochemical methods offer a unique approach to synthesis. For example, electrochemical methods use a metallic anode dissolved in an aprotic solvent to fabricate metal NPs.
5. **Biological methods:** Exploiting the activity of microbes and plants [61], biological methods provide an environmentally friendly approach to nanomaterial synthesis. While biological methods reduce the environmental impact and cost of synthesis, the resulting NPs have lower stability than those synthesized with the help of chemical and physical methods.

8.5 MICROSTRIP PATCH ANTENNA

Planar antennas, which include patch and slot antennas, are a subset of microstrip antennas. A narrow slot in the conductive surface constitutes a slot antenna, whereas microstrip patch antennas have a radiating patch element that is printed on one side of a dielectric substrate and a conductive ground plane on the other side [62]. They are characterized by unique qualities such as low-profile design, lightweight, easy integration, and compatibility with ICs, which have contributed to their widespread use in communication systems, satellites, radar systems and sensing applications. The design has significantly evolved from simple structures to complex high-performance designs within wireless technology. Microwave frequencies to millimeter wave bands are the range across which these microstrip antennas operate so efficiently, making them highly adaptable for various communication settings. Compact, multifunctional systems have become a reality because their planar configuration facilitates integration with other electronic components on the same substrate [63]. However, in the early days, microstrip antenna designs were characterized by a lack of good bandwidths, low gain, and poor radiation efficiency. As such, these constraints have hindered their wider use in practical applications. Nevertheless, numerous research efforts have been conducted to overcome these limitations and extend the performance range of microstrip antennas.

Many key areas in this field of science and technology have seen robust research and innovations leading to transformations in microstrip antennas. It has undergone extensive developments in many crucial areas due to intensive research efforts [64]. The choice of dielectric substrate material plays an important role in the performance of an antenna, including its radiation efficiency, gain, and bandwidth. The use of low-loss dielectrics, metamaterials, and nanostructured materials has led to significant improvements in the performance parameters of microstrip antennas. Moreover, novel feeding techniques like aperture-coupling, proximity-coupling, and stacked-patch feeding have been employed by researchers to enhance radiation characteristics and gain for an antenna. It is these feeding mechanisms that have made multi-band and wideband microstrip antennas possible and more suitable to meet the demanding requirements of modern communication systems.

Over the years, there has been great interest in the amalgamation of microstrip antennas with other state-of-the-art technologies. These include frequency-selective surfaces and reconfigurable structures to create adaptive and tunable microstrip antennas [65]. These dynamic antennas are capable of adjusting their operating frequency, polarization, and radiation patterns in real-time thereby improving their performance characteristics and flexibility in different wireless communication scenarios.

8.6 STRUCTURE OF MICROSTRIP PATCH ANTENNA

Patch antennas, also known as microstrip antennas, are easily recognizable by their simple planar structure. They consist of a radiating patch element printed on one side of a dielectric substrate, with a conductive ground plane on the opposite side [62,65] as shown in Figure 8.4(a). This basic arrangement, combined with their ease of production and integration with electronic circuits, continues to make these antennas popular in a wide range of applications, including wireless communication systems.

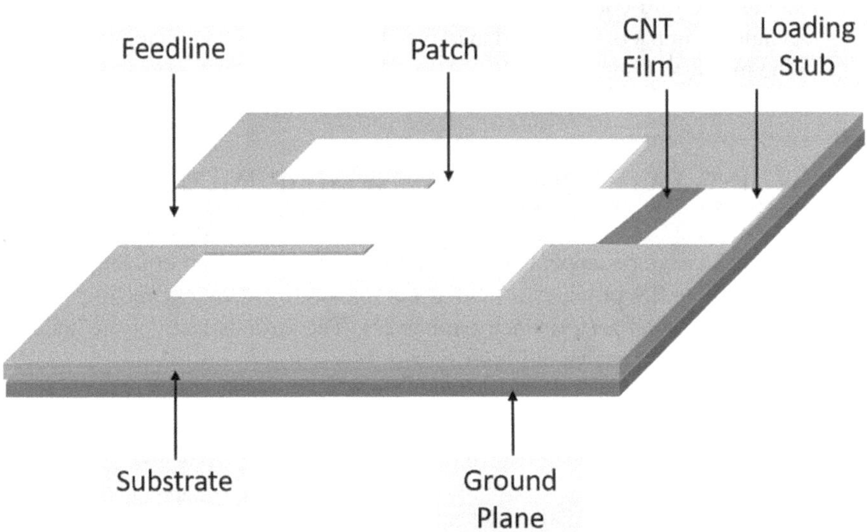

FIGURE 8.4 (a) Microstrip antenna's structural elements (b) Common shapes of patches [62].

The basic structure of a microstrip antenna can be described as follows:

1. **Radiating patch element:** The radiating patch is the most crucial part of the microstrip antenna, responsible for emitting and receiving electromagnetic waves. It is a metallic element (often copper) printed on the top side of the dielectric substrate. The patch can take various shapes, such as square, rectangular, circular, or triangular (Figure 8.4(b)), based on the desired radiation characteristics and frequency response.
2. **Dielectric substrate:** The dielectric substrate provides mechanical support to the radiating patch and helps determine the antenna's radiation properties. It is usually made of a low-loss dielectric material, such as Flame Retardant 4 (FR-4), ceramic, or Teflon, with a relatively high permittivity to achieve a compact antenna size. The thickness of the substrate plays a crucial role in determining the operating frequency and radiation efficiency of the microstrip antenna.
3. **Conductive ground plane:** The conductive ground plane is placed on the bottom side of the dielectric substrate, opposite the radiating patch. It acts as a mirror for the electromagnetic waves generated by the patch, reflecting and directing them in the desired direction. The ground plane also provides a return path for the current flowing through the patch, completing the antenna's circuit.
4. **Feeding mechanism:** To excite the radiating patch and generate electromagnetic waves, a feeding mechanism is employed. The feeding point is usually connected to the radiating patch through a microstrip transmission line. Common feeding techniques include coaxial feeding, microstrip line feeding, aperture-coupling, and proximity-coupling. The choice of feeding mechanism impacts the impedance matching and radiation characteristics of the microstrip antenna.

8.7 FEEDING TECHNIQUES FOR MICROSTRIP PATCH ANTENNA

The feeding technique [64] is an essential aspect of the design of a microstrip antenna as it determines the impedance, bandwidth, radiation pattern, and efficiency of the antenna [64]. Table 8.9 provides a tabular comparison of different feeding methods in terms of various performance parameters. The various feeding methods of microstrip antennas are shown in Figure 8.5:

1. **Microstrip line feed:** In this method, the antenna is fed by a microstrip transmission line that is connected to the metal patch at a particular location, as shown in Figure 8.6a. This method is simple and easy to implement, but it can cause unwanted radiation and can limit the bandwidth of the antenna.
2. **Coaxial probe feed:** In this method, as illustrated in Figure 8.6b, a coaxial probe is inserted through the ground plane and connected to the metal patch. The advantage of this method is that it provides a good impedance match and can be used to achieve wide bandwidths.

TABLE 8.9
Comparison of Different Feeding Methods [64]

Parameter	Microstrip Line Feed	Coaxial Probe Feed	Aperture-Coupled Feed	Proximity-Coupled Feed
Spurious feed radiation	High	High	Less	Minimum
Ease of fabrication	Easy	Soldering and drilling required	Alignment required	Alignment required
Bandwidth	2-5%	2-5%	21%	13%
reliability	Better	Poor	Good	Good
Impedance matching	Easy	Best	Easy	Easy
Advantage	Simple and easy to implement	Provides good impedance match	Offers good isolation and reduces unwanted radiation	Offers good isolation, reduces radiation losses, and can be used to achieve wide bandwidths
Disadvantage	Can cause unwanted radiation	Can be difficult to integrate with the ground plane	Can be challenging to design and implement	Difficult to optimize the coupling distance

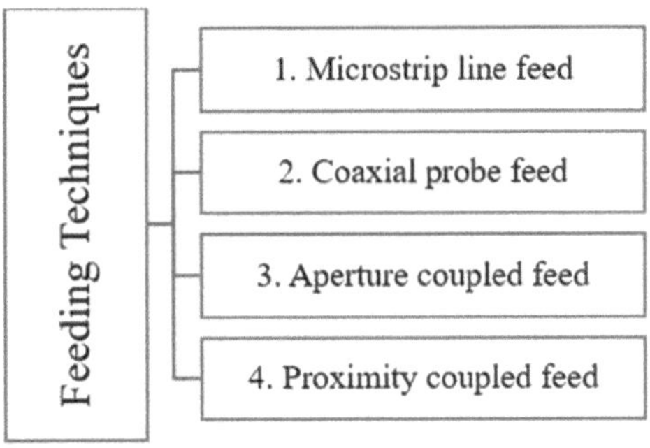

FIGURE 8.5 Various feeding techniques.

3. **Aperture-coupled feed:** In this method, the antenna is fed through an aperture in the ground plane using a microstrip transmission line, as illustrated in Figure 8.6c. This method offers good isolation between the feed line and the radiating element, which reduces unwanted radiation and increases efficiency.
4. **Proximity-coupled feed:** In this method, as depicted in Figure 8.6d, the antenna is fed through capacitive coupling between the metal patch and a feeding element located close to the patch. This method offers good isolation, reduces radiation losses, and can be used to achieve wide bandwidths.

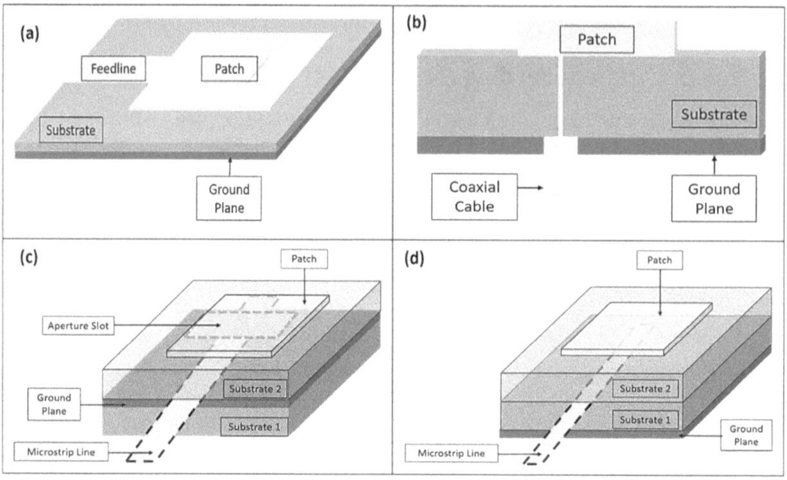

FIGURE 8.6 (a) Microstrip line feed, (b) Coaxial probe feed, (c) Aperture-coupled feed, (d) Proximity-coupled feed.

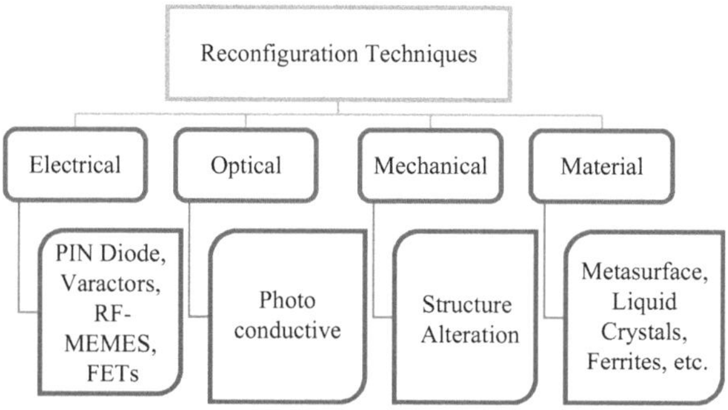

FIGURE 8.7 Various reconfiguration techniques.

8.8 RECONFIGURATION TECHNIQUES FOR MICROSTRIP PATCH ANTENNA

Reconfigurable microstrip antennas can be designed to change their operating characteristics using different techniques [65,66], which can be broadly categorized into the following:

- Electrical reconfiguration techniques
- Optical reconfiguration techniques
- Mechanical reconfiguration techniques
- Material reconfiguration techniques

Table 8.10 provides a tabular comparison of different reconfiguration techniques along with their merits and demerits.

TABLE 8.10
Comparison of Different Reconfiguration Techniques

Reconfiguration Technique	Definition	Advantages	Disadvantages
Electrical reconfiguration	Changing the electrical properties of the antenna using electrical means, such as switches, varactors, or phase shifters	Low power consumption, fast switching speed, and compatibility with existing electronic circuits	Limited range of reconfiguration and potential interference with other electronic devices
Optical reconfiguration	Changing the optical properties of the antenna using optoelectronic materials or devices, such as liquid crystals or electro-optic modulators	Wide range of reconfiguration, high precision, and immunity to electromagnetic interference	Limited bandwidth, high power consumption, and complex fabrication process
Mechanical reconfiguration	Changing the physical shape or position of the antenna using mechanical means, such as MEMS or piezoelectric actuators	High range of reconfiguration, low power consumption, and compatibility with existing mechanical systems	Limited switching speed, mechanical wear and tear, and potential mechanical failure
Material reconfiguration	Changing the material properties of the antenna using materials with tunable properties, such as ferroelectric or magnetic materials	Wide range of reconfiguration, low power consumption, and compatibility with existing fabrication processes	Limited bandwidth, complex material design, and potential interference with other electronic devices

1. **Electrical reconfiguration techniques:** Electrically reconfigurable antennas (RAs) utilize electronic components for switching, such as RF-MEMS, PIN diodes, varactor diodes, and FETs. These elements adjust the surface current distribution in the antenna's radiating structures or edges. By integrating switches into the antenna design, it becomes simpler for engineers to achieve desired reconfigurable functions.
2. **Optical reconfiguration techniques:** Optically RAs change their radiation properties through switches as well. These can be activated optically using silicon switches or other reactive elements. Optically controlled devices have the advantage of eliminating metal wires that might disturb the antenna's radiation qualities. However, when additional metallic microstrip or wired bias lines are involved, the complexity increases. Interference with required radiation patterns becomes a significant challenge for DC-controlled microstrip antennas. This issue can be effectively addressed by using an optically controlled RA [68].
3. **Mechanical reconfiguration techniques:** Antennas can undergo reconfiguration by physically altering their radiating structure. The tuning process involves modifying the radiating parts of the antenna itself. This technique provides the additional benefit of not requiring switch mechanisms, bias lines, or any optical fiber or laser diode integration. However, it is limited by the physical constraints of the device being modified.
4. **Material reconfiguration techniques:** Changing substrate characteristics can also enable reconfiguration in antennas. Materials like liquid crystals, dielectric fluids, ferrites, or metasurfaces play key roles in this process. They allow changes in relative electric permittivity or magnetic permeability. For instance, liquid crystals are nonlinear materials; their dielectric constant can shift when you adjust voltage levels to reorient the molecules. Even a static electric or magnetic field applied to a ferrite material modifies its relative permittivity and permeability significantly.

8.9 DESIGN REQUIREMENTS FOR RAs [69]

Prior to embarking on the design of an RA, RF engineers must confront three challenging questions:

1. **Selection of reconfigurable factor:** What specific aspect of the antenna's behavior needs modification, such as polarization, radiation pattern, or frequency?
2. **Reconfiguration of radiating components:** How can the various radiating components within the antenna's structure be effectively reconfigured to attain the desired outcome?
3. **Minimizing undesirable effects:** Which reconfiguration method can be employed to minimize adverse impacts on the antenna's radiation and impedance characteristics?

8.10 PROCEDURE TO DESIGN A MICROSTRIP ANTENNA

Designing a microstrip antenna involves several steps to ensure its optimal performance and desired characteristics [62,67–69]. Here is a step-by-step procedure to design a microstrip antenna:

8.10.1 Step 1: Define Specifications

Clearly outline the specifications and requirements of the microstrip antenna. This includes the operating frequency range, desired gain, radiation pattern, impedance, polarization, and any other specific application requirements.

8.10.2 Step 2: Choose Substrate Material

Select a suitable dielectric substrate material based on the desired frequency range and performance. The choice of substrate impacts the antenna's size, bandwidth, and radiation efficiency.

8.10.3 Step 3: Calculate Substrate Dimensions

The substrate thickness (h) is determined based on the desired resonant frequency and effective permittivity of the ground plane and patch, using the following equation:

$$h \geq 0.06 \frac{\lambda_{\text{air}}}{\varepsilon_r} \tag{8.1}$$

8.10.4 Step 4: Determine Patch Dimensions

Based on the operating frequency and the desired radiation pattern, calculate the dimensions of the radiating patch. Patch size and shape will impact the antenna's impedance and radiation attributes. For a rectangular patch, the dimensions are calculated according to the following equations:

8.10.4.1 Patch Width (W_p)

The microstrip patch's width is determined by

$$W_p = \frac{c}{2 \times f_r} \sqrt{\frac{2}{\varepsilon_r + 1}} \tag{8.2}$$

where

c = velocity of light in free space
f_r = resonant frequency

8.10.4.2 Effective Dielectric Constant ($\varepsilon_{r_{\text{eff}}}$)

The effective dielectric constant is calculated by

$$\varepsilon_{r_{\text{eff}}} = \frac{\varepsilon_r + 1}{2} + \frac{\varepsilon_r - 1}{2}\left(1 + 12 \times \frac{h}{w}\right)^{-\frac{1}{2}} \tag{8.3}$$

8.10.4.3 Patch Length (L_p)

The microstrip patch's length is determined by

$$L_p = L_{\text{eff}} - 2 \times \Delta L \tag{8.4}$$

where

L_{eff} = effective length
ΔL = extension in length

$$L_{\text{eff}} = \frac{c}{2 \times f_r \times \sqrt{\varepsilon_{r_{\text{eff}}}}} \tag{8.5}$$

$$\Delta L = 0.412 \times h \times \left(\frac{\varepsilon_{r_{\text{eff}}} + 0.3}{\varepsilon_{r_{\text{eff}}} - 0.258}\right) \times \left(\frac{\frac{W_p}{h} + 0.264}{\frac{W_p}{h} + 0.8}\right) \tag{8.6}$$

8.10.5 Step 5: Determine Ground Plane Dimensions

Calculate the dimensions of the conductive ground plane based on the patch dimensions. The ground plane should be larger than the patch to minimize edge effects and enhance radiation efficiency.

8.10.5.1 Ground Width (W_g)

$$W_g \geq W_p + 6h \tag{8.7}$$

8.10.5.2 Ground Length (L_g)

$$L_g \geq L_p + 6h \tag{8.8}$$

8.10.6 Step 6: Determine Feeding Mechanism

Select a feeding method for the antenna. Common feeding methods include coaxial feeding, aperture-coupling, microstrip line feeding and proximity-coupling. The feeding point and method will impact impedance matching and radiation characteristics.

For inset-fed microstrip line, the dimensions are calculated as per the equations mentioned below:

8.10.6.1 Width of Feedline (W_f)

$$W_f = \frac{2\times h}{\pi}\left\{B-1-\ln(2\times B-1)+\frac{\varepsilon_r-1}{2\times\varepsilon_r}\times\left[\ln(B-1)+0.39-\frac{0.61}{\varepsilon_r}\right]\right\} \tag{8.9}$$

where

Z_o = characteristic impedance of feedline

$$B = \frac{377\times\pi}{2\times Z_o\times\sqrt{\varepsilon_r}} \tag{8.10}$$

8.10.6.2 Inset Dimensions

8.10.6.2.1 Inset Length (d)

The proposed microstrip patch antenna uses an inset feeding technique to achieve good impedance matching.

$$R_{\text{in}}\left(y = y_o\right) = \frac{1}{\left(2G_1+G_{12}\right)}\cos^2\left(\frac{\pi}{L}y_o\right) \tag{8.11}$$

where

R_{in} = input impedance of patch
$d = y_o$ (inset length)

$$G_{12} = \frac{1}{120\pi^2}\int_0^{\pi}\left[\frac{\sin\left(\frac{k_oW_p}{2}\cos\theta\right)}{\cos\theta}\right]^2 J_o\left(k_oL\sin\theta\right)\sin^3\theta\, d\theta \tag{8.12}$$

$$G_1 = \frac{1}{120\pi^2}\int_0^{\pi}\left[\frac{\sin\left(\frac{k_oW_p}{2}\cos\theta\right)}{\cos\theta}\right]^2 \sin^3\theta\, d\theta \tag{8.13}$$

8.10.6.2.2 Inset Width (g)

$$g = \frac{c}{\sqrt{2\times\varepsilon_{r\text{eff}}}}\,\frac{4.65\times10^{-9}}{f} \tag{8.14}$$

8.10.7 Step 8: Simulate and Optimize

Use electromagnetic simulation software, such as CST Studio Suite, HFSS, or Agilent Advanced Design System, to model the microstrip antenna and simulate its performance. Optimize the design parameters to achieve the desired specifications, such as resonance frequency, bandwidth, and radiation pattern.

8.11 FUNDAMENTAL PARAMETERS OF MICROSTRIP ANTENNA

There are several fundamental parameters of an antenna, as presented in [62,70,71], that define its performance and behavior. Here are some of the most important parameters:

1. Radiation Pattern
2. Radiation Power Density
3. Radiation Intensity
4. Directivity
5. Gain
6. Antenna Efficiency
7. Beamwidth
8. Beam Efficiency
9. Bandwidth
10. Polarization
11. Input Impedance
12. Antenna Radiation Efficiency

Radiation pattern: The radiation pattern of an antenna is the 3D graphical representation of the intensity of the electromagnetic field radiated by the radiator. It describes the directional properties of the radiator and shows the variation of the electromagnetic field intensity in space as a function of direction.

Radiation power density: Radiation power density is defined as the power per unit area of the electromagnetic wave that is radiated by the radiator. It is a measure of the power density of the radiated energy.

Total power radiated is given by

$$P_{\text{rad}} = \oiint_s W_0 \cdot ds = \int_0^{2\pi}\int_0^{2\pi}\left[\hat{a_r} W_0(r)\right]\cdot\left[\hat{a_r} r^2 \sin\sin\theta d\theta\; d\varnothing\right] = 4\pi r^2 W_0 \qquad (8.15)$$

And the power density is given by

$$W_0 = \hat{a_r}\, W_0 = \hat{a_r}\left(\frac{P_{\text{rad}}}{4\pi r^2}\right)\left(\text{W/m}^2\right) \qquad (8.16)$$

where

P_{rad} = Total radiated power
W_0 = power density
r = radial distance

$\hat{a_r}$ = unit vector in the radial direction
θ = polar angle
$\varnothing$ = azimuthal angle

Radiation intensity: It is the power per unit solid angle of the electromagnetic wave that is radiated by the radiator. It is a measure of the power density of the radiated energy in a particular direction.

It is given by

$$U = r^2 W_{\text{rad}} \tag{8.17}$$

where

U = radiation intensity (W/unit solid angle)
W_{rad} = radiation density (W/m^2)

Directivity: It is the ability of an antenna to focus electromagnetic energy in a particular direction. It is a measure of the efficiency of the antenna in radiating or receiving energy in a specific direction compared to an isotropic radiator.

It is given by

$$D = \frac{U}{U_0} = \frac{4\pi U}{P_{\text{rad}}} \tag{8.18}$$

where

P_{rad} = total radiated power (W)
U = radiation intensity (W/unit solid angle)
U_0 = radiation intensity of isotropic source (W/unit solid angle)

Gain: It is the measure of the increase in power of the electromagnetic wave in a specific direction relative to an isotropic radiator. It is the product of the directivity and the radiation efficiency of the antenna.

It is given by

$$G = 4\pi \frac{U(\theta,\phi)}{P_{\text{in}}} \tag{8.19}$$

where

G = gain of the antenna
$U(\theta,\varphi)$ = radiation intensity
P_{in} = total input (accepted) power

Antenna efficiency: Antenna efficiency is the ratio of the power radiated by the antenna to the power supplied to the antenna. It is a measure of the effectiveness of the antenna in converting the input power into radiated power.

It is given by

$$e_0 = e_r e_c e_d \quad (8.20)$$

where

e_0 = total efficiency (dimensionless)
e_r = reflection efficiency (dimensionless)
e_c = conduction efficiency (dimensionless)
e_d = dielectric efficiency (dimensionless)

Beamwidth: Beamwidth is the angular separation between the two points on the main lobe of the radiation pattern where the power is half of the maximum value.

It is given by

$$\text{HPBW} = 2 * \left(\lambda/(2 * D)\right) \quad (8.21)$$

$$\text{FNBW} = 1.02 * \lambda/D \quad (8.22)$$

where

FNBW = first null beamwidth
HPBW = half power beamwidth
λ = wavelength
D = diameter

Beam efficiency: Beam efficiency is the ratio of the power in the main beam of the radiation pattern to the total radiated power.

It is given by

$$\text{BE} = \frac{\int_0^{2\pi}\int_0^{\theta 1} U(\theta,\varphi)\sin\sin\theta\, d\theta\, d\varphi}{\int_0^{2\pi}\int_0^{\pi} U(\theta,\varphi)\sin\sin\theta\, d\theta\, d\varphi} \quad (8.23)$$

where

BE = beam efficiency of antenna

$\int_0^{2\pi}\int_0^{\theta 1} U(\theta,\varphi)\sin\sin\theta\, d\theta\, d\varphi$ = power transmitted (received) within cone angle θ_1

$\int_0^{2\pi}\int_0^{\pi} U(\theta,\varphi)\sin\sin\theta\, d\theta\, d\varphi$ = power transmitted (received) by antenna

Bandwidth: Without significant deduction of performance, the range of frequencies for which the antenna can operate accounts for bandwidth.

It is given by

$$\mathrm{BW} = F_H - F_L \tag{8.24}$$

where

BW = bandwidth of the antenna
F_H = upper frequency limit of the bandwidth
F_L = lower frequency limit of the bandwidth

Polarization: Polarization refers to the orientation of the electric field of the electromagnetic wave radiated by the antenna. It can be linear, circular, or elliptical.

Input impedance: At the input terminals of the antenna, it is the ratio of voltage to current. It is important for matching the antenna to the transmission line and minimizing the reflection of energy.

It is given by

$$Z_a = R_a + JX_a \tag{8.25}$$

where

Z_a = input impedance of antenna
R_a & X_a = antenna resistances

Antenna radiation efficiency: It is a term that gives the ratio of power that the antenna has been radiating to the power that has been provided as input to the antenna.

It is given by

$$e_{\mathrm{cd}} = \left[\frac{R_r}{R_L + R_r}\right] \tag{8.26}$$

where

e_{cd} = radiation efficiency (conduction dielectric efficiency)
R_r = radiation resistance
R_L = conduction dielectric losses resistance

8.12 IMPACT OF NANOTECHNOLOGY ON MICROSTRIP ANTENNA

Thabet et al. [72] have enhanced the dimensions of the microstrip antenna. They focused on two main aspects: the height of the microstrip antenna substrate and the width of the patch microstrip antenna. This improvement was achieved by using nanocomposite materials, which combine RT Duriod 5880 with nano-fillers such as fumed silica and aluminum oxide. The introduction of these novel nano-fillers results in nanocomposed materials suitable for microstrip antenna substrates. Notably, the height and width of the recommended microstrip antenna model are less than those seen in traditional rectangular designs. This reduction in size is made possible by incorporating nanocomposed materials that possess high relative permittivity. When utilizing this innovative nanocomposite material, an efficient and compact microstrip antenna emerges. Additionally, the advantages include a smaller size, simple fabrication process, improved radiation, increased bandwidth, and better surface wave power for the microstrip antenna. These outcomes are clearly illustrated in Table 8.11.

Urbani et al. [73] focus on the experimental characterization of microstrip antennas composed of various nanofilms, particularly aluminum, and their radiation performance and ultrawideband response, as shown in Figure 8.8. The aluminum patch structure exhibited exceptional radiation performance and an ultrawideband response, indicating potential applications in system-on-chip, space, and specialized fields. The methodology involved indirect excitation through an aperture coupled-feeding mechanism. The findings suggest that microstrip antennas made of nanofilms, especially those with aluminum patches, demonstrate outstanding performance, highlighting their versatility for diverse applications.

Naishadham et al. [74] explore the potential of CNTs for designing antennas, focusing on the tunability of a patch antenna using a CNT thin film, as depicted in figure 8.9. The study shows that by confining the CNT film to a small area outside the radiator, it is possible to achieve about 500 MHz tunability in a patch antenna resonant at 6.6 GHz while improving radiation efficiency. The methodology

TABLE 8.11
Comparison of Parameters of Conventional and Suggested Model [72]

Substrate Material	ε	h (mm)	w (mm)	L (mm)	f (MHz)	Radiation Efficiency
RT Duriod 5880	2.2	5.5146	74.653	87.578	1800	0.5099
RT Duriod 5880 with 0.3 aluminum oxide	2.8	3	71	83	1616.4	0.5322

Component	Dimension (mm)
Patch	4.3 × 5.9
Silicon substrate	10 × 10 × 0.38
RT Duroid substrate	25 × 20 × 0.508
Feeding Line	15 × 0.972
Through-hole diameter	0.8

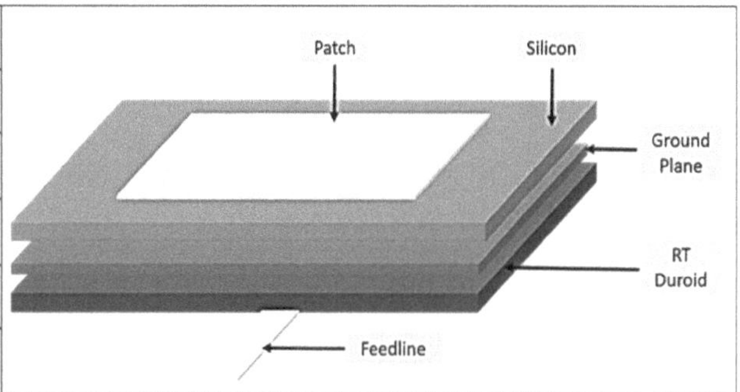

FIGURE 8.8 Structure of the suggested model.

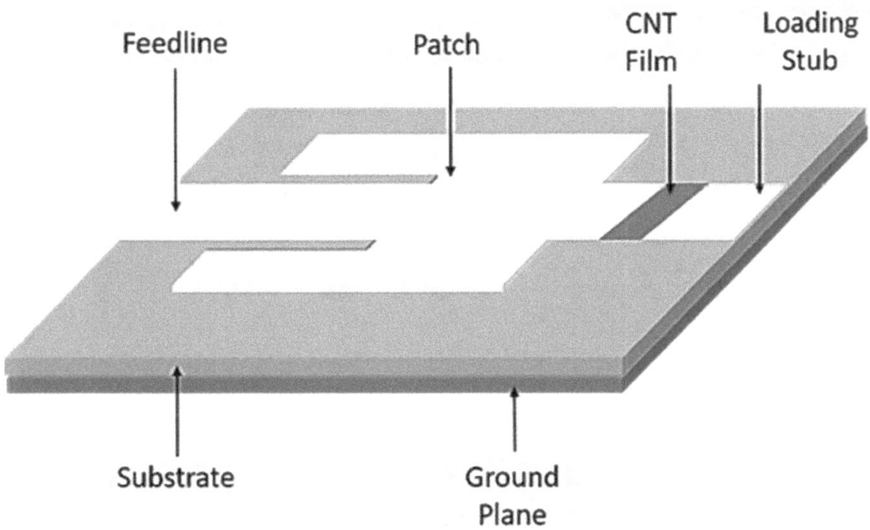

FIGURE 8.9 Structure of the suggested antenna design with CNT film.

involves investigating the effect of varying surface impedance on antenna performance through simulations. Despite CNTs' potential for antenna design, they typically exhibit poor radiation efficiency at microwave frequencies. However, this study shows that by using a CNT thin film and electrodes, significant tunability can be achieved, demonstrating the feasibility of using CNTs in antenna design to achieve both tunability and minimal losses.

Matyáš et al. [75] describe the creation of a flexible microstrip antenna using inkjet printing technology on PET foil with silver NPs, as shown in Figure 8.10, resulting in a lightweight antenna weighing only 0.208 g. The antenna has two frequency bands, 2.02 and 2.3 GHz, making it suitable for wearable electronic devices. The methodology involved synthesizing silver NPs via solvothermal precipitation, creating ink with stabilized silver NPs, and forming antennas with an inkjet printer. The antenna's performance at these frequencies enables it to be used in a variety of applications, potentially replacing several traditional antennas. The paper discusses the antenna's commercial potential for mobile devices and software-defined radio, emphasizing its novel fabrication method and practical implications.

Urbani et al. [77] describe the construction of a microstrip antenna with an ultra-thin iron film patch, which shows excellent performance, reliability, and potential for ultra-wideband applications. The methodology included building the antenna, testing its performance and reliability, and evaluating its suitability for ultra-wideband applications. The antenna demonstrated significant promise for system-on-chip, space applications, and other specialized fields, indicating that it has a wide range of applications.

Song et al. [78] describe the creation of stretchable and mechanically tunable microstrip patch antennas, as shown in Figure 8.11, using silver nanowires to demonstrate adjustable resonant frequency under strain. These antennas are suitable for wireless strain sensing applications and may be applicable to other antenna types.

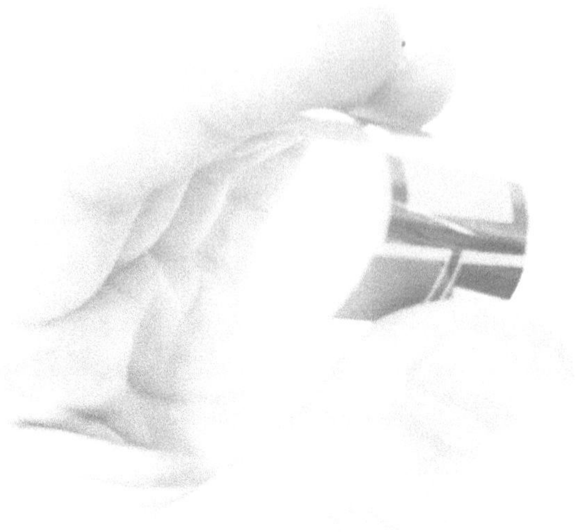

FIGURE 8.10 Demonstration of the flexibility of inkjet [76].

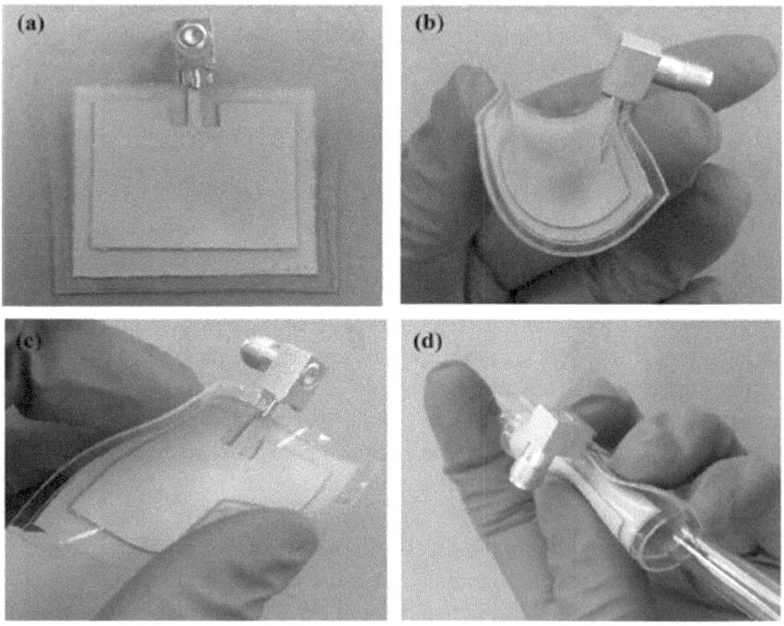

FIGURE 8.11 Photographs of a stretchable microstrip patch antenna composed of AgNW/PDMS flexible conductor: (a) relaxed, (b) bent, (c) twisted, and (d) rolled [78].

The methodology involved creating antennas out of silver nanowires, characterizing their properties under tensile strain, and investigating the relationship between resonant frequency and strain. The study introduces a new type of RA with the potential for more complex patterns and structures.

Kulkarni et al. [79] introduce a unique microstrip antenna. It is compact, operates on multiple bands, and offers high gain and directivity. The design is based on metamaterial technology. The study investigates a microstrip patch antenna that utilizes metamaterials to enhance its properties and achieve multi-band functionality. By modifying the capacitance of the metamaterial structure through metallic loadings placed in each unit cell, we can effectively enable operation at multiple frequencies. This means we can replace several antennas that work at different frequencies with a single one that manages them all. The benefits of this structure include miniaturization, lower costs, and reduced power usage. This patch antenna is set to function across L, S, and C bands. For the analysis, the proposed structure was modeled using Agilent Advanced Design System version 2011.05. Following simulations, the author fabricated it on an FR4 substrate and then utilized a vector network analyzer to measure its performance accurately.

Mahesh et al. [80] discuss the design and fabrication of a rectangular microstrip antenna loaded with zinc NPs to enhance bandwidth and enable multiband operation. Compared to conventional antennas, the antenna loaded with zinc NPs showed a bandwidth improvement of up to 16.66%. The methodology involved designing the antenna, measuring impedance bandwidths, and calculating gain. The results indicate that the zinc NP-loaded antenna resonates at eight bands of frequencies with bandwidths ranging from 1.94% to 16.66%, as demonstrated in Table 8.12. Overall, the use of zinc NPs shows promise for improving the performance of microstrip antennas in wireless communication systems.

Ambalgi et al. [81] focus on developing a new antenna configuration for improved performance in super high-frequency applications. This configuration utilizes nanomaterials and a flexible substrate, aiming to enhance performance, particularly in multi-band applications. The methodology involves studying the antenna's performance using a patch plane and strip line feed technique on a flexible substrate with nanomaterials. Overall, this research explores innovative approaches to antenna design for enhanced performance in high-frequency communication systems and is also used in hybrid renewable energy systems [82].

TABLE 8.12
Comparison of Frequency, Return Loss, Gin, and VSWR of the Suggested Two Antenna Structures [79]

Antennas	Frequency (GHz)	Return Loss (dB)	Gain (dB)	VSWR
RMSA	6.2	−11	2.25	1.50
ZNRMSA	5.1	−18.5	5.42	1.12

8.13 CONCLUSION

Communication engineering has been transformed by nanotechnology. Through manipulating, measuring, modeling, and manufacturing structures, devices, and systems at the nanoscale, new materials, nanomachines, and nano-devices have been produced that possess unique and superior properties. For example, these advancements have found multiple use cases in industries such as electronics, medicine, diagnostics, military, and the food industry. Nanotechnology's historical trajectory from ancient civilizations to recent discoveries highlights how important it has been for human progress. Milestones such as the invention of the scanning tunneling microscope and the discovery of CNTs have opened doors to cutting-edge research in nanotech. Understanding the origin and classification of different nanomaterials is essential in realizing their broad range of applications. Over time, microstrip patch antennas have considerably changed in communication engineering, providing notable benefits such as a flat design, lightweight construction, and ease of integration. Meanwhile, new materials and reconfiguration techniques have improved the performance of microstrip antennas through advances in nanotechnology. Overall, nanotechnology has been transforming communication engineering, offering prospects for developing advanced communication systems and technologies. Exploring the world of nanostructures reveals endless possibilities for new inventions and findings in communication engineering.

REFERENCES

1. Vasantha, S., S. Deepak. (2023). The nano revolution: shaping the future with nano technology. *International Research Journal of Computer Science*, https://doi.org/10.26562/irjcs.2023.v1005.17
2. Litter, M., A. Ahmad. (2023). The world of nanotechnology, Marta Litter's lab, In *Applications of Nanoparticles*, https://doi.org/10.1201/9781003183525-1
3. Umile, G. S., F. Aiello, G. Carullo, A. Facente, D. Restuccia. (2021). Nanotechnologies: an innovative tool to release natural extracts with antimicrobial properties. *Pharmaceutics*, 13 no. 2: 230, https://doi.org/10.3390/pharmaceutics13020230
4. Barhoum, A., M. L. García-Betancourt, J. Jeevanandam, E. A. Hussien., S. A. Mekkawy, M. Mostafa, M. M. Omran, S. Abdalla, M. Bechelany. (2022). Review on natural, incidental, bioinspired, and engineered nanomaterials: history, definitions, classifications, synthesis, properties, market, toxicities, risks, and regulations. *Nanomaterials*, https://doi.org/10.3390/nano12020177
5. Bayda, S., M. Adeel, T. Tuccinardi, M. Cordani, F. Rizzolio. (2019). The history of nanoscience and nanotechnology: from chemical–physical applications to nanomedicine. *Molecules*, https://doi.org/10.3390/MOLECULES25010112
6. Malik, S., K. Muhammad, Y. Waheed. (2023). Nanotechnology: a revolution in modern industry. *Molecules*, https://doi.org/10.3390/molecules28020661
7. Aleya, L., M. M. Abdel-Daim. (2020). Advances in nanotechnology, nanopollution, nanotoxicology, and nanomedicine. *Environmental Science and Pollution Research*, https://doi.org/10.1007/S11356-020-08800-6
8. Binnig, G., H. Rohrer (1983). Scanning tunneling microscopy. *IBM Journal of Research and Development* 30, no.4: 355–369.
9. Iijima, S. (1991). Helical microtubules of graphitic carbon. *Nature* 354, no. 6348: 56–58.
10. Roco, M. C., W. S. Bainbridge. (2001). *Converging Technologies for Improving Human Performance: Nanotechnology, Biotechnology, Information Technology and Cognitive Science*. Alexandria, VA: National Science Foundation.

11. Zhao, Q., N. Cheng, X. Sun, W.-L. Li. (2023). The application of nanomedicine in clinical settings. *Frontiers in Bioengineering and Biotechnology*, https://doi.org/10.3389/fbioe.2023.1219054
12. Fichera, L., G. Li-Destri, N. Tuccitto. (2020). Nanoparticles as suitable messengers for molecular communication. *Nanoscale*, https://doi.org/10.1039/D0NR06999J
13. Krishnaswamy, D., A. S. Helmy, D. D. Wentzloff. (2010). Applications of nanontechnologies in communications. *IEEE Communications Magazine*, https://doi.org/10.1109/MCOM.2010.5473871
14. Boholm, Å. (2011). Nanotechnology, risk and communication. *Journal of Risk Research*, https://doi.org/10.1080/13669877.2011.619774
15. Kim, J. H., J. Lee, J. Lee, A. Facchetti, T. J. Marks, S. K. Park. (2023). Recent advances in low-dimensional nanomaterials for photodetectors. *Small Methods*, https://doi.org/10.1002/smtd.202300246
16. Farrukh, S., W. Hatcher. (2022). Future advances and challenges of nanomaterial-based technologies for electromagnetic interference-based technologies: a review. *Environmental Research*, https://doi.org/10.1016/j.envres.2021.112402
17. Krishnaswamy, D., A. S. Helmy, D. D. Wentzloff. (2010). Applications of nanontechnologies in communications. *IEEE Communications Magazine*, https://doi.org/10.1109/MCOM.2010.5473871
18. Grillot, F., F. Grillot, J. Duan, J. Duan, B. Dong, H. Huang. (2021). Uncovering recent progress in nanostructured light-emitters for information and communication technologies. *Light-Science & Applications*, https://doi.org/10.1038/S41377-021-00598-3
19. Rahman, M.-U. (2021). Sources of nanomaterials, https://doi.org/10.1016/B978-0-12-823823-3.00007-0
20. Patyal, U., V. Kumar, V. Datta, M. Singh, N. Srivastav, Y. Muralidaran, ... M. Yadav. (2022). Manufacturing of nanomaterials from microbial sources and their roles in industrial innovations. *Microbial Nanotechnology: Advances in Agriculture, Industry and Health Sectors*, 169, https://doi.org/10.1515/9783110754476-008
21. Rofeal, M., F. A. Abdelmalek, A. Steinbüchel. (2022). Naturally-sourced antibacterial polymeric nanomaterials with special reference to modified polymer variants. *International Journal of Molecular Sciences*, https://doi.org/10.3390/ijms23084101
22. Chen, H., Y. Zheng, J. Li, L. Li, X. Wang. (2023). AI for nanomaterials development in clean energy and carbon capture, utilization and storage (CCUS). *ACS Nano*, https://doi.org/10.1021/acsnano.3c01062
23. Kulkarni, D., R. Sonwane, S. Shelke, M. P., More, S. R. Pardeshi, V. Junnuthula, S. Dyawanapelly. (2023). Biofabrication of nanoparticles: sources, synthesis, and biomedical applications. *Frontiers in Bioengineering and Biotechnology*, https://doi.org/10.3389/fbioe.2023.1159193
24. Patyal, U., V. Kumar, V. Datta, M. Singh, N. Srivastav, Y. Muralidaran, ... M. Yadav. (2022). Manufacturing of nanomaterials from microbial sources and their roles in industrial innovations. *Microbial Nanotechnology: Advances in Agriculture, Industry and Health Sectors*, 169, https://doi.org/10.1515/9783110754476-008
25. Strambeanu, N., L. Demetrovici, D. Dragos. (2015). Natural sources of nanoparticles, https://doi.org/10.1007/978-3-319-11728-7_2
26. Daulton, T. L. (2005). Nanodiamonds in the cosmos, https://doi.org/10.1007/1-4020-3322-2_5
27. Київ-КНУБА (2023). Classification of nanomaterials and the oppourtunities of their use in civil engineering. Містобудування та територіальне планування, https://doi.org/10.32347/2076-815x.2023.82.95-113
28. Dave, V. (2022). Fundamentals of nanomaterials, https://doi.org/10.1201/9781003216308-3
29. Dena, A. S. A., O. A. Abdelaziz, I. M. El-Sherbiny. (2022). Nanomaterials: classification, composition, and recent advances in synthesis. *Immunomodulatory Effects of Nanomaterials*, 1–19, https://doi.org/10.1016/b978-0-323-90604-3.00007-3

30. Selvaraj, R., D. Nagarajan. (2022). Dimensionality nanostructures classification on scanning electron microscopy images using two-stage hybrid classification algorithm. *Nano*, https://doi.org/10.1142/s1793292022500576
31. Yadav, S., S. Daniel. (2023). Green synthesis of zero-dimensional carbon nanostructures in energy storage applications – a review. *Energy Storage*, https://doi.org/10.1002/est2.500
32. Saha, S., S. Bansal, M. Khanuja. (2022). Classification of nanomaterials and their physical and chemical nature. In *Nano-Enabled Agrochemicals in Agriculture*, Elsevier, pp. 7–34, https://doi.org/10.1016/B978-0-323-91009-5.00001-X.
33. Guari, Y. (2024). Advanced porous nanomaterials: synthesis, properties, and applications. *Nanomaterials,* 14: 1602, https://doi.org/10.3390/nano14191602
34. Abdel-Karim, R. (2023). Nanoporous materials as versatile nanoplatforms for drug delivery applications: properties, recent progress, and challenges. In: Malviya, R., S. Sundram (eds.), *Engineered Biomaterials*, Singapore: Springer Nature Singapore, pp. 495–519. https://doi.org/10.1007/978-981-99-6698-1_16
35. Zakharov, V. N., I. K. Kudryavtsev, S. F. Dunaev, K. A. Paseshnichenko, L. A. Aslanov. (2022).Triazine 2D nanosheets as a new class of nanomaterials: crystallinity, properties and applications. *Colloids Interfaces*, 6: 20, https://doi.org/10.3390/colloids6020020.
36. Barhoum, A., M. L. García-Betancourt, J. Jeevanandam, E. A. Hussien, S. A. Mekkawy, M. Mostafa, M. M. Omran, S. M. Abdalla, M. Bechelany. (2022). Review on natural, incidental, bioinspired, and engineered nanomaterials: history, definitions, classifications, synthesis, properties, market, toxicities, risks, and regulations. *Nanomaterials*, 12: 177, https://doi.org/10.3390/nano12020177.
37. Hussain, C. M., M. B. Ahamed, editors. (2024). *Functionalized nanomaterials based supercapacitor: design, performance and industrial applications.* 1st ed. Singapore: Springer Nature Singapore, https://doi.org/10.1007/978-981-99-3021-0
38. Chen, J. (2023). Crystallization and assembly-driven nanostructures for energy, electronics, environment, and emerging applications. *Nanomaterials*, 13: 637, https://doi.org/10.3390/nano13040637
39. Doa, M. J. (2021). Categorization of next generation nanomaterials: international cooperation on the categorization of nanomaterials in the regulatory chemical context, https://doi.org/10.48550/ARXIV.2101.03870
40. Brown, S. C. (2021). A perspective on the grouping and categorization of nanomaterials, https://doi.org/10.48550/ARXIV.2101.03869
41. Yamamoto, T., Y. Takahashi, N. Toyoda. (2021). Dispersion of nano-materials in polymer composite materials. *MATEC Web Conference*, 333: 11003, https://doi.org/10.1051/matecconf/202133311003
42. Mabrouk, M, Das DB, Salem ZA, Beherei HH. (2021). Nanomaterials for Biomedical Applications: Production, Characterisations, Recent Trends and Difficulties. *Molecules*, 26: 1077, https://doi.org/10.3390/molecules26041077.
43. Monika, V., S. Gaurav, V. Sharma, V. Rawat, K. Behera, A. Chhabra. (2021). Metal-based nanomaterials: a new arena for catalysis. In: Hussain, C. M., S. K. Shukla, B. Mangla (eds.), *Functionalized Nanomaterials for Catalytic Application*. 1st ed., Wiley, pp. 329–353, https://doi.org/10.1002/9781119809036.ch12
44. Gómez-López, P., A. Puente-Santiago, A. Castro-Beltrán, L. A. S. do Nascimento, A. M. Balu, R. Luque, C. G. Alvarado-Beltrán. (2020). Nanomaterials and catalysis for green chemistry. Current Opinion in Green and Sustainable Chemistry, 24: 48–55, https://doi.org/10.1016/j.cogsc.2020.03.001
45. Delogu, F., E. Arca, I. V. Shvets. (2010). The synthesis of nanostructured and nanometer-sized systems. *Key Engineering Materials*, 444: 99–131, https://doi.org/10.4028/www.scientific.net/KEM.444.99.

46. Singh, A. K., R. Kumar, H. Maity, P. Singh, S. Singh. (2024). Analog and linearity performance analysis of ferroelectric vertical tunnel field effect transistor with and without source pocket. International Journal of Numerical Modelling: Electronic Networks, Devices and Fields, 37 no. 4: e3274.
47. Dhand, C., N. Dwivedi, X. J. Loh, A. N. Jie Ying, N. K. Verma, R. W. Beuerman, R. Lakshminarayanan, S. Ramakrishna. (2015). Methods and strategies for the synthesis of diverse nanoparticles and their applications: a comprehensive overview, *RSC Advances*, 5: 105003–105037, https://doi.org/10.1039/C5RA19388E
48. Herlekar, M., S. Barve, R. Kumar. (2014). Plant-mediated green synthesis of iron nanoparticles. *Journal of Nanoparticles*, 2014: 1–9, https://doi.org/10.1155/2014/140614
49. Li, G.-R., H. Xu, X.-F. Lu, J.-X. Feng, Y.-X. Tong, C.-Y. Su. (2013). Electrochemical synthesis of nanostructured materials for electrochemical energy conversion and storage. *Nanoscale*, 5: 4056, https://doi.org/10.1039/c3nr00607g
50. Swiatkowska-Warkocka, Z., A. Pyatenko, Y. Shimizu, M. Perzanowski, A. Zarzycki, B. R. Jany, M. Marszalek. (2018). Tailoring of magnetic properties of nio/ni composite particles fabricated by pulsed laser irradiation. *Nanomaterials*, 8: 790, https://doi.org/10.3390/nano8100790
51. Singh, A. K., R. Kumar, S. Jit. (2024). Dielectric modulated TFET on SELBOX substrate as a label-free biosensor applications: analytical modeling study and sensitivity analysis. *Physica Scripta*, 99 no. 10: 105413.
52. Ratnesh, R. K., M. K. Singh, V. Kumar, S. Singh, R. Chandra, M. Singh, J. Singh. (2024). Mango leaves (Mangifera Indica) derived highly florescent green graphene quantum dot nanoprobes for enhanced on-off dual detection of cholesterol and Fe^{2+} ions based on molecular logic operation. *Biomaterials Science & Engineering, American Chemical Society* 7: 4417–4426, https://doi.org/10.1021/acsabm.4c00292
53. Lee, S. H., S. P. Woo, N. Kakati, D.-J. Kim, Y. S. Yoon. (2018). A comprehensive review of nanomaterials developed using electrophoresis process for high-efficiency energy conversion and storage systems. *Energies*, 11: 3122, https://doi.org/10.3390/en11113122
54. Yadav, M. K., R. Kumar, R. K. Ratnesh, J. Singh, R. Chandra, A. Kumar, A. K. Singh. (2024). Revolutionizing technology with spintronics: devices and their transformative applications. *Materials Science and Engineering: B*, 303: 117293.
55. Marin Rivera, R., G. Zante, J. M. Hartley, K. S. Ryder, A. P. Abbott. (2022). Catalytic dissolution of metals from printed circuit boards using a calcium chloride–based deep eutectic solvent. *Green Chemistry*, 24: 3023–3034, https://doi.org/10.1039/D1GC04694B
56. Kumar, R., R. K. Ratnesh, J. Singh, R. Chandra, G. Singh, V. Vishnoi. (2023). Recent prospects of medical imaging and sensing technologies based on electrical impedance data acquisition system. *Journal of The Electrochemical Society,* 170: 117507, https://doi.org/10.1149/1945-7111/ad050f
57. Kothiyal, S. R., R. K. Ratnesh, A. K. Shukla. (2023). Enhancing sensing with surface plasmons resonance: theoretical insights and design strategies in photonic crystal fiber, *International Conference Artificial Intelligence and Information Technologies, (ICAIIT)-2024*, Taylor & Francis, 279–285, https://doi.org/10.1201/9781003510833-46
58. Ratnesh, R. K., M. Singh, S. Pathak, V. Dakulagi. (2020). Reactive magnetron sputtered-assisted deposition of nanocomposite thin films with tunable magnetic, electrical and interfacial properties, *Journal of Nanoparticle Research*, 22: 290, https://doi.org/10.1007/s11051-020-05017-z
59. Riyadh Mohammed, H., A. O. Ibrahim. (2022). Preparation of hybrid junction light emitting device from cadmium sulphite quantum dot/ poly-tpd junction. *University of Thi-Qar Journal of Science*, 8: 100–103, https://doi.org/10.32792/utq/utjsci.v8i2.870

60. Ratnesh, R. K. (2019). Hot injection blended tunable CdS quantum dots for production of blue LED and a selective detection of Cu^{2+} ions in aqueous medium, *Journal of Optics and Laser Technology*, 116: 103–111, https://doi.org/10.1016/j.optlastec.2019.03.001
61. Holyk, J., D. Kajnts, D. Vantyukh. (2023). Classification of nanomaterials and the oppourtunities of their use in civil engineering. *Urban Development Spat Plan*, 95–113, https://doi.org/10.32347/2076-815x.2023.82.95-113
62. Kumar, R., R. K. Ratnesh, J. Singh, R. Chandra, G. Singh, and V. Vishnoi. (2023). Recent prospects of medical imaging and sensing technologies based on electrical impedance data acquisition system. *Journal of the Electro Chemical Society*, 170: 117507, https://doi.org/ 10.1149/1945-7111/ad050f
63. Litter, M. I., A. Ahmad. (2023). The world of nanotechnology. In *Industrial Applications of Nanoparticles*. 1st ed., Boca Raton: CRC Press, pp. 1–15, https://doi.org/10.1201/9781003183525-1
64. Reddy, M. H., R. M. Joany, G. Manikandan, A. S. A. Nisha. (2017). Design of microstrip patch antenna with multiple slots for satellite communication. *2017 International Conference on Communication and Signal Processing (ICCSP)*, Chennai, India, 0830–0834, https://doi.org/10.1109/ICCSP.2017.8286482
65. Ratnesh, R. K., M. K. Singh, J. Singh. (2024). Enhancing ZnO/Si heterojunction photodetector performance for ultra high responsivity across wide spectral range. *Journal of Material Science: Materials in Electronics* 35: 756, https://doi.org/10.1007/s10854-024-12516-5
66. Srinivasan, K., N. Vijayalakshmi, B. Jeyanth, J. M. Jahith, T. C. Raghul. (2021). Design and analysis of microstrip slot antenna, *2021 7th International Conference on Advanced Computing and Communication Systems (ICACCS)*, Coimbatore, India, 868–873, https://doi.org/10.1109/ICACCS51430.2021.9441752
67. Kumar, R., R. K. Ratnesh, J. Singh, A. Kumar, R. Chandra. (2024). IoT-driven experimental framework for advancing electrical impedance tomography. *Journal of Solid-State Science and Technology* 13: 027002, https://doi.org/10.1149/2162-8777/ad2331
68. Costantine, J., R.Kanj, Z. Ghorayeb, et al. (2016). A radiation pattern reconfigurable antenna for WLAN access. *2016 United States National Committee of URSI National Radio Science Meeting (USNC-URSI NRSM)*, Boulder, CO, USA, 1–2, https://doi.org/10.1109/USNC-URSI-NRSM.2016.7436227
69. Garg, A., R. K. Ratnesh. (2022). Solar cell trends, and the future: a review. *Journal of Pharmaceutical Negative Results*, 13: 2051–2060. https://doi.org/10.47750/pnr.2022.13.S06.268
70. Kothiyal, S. R., R. K. Ratnesh, A. Kumar. (2023). Field effect transistor (FET)-sensor for biological applications. *International Conference on Device Intelligence, Computing and Communication Technologies, (DICCT)-2023*, IEEE, 433–438, https://doi.org/10.1109/DICCT56244.2023.10110155
71. Garg, A., R. K. Ratnesh, R. K. Chauhan, N. Mittal, H. Shankar. (2022). Current advancement and progress in BioFET: a review. *International Conference on Signal and Information Processing (IConSIP)*, IEEE, 1–7, https://doi.org/10.1109/IConSIP49665.2022.10007517
72. Thabet, A., A. Z. El Dein, A. Hassan. (2011). Design of compact microstrip antenna by using new nano-composite materials, *The 4th IEEE International NanoElectronics Conference*, Taoyuan, Taiwan,, 1–2.
73. Urbani, F., D. W. Stollberg, A. Verma. (2012). Experimental characterization of nano-film microstrip antennas. *IEEE Transactions on Nanotechnology* 11: 406–411
74. Naishadham, K. (2014). An investigation on the tuning of a microstrip patch antenna using carbon nanotube thin films. *2014 IEEE Antennas and Propagation Society International Symposium (APSURSI)*, Memphis, TN, USA, 900–901.

75. Matyáš, J., P. Slobodian, L. Munster, R. Olejník, P. Urbánek. (2017). Microstrip antenna from silver nanoparticles printed on a flexible polymer substrate. *Materials Today: Proceedings* 4: 5030–5038.
76. Haider, R., M. Hussain, I. Ayman. (2013). Design, fabrication, and testing of flexible antennas. *Advancement in Microstrip Antennas with Recent Applications,* InTech. https://doi.org/10.5772/50841.
77. Urbani, F., D. W. Stollberg, A. Verma. (2010). Outstanding performance of a nanofilm microstrip antenna, *2010 IEEE Nanotechnology Materials and Devices Conference*, Monterey, CA, USA, 160–163.
78. Song, L., A. Myers, J. J. Adams, Y. Zhu. (2014). Stretchable and reversibly deformable radio frequency antennas based on silver nanowires. *ACS applied Materials & Interfaces* 6 no. 6: 4248–4253.
79. Kulkarni, N., G. Lohiya. (2016). A compact microstrip patch antenna using metamaterial. *International Journal of Engineering Trends and Technology* 42: 365–369.
80. Mahesh, P. (2018). Zinc nanoparticles loaded rectangular microstrip antenna for multiband operation. *International Journal for Research in Applied Science and Engineering Technology* 6: 261–264, https://doi.org/10.22214/ijraset.2018.5043
81. Ambalgi, P. (2020). Nanomaterial fabricated patch antenna for super high frequency applications engineering & technology. *International Journal of Science and Technology*, 1: 1–7.
82. Pandey, P. K., R. Kumar, V. Gupta, et al. (2024). An innovative hybrid controller-based combined grid-connected hybrid renewable energy system. *Electrical Engineering*, https://doi.org/10.1007/s00202-024-02363-2

9 Nanotechnology Mechanistic Techniques and Ideas for Mechatronics

Aditi Singhal, Mansi Verma, and Sanjay Kumar Singh

9.1 INTRODUCTION

Metal nanoparticles (NPs) have garnered substantial scientific interest due to their exceptional properties, prompting extensive research efforts [1–3]. The precise control of NP size and distribution is pivotal for optimizing their physical and chemical characteristics, which are crucial for their wide-ranging applications in catalysis, energy systems, and biomedical technologies [4,5].

Various techniques, such as laser ablation, vapor deposition, and capping ligands, have emerged to manipulate nanomaterial size [6,7]. However, recent investigations, including our group's research, underscore the profound impact of controlling transport phenomena, particularly in the nascent stages of synthesis, on final NP properties. Parameters like stirring speed, reaction volume, and reactor design in batch configurations, as well as reactor and mixer design in flow setups, intricately influence mixing efficiency during synthesis. Computational fluid dynamics (CFD) and an integrated population balance model (PBM) provide a robust framework to comprehend the synthesis of NPs amidst complex fluid dynamics and polydisperse entity formation. Our study unveils an experimentally validated PBM-CFD model, focusing on the chemical reduction synthesis of silver NPs [8]. This model offers precise predictions regarding the influence of flow parameters and reaction conditions on size and distribution evolution. Such predictive capabilities are pivotal for engineering optimized flow reactors, obviating the need for empirical trial-and-error approaches.

9.2 METHOD

This research delves into the chemical reduction process of silver nitrate to produce silver NPs, acting as a potent reducing agent as per the reaction:

$$8Ag^{+} + BH_4^{-} + 8OH^{-} \rightarrow 8AgO + H_2BO_3^{-} + 5H_2O \quad (9.1)$$

DOI: 10.1201/9781003634553-9

To precisely model AgNP production in microreactors, an integrated computational tool was developed [9]. $AgNO_3$ and $NaBH_4$ aqueous solutions, the reactants, are combined in a T-mixer of a certain size. A defined concentration of $AgNO_3$ is injected along the mixer tube, while $NaBH_4$ is introduced perpendicular to the reactor [10].

Two reactor configurations were studied: a three-dimensional helical flow reactor (HFR) and a straight flow reactor [11]. The HFR consists of a bent reaction tube around a cylindrical support, with different helical diameters modeled for the HFR [12]. The primary liquid phase includes the inflow stream with reactants, while the silver NPs generated in situ make up the solid phase. The simulation involves solving three separate models concurrently [13].

1. The multiphase model, derived from the mixture model, computes pressure and velocity fields for both phases within the system [14].
2. A model for primary phase species transfer is employed to analyze the transfer of species within the primary phase of the system [15].
3. The production and growth dynamics of NPs in the system are simulated by means of a population balance model that uses the quadrature method of moments [16].

9.2.1 Population Balance Model

The Finke–Watzky pseudo elementary two-step mechanism simulates the creation of silver NPs via chemical reduction [17]. This mechanism operates under the assumption of a pseudo-first-order reaction, enabling the calculation of localized parameters like nuclei size, NP formation rate, and surface growth within each differential fluid cell of the reactor [18]. The two-step methodology considers growth beyond the initial nucleation stage [19]. While nucleation is typically considered a pseudo-first-order process, in this scenario, in this scenario, the concentration of the reducing agent is factored into the nucleation rate equation to account for mixing effects [20,21].

Stated differently, nucleation occurs exclusively when a reducing agent and a metal precursor are present.

$$\text{Nucleation} \quad BH_4^- + Ag^+ \rightarrow AgSNR = K_{nuc}[BH^{-4}][Ag^+] \tag{9.2}$$

The rate of growth is expressed as

$$\text{Growth} \quad Ag^+ + AgS \rightarrow AgSGR = K_{growth}[Ag^+][Ag]s \tag{9.3}$$

where K_{nucl} (particles $m^{-3}M\text{-}2s^{-1}$) and K_{growth} ($mM\text{-}2s^{-1}$) are kinetic constants for nucleation and apparent growth, respectively.

The multiphase model employed is ANSYS Fluent's multiphase modeling capabilities. The mixture model is selected because, as shown in earlier studies for systems of a similar nature, it achieves a modest computational cost [22]. This model is capable of capturing the evolving distribution of particle sizes within the main flow field, specifically focusing on the dispersed phase. It establishes a relative velocity for each phase (liquid and solid) and solves a unified momentum equation for the multiple phases being transported.

Global resistance to flows in a cell is derived from the mean diameter of the NP distribution, not from the influence of individual solid particles on the flow [23]. Using the Manninen et al. method, the algebraic link for the relative velocity was explained for the slip velocity. Solid stress resulting from collisions and friction is disregarded. It is believed that NPs are spherical in shape. The Sauter mean diameter (d_{32}) is derived from the ratio of the third moment (m^3, representing particle volume, m^3) to the second moment (m^2, representing particle surface area, m^2). This parameter is instrumental in calculating the velocity field pertaining to the secondary solid phase when integrating the PBM with multiphase fluid dynamics computations.

$$d_{32} = m^3 \quad m^2 \tag{9.4}$$

The generation of the secondary solid phase will involve the mass transfer coupling rates (Mt, expressed in kmolAg,s $m^{-3}s^{-1}$) between secondary (solid) and primary (liquid) phases. This connection will once again link the PBM with the multiphase fluid dynamic model, as detailed in the provided equations [24,25].

$$\text{MtAg,s}(\text{nucleation}) = \rho\text{Ags MwAgs } \pi\text{6d3nucNR} \tag{9.5}$$

$$\text{MtAg,s}(\text{growth}) = \rho\text{Ags MwAgs } \pi\text{6 3GRm2} \tag{9.6}$$

where ρAgs and MwAgs are the secondary phase's density and molecular weight of silver, and $\pi/6$ is the sphericity constant.

9.2.2 Photodynamic Therapy (PDT)

For more than a century, PDT has found applications in medicine, primarily targeting undesired biological cells like those found in skin cancer [26]. The term "photodynamic action" originated from the observations of a medical student who noticed that paramecia cultivated in darkness survived, while those cultured with the fluorescent dye acridine orange perished upon exposure to light [27]. Later research elucidated the method of leveraging PDT's deadly impact to address skin malignancies, confirming the crucial role of oxygen in the photosensitization process.

Currently, PDT stands out as a promising therapeutic choice for various ailments, with cancer being a prominent focus [28]. Figure 9.1 represents significant potential either independently or in combination with traditional treatment methods like chemotherapy and radiation therapy. PDT has some unique benefits, such as being relatively less intrusive, having lower systemic toxicity, and offering a wide spectrum of cellular targets [29]. PDT works by stimulating a photosensitizer (PS), a light-sensitized tumor-targeting substance, with appropriate light irradiation. Reactive oxygen species (ROS) are highly harmful because PS molecules have the capacity to take up and provide electrons to molecules of oxygen. The production of these ROS leads to irreversible harm in targeted tissues, triggering a cascade of signals that signal both survival and cellular death pathways. When cells are unable to cleanse and repair

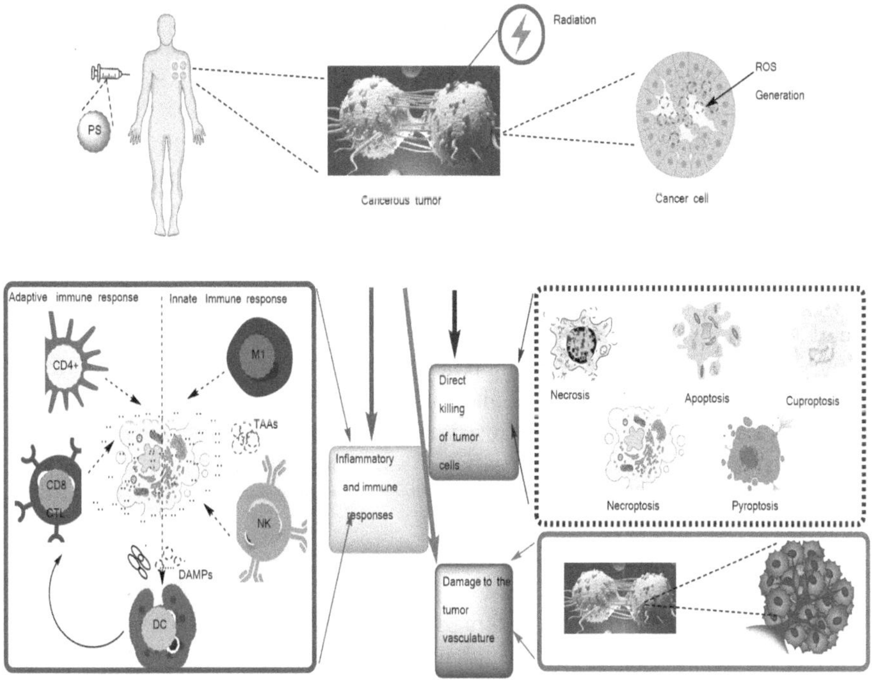

FIGURE 9.1 Photodynamic therapy for cancer treatment.

the damage they have caused, high levels of oxidative stress set off processes that ultimately lead to cell death [30].

9.3 MECHANISM OF PDT

PDT results from the synergistic interaction of molecular oxygen, a specific wavelength of light, and a non-toxic photosensitizing drug, which produces damaging ROS. The PS medications can be injected intravenously, topically, orally, intraperitoneally, intra-arterially, or intratumorally. Using any of these delivery methods will cause the drugs to accumulate selectively inside the tumor [31]. The PS is activated by further exposure to light, and a photochemical reaction is triggered by the presence of molecular oxygen.

The PS is first excited from its ground state (singlet state, 1PS) to a transient excited singlet state (1PS*) [32]. A fraction of this energy propels the PS into a stable electronically excited state (triplet state, 3PS*) through intersystem crossing, while the remainder is released as fluorescence by the excited 1PS*. The triplet state can transfer energy to biological molecules, initiating two key reactions:

1. In the I Type reaction, hydroxyl radical (OH•), superoxide anion radical (O^{2}-•), and hydrogen peroxide (H_2O_2) donate protons and electrons to nearby oxygen and water molecules, leading to the generation of free radicals.

2. In the II Type reaction, energy is directly transferred to adjacent oxygen molecules within the cytoplasm, resulting in the formation of singlet oxygen ($1O_2$) [33,34].

Because PSs can both absorb light of specific wavelengths and initiate the photophysical or photochemical reactions, they are crucial to PDT. The initial PS utilized in therapy was the "hematoporphyrin derivative" (HpD). Subsequently, over a thousand synthetic and natural PSs have been discovered. PDT has utilized diverse classes of PSs, beginning with the initial generation of PSs that included various forms of HpD, such as Photofrin®. Photofrin® was the first PS molecule to receive approval from the United States Food and Drug Administration (USFDA) for use in cancer treatment in 1995 [35]. The efficacy of treatment relies on the PSs' capacity to selectively target cancerous cells and the ensuing cellular interactions. Metal-based NPs have shown promise as a method to improve the distribution of PSs and enhance the outcomes of PDT in this aspect [36]. PDT's three separate anti-tumor mechanisms provide it with a unique therapeutic option for cancer:

1. Direct lethal effect on tumor cells resulting from reactive species
2. Injury to the tumor's blood vessels
3. Stimulation of the immune system through improved presentation of cancer cell antigens to T cells and the phagocytic uptake of dead cancer cells by dendritic cells.

The efficacy of PDT in causing cell death directly depends on where the PS is located within the cell. The resulting ROS have the potential to harm cellular compartments such as the endoplasmic reticulum, lysosomes, or mitochondria [37]. Depending on the severity of the ROS-induced damage, it may be irreparable or irreversible. Interestingly, mild oxidative stress triggers an adaptive response aimed at maintaining cellular balance and repairing the damage. These adaptive mechanisms primarily involve removing oxidized biomolecules and inhibiting the synthesis of new proteins. Examples of these survival strategies include the nuclear factor E2-related factor 2 (Nrf2) pathways and the integrated stress response, as well as the unfolded protein response [38,39].

9.4 MAGNETIC NPs

Using magnetic NPs and localized heat creation, magnetic hyperthermia cancer therapy offers a possible means of selectively treating cancer cells [40,41]. As a component of the therapy, biocompatible magnetic NPs are injected into the body, where they accumulate at the site of the tumor. These NPs typically consist of soft magnetic materials, such as iron oxide, characterized by small hysteresis curves [42]. To enhance their specificity for cancer cells, these NPs can be modified with specific chemicals. Once they have amassed within the tumor, exposure to an alternating magnetic field induces rapid oscillations in the NPs. This process produces heat through mechanisms such as Brownian relaxation, a hysteresis loss, and Néel relaxation [43,44].

Hysteresis loss refers to the conversion of energy into heat due to magnetic materials reacting sluggishly to alterations in the magnetic field they experience. The production of heat by magnetic NPs leads to localized hyperthermia, elevating the temperature specifically within the tumor region [45–47]. The goal of this localized hyperthermia is to cause apoptotic cell death by raising the temperature to between 42°C and 45°C. Additionally, hyperthermia can improve the effectiveness of other cancer therapies, such as radiation therapy or chemotherapy, by heightening the sensitivity of cancer cells [48]. A significant advantage of magnetic hyperthermia treatment is its precise targeting of tumor cells while minimizing harm to adjacent healthy tissues. Moreover, the risk of off-target effects can be diminished through the functionalization of NPs with specific ligands, facilitating their recognition and binding to cancer cells [49–51].

This therapeutic strategy is non-invasive, targeted, and can be used in concert with other therapies for optimal outcomes because the magnetic field can be carefully controlled and concentrated to generate localized heat within the tumor area. Because they function as heat-generating transducers, magnetic NPs—produced through various techniques and functionalized to suit biomedical applications—are a crucial component of hyperthermia therapy [52]. $La_{1-x}Sr_xMnO_3$, a magnetic perovskite based on rare earth metals, has emerged as a potential substitute for iron oxide-based magnetic NPs commonly employed in hyperthermia applications. The ability to modify the doping ratio and thus adjust the NPs' Curie temperature for medical hyperthermia makes them a favorable option.

Replacing La with Sr in $LaMnO_3$ transforms a portion of Mn^{3+} ions into Mn^{4+} ions, thereby altering its antiferromagnetic properties. Moreover, a paramagnetic state can emerge, contingent on the extent of substitution [53]. It has been demonstrated that $La_{0.7}Sr0_3MnO_3$, with its superior Curie temperature and magnetic properties, is the optimal Sr replacement ratio for applications involving magnetic hyperthermia. Nevertheless, ensuring biocompatibility remains crucial. Silica surface modification is appealing as it allows for further functionalization and enhances biocompatibility. This paper presents an improved method for creating silica-coated LSMO magnetic cores with the goal of improving biocompatibility and appropriateness for magnetic hyperthermia therapy.

The coercivity and remanence values of our fabricated nanosystem are notably lower than those reported for LSMO nanosystems in prior studies. Additionally, for the first time, we have demonstrated the impact of a silica coating on the surface chemistry of bare LSMO NPs through comprehensive XPS analysis, particularly in the context of hyperthermia therapy. This discovery may open up new avenues for investigating the surface chemistry of nanosystems and ligand functionalization for tailored applications. Our aim with this innovative approach is to advance the development of precise and efficient cancer treatments [54,55].

9.5 LATENT FINGERPRINT ENHANCEMENT BY AG NPs

An individual's fingertips and palms have papillary ridges that can come into accidental contact with surfaces to form fingermarks [56]. A fingermark is created when the original ridge pattern is partially or completely replicated and transferred onto the surface of skin secretions, such as sweat and oily substances. The predominant form of evidence encountered at crime scenes is often latent fingerprints, which are imperceptible to the naked eye [57]. Making them visible therefore requires applying the proper physicochemical development processes to produce a visual contrast between the underlying substrate and the fingermark residues [58].

Most of the visual, physical, and chemical methods for latent fingerprint formation that have been studied include the application of silver nitrate solution, cyanoacrylate or iodine fuming, ninhydrin, or colored granules. Novel techniques for latent fingermark visualization have recently been made possible by the ability to alter features of nanoscale materials and to functionalize them using chemical components that leave a fingermark trail behind [59].

There is considerable interest in the study and application of NPs for latent fingerprint enhancement because physical size and chemical composition both affect electrical and optical capabilities at the nanoscale. Examples of these NPs include carbon dots, quantum dots (QDs), and NPs of gold and silver (AgNPs and AuNPs) [60].

Another benefit of possessing an extensive surface area is the increased density of molecular interactions, resulting in a higher population of binding sites for specific components like fatty acids, amino acids, or metabolites commonly found in sweat at both the cellular and molecular levels [61–64]. An additional advantage is the great spatial resolution of the generated image. Despite the benefits mentioned above, there are still certain drawbacks to nanomaterial-based techniques for latent fingerprint creation. These include issues with synthesis, stability, and health and safety. By investigating the creation of composite materials in the form of SiO_2, Fe_2O_3, or zeolites in core–shell structures, these issues can be avoided. Studying the manufacturing of NPs presents an alternative approach that can interact with the surface rather than the fingerprint residue. For instance, compounds based on silver NPs (AgNPs) can yield a negative or reversed impression [65–68].

In this regard, an efficient technique that has evolved is the electrochemical deposition of species on metallic surfaces indicated by fingers, such as conjugated polymers and AgNPs [69]. Their preferred contact with the surface over the fingermark residue is responsible for the higher success rates. One approach that is seen as extremely promising is the use of electrochemical methods to create latent fingerprints on conducting surfaces. Examples that have been reported include the scanning Kelvin probe, electrolysis, and electrostatic enhancement. Nevertheless, these techniques are frequently costly and require technological advancements that are not always feasible for use in forensic labs. One effective and simple technique for creating metal NPs is electrochemical deposition [70,71].

To improve latent fingerprints, Qin et al. suggested using AgNPs or AuNPs that are spatially selectively electrodeposited on conducting surfaces, such as stainless steel coins, Au, Pt, and glass covered with indium/tin oxide (ITO). Utilizing this method, a distinct negative representation of the fingerprint is generated by employing the fingermark residue as an insulating agent during the electrodeposition procedure. This ensures that the finger's "furrows" serve as a barrier to prevent residue from growing on the metal NPs. Essentially, this process involves through-mask electrodeposition, with the organic matter, such as fatty acids from the finger's contact print acting as a deposition-blocking overlayer. This technique resembles the creation of blocking "monolayer-like" films through contact printing [72].

In terms of materials, this involves applying polymer films through electrodeposition to the exposed metal areas of fingerprinted surfaces, ensuring specific spatial orientation. Here, we develop this idea by electrodepositing AgNPs on forensically significant substrates, such as brass, copper, and stainless steel. A rigorous, regulated, and effective methodology for enhancing latent fingermarks on metallic surfaces is the ultimate objective. This study aims to achieve the following specific goals [73].

i. Determine and optimize the AgNP deposition electrochemical control function;
ii. Understand how the final produced mark is affected by latent mark aging
iii. Analyze the quality of the resulting images.

9.6 MECHATRONICS TECHNOLOGIES AS AN INNOVATIVE DETECTING APPROACH

The word "mechatronics," which combines the words "mechanical" and "electronics," was coined and first published in the 1980s. Specifically, this concept was first proposed by Yasakawa Electric Company engineer Tetsuro Mori, who worked in the unique neighborhood of Chiyoda-Ku in Tokyo, Japan [1969]. In the discipline of mechatronics, the principles of computing, electric/electronic engineering, mechanics, and multidisciplinary engineering are combined with the goal of developing and refining new artificial intelligence-containing technologies [74].

These days, it is utilized in a wide range of technical fields and industries, including anti-lock brakes, robotic-assisted surgery systems, medical robots, self-driving cars, smart biosensors, calculators, the agriculture and food industries, Roomba® (iRobot Corporation), NASA's Curiosity rover, and anti-lock brakes. Figure 9.2 represents the various applications of mechatronics technologies [75]. The mechatronics system is composed of three components: informatics, electronics, and mechanics [76] introducing "Industry 4.0," a next generation of the industrial revolution resulting from mechatronics that integrates autonomous systems, Internet of Things, machine-to-machine interfaces, and smart factories. Currently, there is a shift in the mechatronic sciences toward intelligent technical systems [77]. Established in 1993, the Mechatronics in Medicine Laboratory (a division of the Computer Aided Systems Engineering Section) [78] aims to conduct research and create mechatronic surgical assistance during the expansion of robotic surgery utilization. Combining robotics, modeling, artificial intelligence, and sensory devices can mimic human function and analysis in a number of experiments [79]. Because of the complexity of the processes involved in microbiologically influenced corrosion, as well as the challenges in detecting, interpreting, and monitoring the results obtained through the aforementioned laboratory techniques, applying mechatronics to the field of biocorrosion may help overcome current obstacles and limitations.

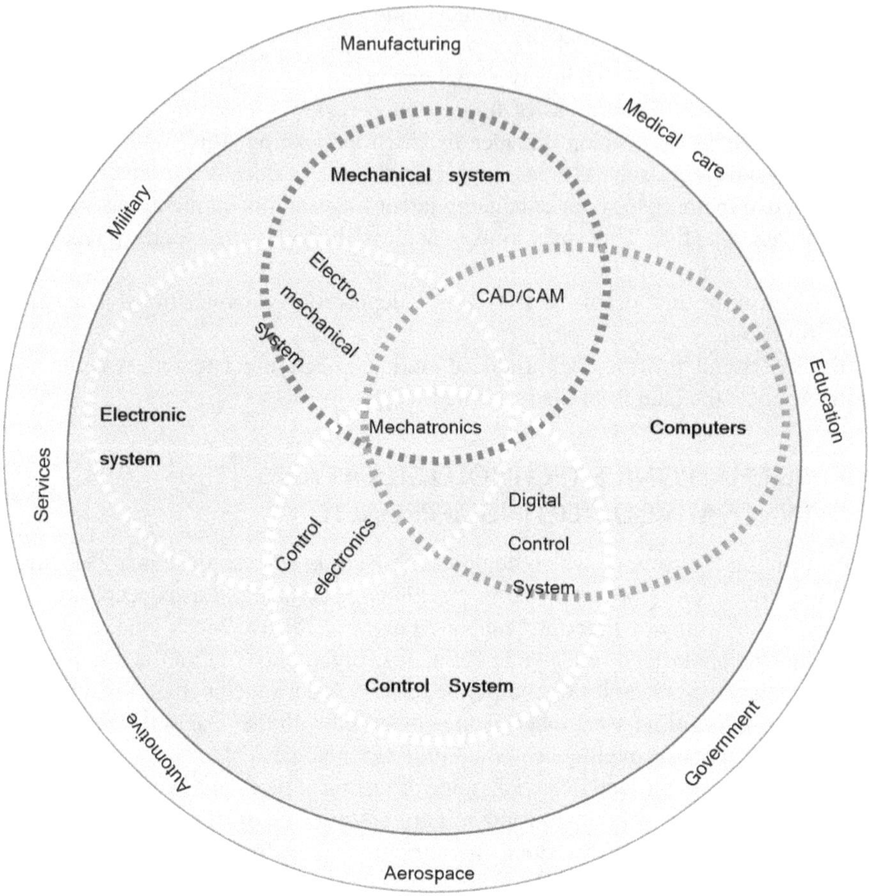

FIGURE 9.2 Various mechatronics applications.

The MAKRO project was completed in 1999 by Feelisch and Stamler [80]. The goal of the project was to build inspection robots that could perform tasks in wastewater collection systems that are inaccessible to humans. These robots are equipped with an infrared scanner, a video/CCTV camera, an ultrasonic sensor, and a structured light source. This may have been a precursor to developments in mechatronics that could evaluate other kinds of corrosion with high sensitivity and specificity [81]. Regarding pipelines, visual data from the internal pipeline, associated with a location stamp, is considered essential for pipe inspection. Hamilton et al. [Egypt, 2018] introduced an autonomous sewage pipeline inspection robot for detecting hidden corrosion and locating cracks [82]. In this research, a novel robot was designed for the online evaluation of irregularities in interior pipelines, and subsequently, a user interface was developed for offline processing and interpretation of the obtained visual information.

9.6.1 Bacterial Biofilm and Bio Corrosion

The biofilm that bacteria create serves as a surrogate for them, as microcolonies grow in biocorrosion locations. Specifically, biofilms are complex surface-attached bacterial communities held together by self-produced polymer matrices, which mostly consist of extracellular DNA (deoxyribonucleic acid), secreted proteins, and exopolysaccharides [83]. The role of bacteria and their biofilms in the MIC process can be explained through three different methods. Firstly, electroactive sessile cells in biofilms, rather than planktonic cells, are able to extract electrons from energetic metallic or nonmetallic surfaces for cellular respiration, thus directing the transport of chemical species toward or away from the surface and directly causing corrosion in these collections of bacteria, whether in mono- or poly-microbial aggregates. Secondly, biofilms may have a high density of pre-existing corrosive substances, such as water with CO_2, in accordance with the inorganic passive layer (often a metal oxide film on a metallic surface) or organic acids beneath abiotic surfaces. Here, biofilms can indirectly accelerate biocorrosion events by contributing to the loss of protective coatings [84]. It is interesting to note that salt film precipitation and oxide formation can both result in thick film passivity. Sometimes, salt layer precipitation either provides sufficient protection on its own or acts as a catalyst for a thin passive film. Magnesium (Mg) exhibits extremely poor corrosion behavior in orthopedic applications, as demonstrated by Chidanand Hegde et al. [85]. For MC3T3 osteoblast cells, coating magnesium alloy with poly(vinyl acetate)/dichloromethane sheets provided exceptional bonding strength. An additional type III biocorrosion can be defined as a microbial attack that results in the creation of extracellular organic material, such as polyurethane and its organic plasticizer, with the intention of organisms extracting organic carbon and energy [86].

9.6.2 Biosensing Techniques

A few papers on the creation of biosensor-based in-situ testing strips for the identification of biocorrosion bacteria in industrial, medical, and/or environmental samples have been published [87]. These sensing techniques still need to be improved upon and commercialized, though. Biosensing techniques are showing impressive confidence for diagnosis because of features like real-time analysis capability, affordability, time-saving, and feasibility [88]. Many types of biosensors have emerged in the modern era, including mechanical, piezoelectric, magnetic, colorimetric, optical, electrochemical, and fluorescent biosensors. Infectious (bacterial and viral) agent detection has evolved into a unique analysis using electrochemical DNA biosensors [89]. Benefits among the different kinds of biosensors include reduced limits of detection (LODs), automation, affordability, high sensitivity and specificity, and the introduction of disposable instruments and diagnostic techniques. Accordingly, Kobra Salimiyan rizi created an electrochemical DNA biosensor to identify the SRB genome [90]. Due to their large pore volumes, strong mechanical strengths, high surface areas, and quick electron transfer, three-dimensional (3D) graphene was integrated into a nanocomposite that was created for the biosensor in question. This enhanced the biosensor's sensitivity. G-Au NPs were fixed on the glassy carbon electrode, which serves as the working electrode, using phenol and Nafion. The redox label was methylene blue (MB) trihydrate. The results of the differential pulse voltammetry measurements were obtained in the presence of MB molecules under optimal experimental conditions [91].

The estimated linear working range and attained LOD of the suggested DNA biosensor were found to be $9.41 \times 10\pi$ 15M (M) and $1.0 \times 10\pi$ 14–$1.0 \times 10\pi$ 7 M, respectively, for the target DNA, which is dsrAB DNA. Additionally, the dsrAB DNA sequence in the fluid taken from a petroleum field is satisfactorily detected by this biosensor. When compared to both quantitative RT-PCR and gel electrophoresis, the performance LOD of this biosensor demonstrated a better LOD. However, only four distinct, synthetic single-strand oligonucleotide sequences: two-mutation target, single-mutation target, and non-complementary sequence, full complementary target were used to test the suggested biosensor's selectivity. However, genuine bacterial genomic samples from sulfate-reducing bacteria other than SRB with high genetic similarity that can exist in biocorrosion sites are thought to be a superior way to analyze it [92].

9.6.3 Microfluidics Technology

The science of carefully controlling and modifying micro- or nano-fluids in very small-scale spaces, like micro-channels, is called microfluidic technology. The fundamental idea behind microfluidics is to include simple micro-sized platform procedures that frequently require an entire lab. A single integrated miniaturized circuit for the integration of several biomedical, chemistry, biotechnology, or medical diagnostic laboratory operations together is referred to as a lab-on-a-chip [93]. The creation of a microfluidic chip involves the molding or engraving of a microchannel platform into materials like poly-methyl methacrylate, silicon, glass, polydimethicone (PDMS), and polyethylene terephthalate glycol. The inputs and outputs in this chip create an interface between the macro- and micro-worlds by connecting the chip's micro-channel network to the external environment. The three types of regular microfluidic systems are digital (DMF), continuous, and droplet-based. Typically, the droplets on the microfluidic sensing devices are moved using the electrowetting-on-dielectric approach. Integrated microfluidic biosensors can now be employed as micromachines for online biocorrosion detection.

9.6.4 Bioremediation

Heavy metals from water and organic pollutants from soil can be removed by using NPs. Silver NPs have the ability to efficiently break down a variety of contaminants like organic dyes and substances present in wastewater. Numerous nanomaterials, such as carbon nanotubes, metal oxides, fibers, and nanoscale zeolites, have been investigated for use in cleanup processes. When utilized in cleanup, nanoscale particles can reach places that larger ones cannot. Before reacting with pollutants, they can be coated to aid in transit and prevent reactivity with nearby soil matrices [94].

Nanoscale zero-valent iron (nZVI) is one commonly utilized nanomaterial for remediation applications. Chlorinated solvents that have contaminated groundwater have been cleaned up at several hazardous waste sites using it. Due to their detrimental impact on the environment and human health, heavy metals including arsenic, thallium, cadmium, mercury, and lead have drawn significant interest regarding the removal of these substances from natural water sources. NPs of superparamagnetic iron oxide work well as a sorbent for these toxic substances. Therefore, because analytical techniques to assess the trace concentration of NPs are lacking, there have been no measurements of designed NPs in the environment available [95].

9.6.5 Electronic Sensors

9.6.5.1 Si-Field Effect Gas Sensor

The vast potential of monolayer-protected NPs for gas sensing applications has been demonstrated by recent studies [96]. The application of gold-NP (Au-NP)-based gas sensors for NOx detection has been the subject of numerous investigations. The main tenet is that NOx has a demonstrated strong affinity for gold surfaces, as shown by Li Han and colleagues [97].

Gold thin film that has been thermally evaporated has been utilized to monitor NOx, with little to no reaction from the sensor for other interfering gases such as CO and H_2 [98]. Recent research on the size-dependent sensor response of Au-NP-based sensors has suggested a simple link between the sensing response and the grain size of the gold catalytic layer [99]. Micro-porous silicon films doped with Au have been used to monitor NOx, and their reactions to NO_2 and NO species are similar [100].

9.6.5.2 Semiconductor-Based Sensor

Porphyrin molecules have been chemically functionalized to III-V compound semiconductors, such as GaAs, and this approach has been proposed as a means of selectively detecting NO [101]. Because of its high binding affinity to NO, hemin (Fe-porphyrin) has been proven to be the most suitable porphyrin compound for NO sensing. Porphyrin compounds are generally widely available, each with specialized features and a range of gas molecules they can target for sensor applications. The combination of an organic substance (porphyrin molecules) with an inorganic material and the potential to create tailored sensor qualities is of scientific interest. Because of their broad range of electrical characteristics and chemical inertness, III-V compounds are particularly intriguing as transduction platforms [102]. It has been demonstrated that when NO is detected using hemin functionalized to GaAs, its surface potential decreases, increasing the current flowing through the GaAs device [103]. Therefore, the sensitivity of these semiconductors to adsorbates is another advantageous feature.

9.7 SEMICONDUCTOR NPs IN TARGETED DRUG DELIVERY

At the nanoscale, semiconductor NPs are regarded as semiconducting materials. Their electrical conductivity is lower than that of an insulator and higher than that of a conductor. The NPs have a high surface area, exhibit quantum effects, and range in size from 1 to 20 nm. NPs of QDs are thought to be semiconductors possessing magnetic and optoelectronic characteristics. These QDs exhibit strong photoluminescence and fluorescence. Owing to these characteristics, trials were conducted for drug administration, and QDs were tried for diagnostics [104].

Subsequently, it was discovered that this helped with radiation, chemotherapy, and imaging with ionizing radiation therapy for cancer treatment. Doxorubicin is delivered to human lung cancer cells using QDs. They used the Ag-In-Zn-S QDs as the foundation for their customized medication delivery method. To target doxorubicin (DOX) to folate receptors on adenocarcinoma human alveolar basal epithelial cells (A549), the QD crystals were modified with 11-mercaptoundeanoic acid (MUA), L-cysteine, and lipoic acid, and adorned with folic acid. The folate receptor-negative control was NIH/3T3 [105]. Subsequently, the attachment of FA to QD nanocrystals and DOX to QD-FA nanocarriers was verified by FT-IR. The amount of FA and DOX anchoring to the QD nanocrystals was measured using UV-Vis analysis. In comparison to other produced QD sample forms, QD-FA-DOX nanoconjugates exhibited increased cytotoxicity upon biological screening. Among other cytotoxins, A549 cells were shown to be most susceptible to the effects of QD-MUA-FA-DOX, while NIH/3T3 cells showed no discernible cytotoxicity. Even the in vitro scratch test shows that using QD-MUA-FA-DOX inhibits the migration of A549 cells. For this reason, QD-MUA-FA-DOX nanoconjugates may be employed as a drug delivery technique to treat adenocarcinoma-developing human alveolar basal epithelial cells [106].

9.8 CARBON-BASED NPs FOR PRECISE MEDICATION ADMINISTRATION (TARGETED DRUG DELIVERY)

Since carbon-based NPs have favorable mechanical, electrical, optical, thermal, and chemical properties, they are employed in medicine for targeted medication delivery. There are many different forms of carbon, including diamonds, graphite, and amorphous carbon [107].

9.9 GRAPHENE OXIDE

Graphene is a two-dimensional material consisting of a homogenous, flat monolayer of carbon atoms tightly packed in a hexagonal or honeycomb configuration. Because of its high surface area, optical qualities, and biocompatibility, grapheme and its variants are used in biomedical applications. The folic acid-targeted method is used to administer DOX and MRI through GO-coated Fe_3O_4 nanocomposites. The stable material employed for targeted drug delivery is FA-Fe_3O_4 @nGO-DOX. The survivability of the FA-Fe3O4@nGO-DOX nanoplatforms was tested using the Cellosaurus cell line/MGC-803 selective uptake. Consequently, the selected tumor cells are killed by pH-mediated drug release and the selective absorption of the NPs through MSC-803 utilizing folic acid receptors [108].

9.10 GO-COATED FE_3O_4 - NANOCOMPOSITE THAT IS STABLE AND BIOCOMPATIBLE

A strong drug delivery vehicle for targeted therapy and concurrent tumor MR imaging. Cisplatin (CDDP) was added to the graphene oxide (GO), chitosan composite (CS) materials, and hydroxyapatite (HAP). These structures promote the formation of osteoblast cells and impede the development of osteosarcoma cells.

The nanocomposite GO/HAP/CS-3/CDDP can be used to replace cancer-affected bones as it demonstrated increased cytotoxicity against cancer cells and great vitality in osteoblast-like cells [109].

9.11 GRAPHENE QDs (GQDs)

GQDs were created by loading sodium salicylate, a model medication, onto cross-linked chitosan and encasing it in carboxymethyl cellulose (CMC). The gastrointestinal tract's consistent and regulated medication release pattern is demonstrated by the nanocomposites. When added to the CMC hydrogel film, GQDs show a long release pattern and a high loading capacity. As demonstrated by DOX-loaded CMC/GQD nanocomposites, these films exhibit minimal toxicity against blood cancer cells (K562) and can be used as effective anticancer materials. GQD was used to treat wound inflammation and the rate of healing. The medication for healing wounds is a combination of sodium salicylate-coated chitosan aerogel and GQD with fluorescence emission capacity [110].

9.12 CONCLUSION

The integration of nanotechnology into mechatronics offers transformative potential that can significantly enhance the performance, efficiency, and capabilities of mechatronic systems. This multidisciplinary approach leverages the unique properties of nanomaterials and nanoscale phenomena to develop innovative solutions that are not feasible with traditional technologies. The convergence of nanotechnology and mechatronics represents a frontier of engineering innovation, with the potential to revolutionize various industries, from healthcare to manufacturing. By harnessing the unique properties of nanomaterials and nanoscale mechanisms, we can develop smarter, more efficient, and highly specialized mechatronic systems. Continued research and interdisciplinary collaboration will be key to overcoming current limitations and unlocking new possibilities for the future.

REFERENCES

1. F.C. Walsh, D.V. Bavykin, L. Torrente-Murciano, A.A. Lapkin, B.A. Cressey, Synthesis of novel composite materials via the deposition of precious metals onto protonated titanate (TiO_2) nanotubes, *Trans. Inst. Met. Finish*, 2006, 293–299. https://doi.org/10.1179/174591906X149077
2. T. Murcian, G.J. Hutchings, D. Kiely, Enhanced Au-Pd activity in the direct synthesis of hydrogen peroxide using nanostructured titanate nanotube supports, *Chem Cat Chem*, 2014, 2531–2534. https://doi.org/10.1002/cctc.201402361
3. T. Murcian, C. Villager, Selective oxidation of salicylic alcohol to aldehyde with O_2/H_2 using Au-Pd on titanate nanotubes catalysts, *Chem Cat Chem*, 2015, 925–927. https://doi.org/10.1002/cctc.201403040
4. C. Kelly, S. Zhao, The optical properties of metal nanoparticles: the influence of size, shape, and dielectric environment, *J Phys. Chem. B*, 107, 2003, 668–677 https://doi.org/10.1021/jp026731y
5. F. Kruis, Synthesis of nanoparticles in the gas phase for electronic, optical and magnetic applications – a review, *J. Aerosol Sci*, 1998, 511–535. https://doi.org/10.1016/S0021-8502(97)10032-5

6. S. Salabas, Magnetic nanoparticles: synthesis, protection, functionalization, and application, *Angew. Chemie Int. Ed.*, 2007, 1222–1244. doi:10.1002/anie.200602866.
7. T. Murciano, B. Lapkin, W. Walsh, Highly selective Pd/titanate nanotube catalysts for the double-bond migration reaction, *J. Catal.*, 2007, 272–278. https://doi.org/10.1016/j.jcat.2006.10.015
8. B. Pareek, J. Gupta, Panwar: synthesis and applications of noble metal nanoparticles: a review, *Adv. Sci. Eng. Med.*, 2017, 527–544. https://doi.org/10.1166/asem.2017.2027
9. S. Nishida, K. Yamamoto, M. Kobayashi, N. Kitagawa, Facile synthesis of size-controlled Rh nanoparticles via microwave-assisted alcohol reduction and their catalysis of CO oxidation, *Chem. Lett.*, 46, 2017, 1257. https://doi.org/10.1246/cl.170440
10. P. Campbell, The effect of size-dependent nanoparticle energetics on catalyst sintering, *Sci. Org.*, 2002, 811–814. https://doi.org/10.1126/science.1075094
11. H. Wang, M. Kafshgari, Optical properties and applications of plasmonic-metal nanoparticles, *Adv. Funct. Mater.*, 2020, 664. https://doi.org/10.1002/adfm.202005400
12. K. Wu, D. Su, J. Liu, R. Saha, J.P. Wang, Magnetic nanoparticles in nanomedicine: a review of recent advances, *Nanotechnology*, 2019, 502003. https://doi.org/10.1002/adfm.202005400
13. J. Kolhatkar, W. Litvinov, T. R. Lee, Tuning the magnetic properties of nanoparticles, *Int. J. Mol. Sci.*, 2013, 15977–16009. https://doi.org/10.3390/ijms140815977
14. M. Asano, S. Inoue, W. Furuta, K. Muranaka, K. Hayashi, K. Mae: Incorporative mixing in microreactors: Influence on reactions and importance of inlet designation, *Chem. Eng. J.*, 2023, 138942. https://doi.org/10.1016/j.cej.2022.138942
15. P. Gao, M. Torrente, Recent progress on the manufacturing of nanoparticles in multi-phase and single-phase flow reactors, *Curr. Opin. Chem. Eng.*, 2020, 26–33. https://doi.org/10.1016/j.coche.2020.03.008
16. P. Gao, M. Torrente, Tailoring the size of silver nanoparticles by controlling mixing in microreactors, *Chem. Eng. J.*, 2022, 134112. https://doi.org/10.1016/j.cej.2021.134112
17. W. Zhou, G. Xu, Impact of mixing for the production of CuO nanoparticles in supercritical hydrothermal synthesis, *Ind. Eng. Chem. Res.*, 2014, 481–493. https://doi.org/10.1021/ie4029413
18. O. Kaspar, H. Koyuncu, V. Hubatova, V. Tokarova, Influence of channel height on mixing efficiency and synthesis of iron oxide nanoparticles using droplet-based microfluidics, *RSC Adv.*, 10, 2020, 15179–15189. https://doi.org/10.1021/ie4029413
19. B. Harries, R. Barrow, A numerical model for segmented flow in a microreactor, *Int. J. Heat Mass Transf.*, 2003, 3313–3322. https://doi.org/10.1039/D0RA02470H
20. Z. Gao, M. Han, Rational design of the inlet configuration of flow systems for enhanced mixing, *J. Flow Chem.*, 2021, 589–598. https://doi.org/10.1007/s41981-021-00184-z
21. P. Dasgupta, S. Torelli, L. Paulson, Computational fluid dynamics modeling and analysis of silica nanoparticle synthesis in a flame spray pyrolysis reactor, *Combust. Flame*, 2022, 11178. https://doi.org/10.1016/j.combustflame.2021.111789
22. L. Liu, O. Sun, W. Odoom, Modeling of silver nanoparticle formation in a microreactor: reaction kinetics coupled with population balance model and fluid dynamics, *Ind. Eng. Chem. Res.*, 2014, 4263–4270. https://doi.org/10.1021/ie4031314
23. M.A. Watzky, R.G. Finke, Transition metal nanocluster formation kinetic and mechanistic studies. a new mechanism when hydrogen is the reductant: Slow, continuous nucleation and fast autocatalytic surface growth, *J. Am. Chem. Soc.*, 1997, 10382–10400. https://doi.org/10.1021/ja9705102
24. T. Manninen, K. Taivassalo, On the mixture model for multiphase flow, *VTT Publ.*, 1996, 3–67. https://doi.org/10.1016/j.combustflame.2021.111789
25. S. Tampa, M. Matei, C. Mitran, Photodynamic therapy: a hot topic in dermato-oncology. *Oncol Lett.*, 2019, 4085–4093. https://doi.org/10.3892/ol.2019.9939

26. M. Ochsner, Photophysical and photobiological processes in the photodynamic therapy of tumours, *J. Photochem. Photobiol. B Biol.*, 1997, 1–18 https://doi.org/10.1016/S1011-1344(96)07428-3
27. G. Niculescu, Photodynamic therapy, *An Up-to-Date Rev. Appl. Sci.*, 11, 3626. https://doi.org/10.3390/app11083626
28. S. Donohoe, G.D.S. Arnaut, Cell death in photodynamic therapy: from oxidative stress to anti-tumor immunity, *Biochim. Biophys. Acta Rev. Cancer*, 1872, 188308. https://doi.org/10.1016/j.bbcan.2019.07.003
29. A. Santos, W. Terra, A. Gomes, B. Ravagnani, Rate determines PDT efficiency in breast cancer cells displaying different GSH levels, *Photochem. Photobiol.*, 2020, 658–667. https://doi.org/10.1111/php.13182
30. C. Duan, C. Guo, W. Han, photodynamic therapy mediated by nontoxic core-shell nanoparticles synergizes with immune checkpoint blockade to elicit antitumor immunity and antimetastatic effect on breast cancer, *J. Am. Chem. Soc.*, 2016, 16686–16695. https://doi.org/10.1021/jacs.6b09538
31. C. Efimova, T. Meeren, H. Hammad, D, Vaccination with early ferroptotic cancer cells induces efficient antitumor immunity, *J. Immunotherap. Cancer*, 2020, e001369. https://doi.org/10.1136/jitc-2020-001369
32. L. Fabian, M. Hock, W. Wiench, Tissue distribution and toxicity of intravenously administered titanium dioxide nanoparticles in rats, *Arch. Toxicol.*, 2008, 151–157. https://doi.org/10.1007/s00204-007-0253-y
33. M.Fakhar-eAlam, S.M.U. Ali, Z.H. Ibupoto, M. Atif, M. Kashif, M. Willander. Sensitivity of A-549 human lung cancer cells to nanoporous zinc oxide conjugated with Photofrin, *Lasers Med. Sci.*, 2012, 607–614. https://doi.org/10.1007/s10103-011-0989-8
34. B. Imanparast, S. Bakhshizadeh, P. Sazgarnia, Pegylated hollow gold-mitoxantrone nanoparticles combining photodynamic therapy and chemotherapy of cancer cells, *Photodiagn. Photodyn. Ther.*, 2018, 295–300. https://doi.org/10.1016/j.pdpdt.2018.07.011
35. M. Jalili, T. Switaj, W. Nowis, C. Massalska, M. Radzikowska, J. Gołab, Effective photoimmunotherapy of murine colon carcinoma induced by the combination of photodynamic therapy and dendritic cells, *Clin. Cancer Res.*, 2004, 4498–4508. https://doi.org/10.1158/1078-0432.CCR-04-0367
36. P. Jang, K. Tung, Gold nano rod-photosensitizer complex for near-infrared fluorescence imaging and photodynamic/photothermal therapy in vivo, *ACS Nano*, 2011, 1086–1094. https://doi.org/10.1021/nn102722z
37. J. Jeevanandam, A. Barhoum, Y.S Chan, A. Dufresne, Review on nanoparticles and nanostructured materials: history, sources, toxicity and regulations, *Beilstein J. Nanotechnol.*, 2018, 1050–1074. https://doi.org/10.3762/bjnano.9.98
38. F. Jin, D. Liu, Nanomaterials-based photodynamic therapy with combined treatment improves antitumor efficacy through boosting immunogenic cell death, *Int. J. Nanomed.*, 2021, 4693–4712. https://doi.org/10.2147/IJN.S314506
39. E.A. Perigo, G. Hemery, O. Sandre, D. Ortega, E. Garaio, F. Plazaola, Fundamentals and advances in magnetic hyperthermia, *Appl. Phys. Rev.*, 2015, 41302. https://doi.org/10.1063/1.4935688
40. D. Chang, M. Lim, J.A.C.M. Goos, R. Qiao, Y.Y. Ng, F.M. Mansfeld, et al., Biologically targeted magnetic hyperthermia: potential and limitations, *Front. Pharmacol.*, 2018, 881–890. https://doi.org/10.3389/fphar.2018.00831
41. R.K. Ratnesh, M. Singh, S. Pathak and V. Dakulagi, Reactive magnetron sputtered-assisted deposition of nanocomposite thin films with tunable magnetic, electrical and interfacial properties, *J. Nanoparticle Res.*, 2020, 22, 290. https://doi.org/10.1007/s11051-020-05017-z

42. T. Neuberger, B. Schopf, H. Hofmann, M. Hofmann, B. von Rechenberg, Superparamagnetic nanoparticles for biomedical applications: possibilities and limitations of a new drug delivery system, *J. Magn. Magn Mater.*, 2005, 483–496. https://doi.org/10.1016/j.jmmm.2005.01.064
43. M.K. Yadav, R. Kumar, R.K. Ratnesh, J. Singh, R. Chandra, A. Kumar, V. Vishnoi, G. Singh, A.K. Singh, Revolutionizing technology with spintronics: devices and their transformative applications, *Mater. Sci. Eng. B*, 2024, 303, 117293. https://doi.org/10.1016/j.mseb.2024.117293A
44. A. Villanueva, M. Canete, A.G. Roca, M. Calero, S. Veintemillas, C.J. Serna, The influence of surface functionalization on the enhanced internalization of magnetic nanoparticles in cancer cells, *Nanotechnology*, 2009, 115103. https://doi.org/10.1088/0957-4484/20/11/115103
45. R. Hao, R. Xing, Z. Xu, Y. Hou, S. Gao, S. Sun, Synthesis, functionalization, and biomedical applications of multifunctional magnetic nanoparticles, *Adv. Mater.*, 2010, 2729–2742. https://doi.org/10.1002/adma.201000260
46. R.K. Ratnesh, M.K. Singh, J. Singh, Enhancing ZnO/Si heterojunction photodetector performance for ultra high responsivity across wide spectral range, *J. Mater. Sci.: Mater. Electron.*, 2024, 35, 756. https://doi.org/10.1007/s10854-024-12516-5
47. I. Obaidat, Magnetic properties of magnetic nanoparticles for efficient hyperthermia, *Nanomaterials*, 2015, 63–89. https://doi.org/10.3390/nano5010063
48. T. Ahmed, T. Kondo, Hyperthermia: an effective strategy to induce apoptosis in cancer cells, *Apoptosis*, 20(11), 2015, 1411–1419. https://doi.org/10.1007/s10495-015-1168-
49. C. Maurici, B. Wylleman, Hyperthermia enhances efficacy of chemotherapeutic agents in pancreatic cancer cell lines, *Biomolecules*, 2022, 3641. https://doi.org/10.1007/s10495-015-1168-3
50. X. Mei, R. ten Cate, C.M. van Leeuwen, H.M. Rodermond, L. de Leeuw, D. Dimitrakopoulou, et al. Radiosensitization by hyperthermia: the effects of temperature, sequence, and time interval in cervical cell lines, *Cancers*, 12(3), 2020, 381–382. https://doi.org/10.3390/cancers12030582
51. S.R. Kothiyal, R.K. Ratnesh, A.K. Shukla, Enhancing Sensing with Surface Plasmons Resonance: Theoretical Insights and Design Strategies in Photonic Crystal Fiber, *International Conference Artificial Intelligence and Information Technologies, (ICAIIT)-2024*, Taylor & Francis, 2023, 279–285. https://doi.org/10.1201/9781003510833-46
52. P. Moroz, S.K. Jones, J. Winter, B.N. Gray, Targeting liver tumors with hyperthermia: ferromagnetic embolization in a rabbit liver tumor model, *J. Surg. Oncol.*, 2001, 22–29. https://doi.org/10.1002/jso.1118
53. W. Jones, W. Gray, Treatment of experimental rabbit liver tumours by selectively targeted hyperthermia, *Int. J. Hyperther.*, 2002, 117–128. https://doi.org/10.1080/02656730110103519
54. G. Veiseh, G. Zhang, Design and fabrication of magnetic nanoparticles for targeted drug delivery and imaging, *Adv. Drug Deliv. Rev.*, 2010, 284–304. https://doi.org/10.1016/j.addr.2009.11.002
55. A. Becue, Emerging fields in fingermark (Meta) detection – a critical review, *Anal. Methods*, 2016, 7983–8003. https://doi.org/10.1039/C6AY02496C
56. M. Wang, A. Yu, Y. Zhu, M. Yang, C. Mao, Fluorescent nanomaterials for the development of latent fingerprints in forensic sciences, *Adv. Funct. Mater.*, 2017,1606243–1606258. https://doi.org/10.1002/adfm.201606243
57. G. Ramotowski, *Advances in Fingerprint Technology*, 3rd ed. CRC Press, Boca Raton, FL, 2013, 28–36. https://doi.org/10.1201/b12882
58. L. Champod, P. Margot, M. Stoilovic, *Fingerprints and Other Ridge Skin Impressions*. CRC Press, Boca Raton, FL, 2016, 22–26. https://doi.org/10.1002/adfm.201606243

59. M. Kanodarwala, L. Spindler, Nanoparticles used for fingermark detection – a comprehensive review, *WIREs Forensic Sci.*, 2019, e1341. https://doi.org/10.1002/wfs2.1341
60. G.S. Sodhi, J. Kaur, Physical developer method for detection of latent fingerprints: a review, *Egypt. J. Forensic Sci.*, 6 2016, 44–47. https://doi.org/10.1016/j.ejfs.2015.05.001
61. E. Prabakaran, P. Kriveshini, Nanomaterials for latent fingerprint detection: a review, *J. Mater. Res. Technol.*, 122, 2021, 1856–1885. https://doi.org/10.1016/j.jmrt.2021.03.110
62. R.K. Ratnesh, Hot injection blended tunable CdS quantum dots for production of blue LED and a selective detection of Cu^{2+} ions in aqueous medium, *J. Opt. Laser Technol.*, 2019, 116, 103–111. https://doi.org/10.1016/j.optlastec.2019.03.001H.S
63. A. Garg, R.K. Ratnesh, Solar cell trends, and the future: a review, *J. Pharm. Negat. Results*, 2022, 13, 2051–2060. https://doi.org/10.47750/pnr.2022.13.S06.268
64. F. Costa, S. Sousa, Dias, A.M.L. Assis, A.C. Jesus, A.S. Ribeiro, M. Navarro, SATS@CdTe hierarchical structures emitting green to red colors developed for latent fingerprint applications, *Dyes Pigments*, 180, 2020, 108483. https://doi.org/10.1016/j.dyepig.2020.108483
65. P. Garg, T. Chandra, R. Ahlawat, N. Mittal, R.K. Ratnesh, S.K. Tripathi, Star Galaxy Image Classification Via Convolutional Neural Network, *ICOSEC-2022*, IEEE, 2022, 1156–1168. https://doi.org/10.1109/ICOSEC54921.2022.9952065
66. C. Assis, M. Alves, V. Oliveira, A. Hillman, A.S. Ribeiro, From nanomaterials to macromolecules: innovative technologies for latent fingerprint development, *WIREs Forensic Sci.*, 2023, e1475. https://doi.org/10.1002/wfs2.1475
67. M. Zhao, D. Song, Y. Liu, Y. Luo, S. Lin, C.-Y. Liu, New luminescent nanoparticles based on carbon dots/SiO2 for the detection of latent fingermarks, *Anal. Methods*, 2017, 4770–4775. https://doi.org/10.1039/C7AY01316G
68. C. Wan, W. Chen, Li, Cui, B. Yuan, Preparation of novel magnetic nanomaterials based on "facile coprecipitation" for developing latent fingerprints in crime scenes, *ACS Omega*, 2022, 1712–1721. https://doi.org/10.1021/acsomega.1c04208
69. C. Zhang, Synthesis of composite particles with Fe3O4 core and Ag shell for the development of fingerprints, *Bull. Korean Chem. Soc.*, 2013, 2378. https://doi.org/10.5012/bkcs.2013.34.5.1457
70. Z. Hu, H. Dai, W. Zhou, J. Wei, H. Zhang, Z. Ye, Y. Qiu, Y. Chen, Z. Duan, J. Wang, W. Zhang, F. Xie, R. Guo, Corrosion resistant solid-state carbon dots@silicalite-1 composite for latent fingerprints detection, *J. Alloys. Compd.*, 2021, 161660. https://doi.org/10.1016/j.ccr.2024.215976
71. L. Yang, Y. Lu, L. Tang, J. Zhang, L. Ge, F. Li, Visualization of latent fingerprints using a simple "silver imaging ink, *Anal. Methods*, 2016, 6293. ,https://doi.org/10.1039/C6AY01811D
72. V. Prasad, L. Prasad, S. Lukose, P. Agarwal, Latent fingerprint development by using silver nanoparticles and silver nitrate – a comparative study, *J. Forensic Sci.*, 66(3), 2021, 1065–1074. https://doi.org/10.1111/1556-4029.14664
73. B. Balamurugan, T. Maruyama, Evidence of an enhanced inter band absorption in Au nanoparticles: size-dependent electronic structure and optical properties, *Appl. Phys. Lett.*, 2005, 143105 https://doi.org/10.1063/1.2077834
74. R.K. Ratnesh, M.S. Mehata, Investigation of biocompatible and protein sensitive highly luminescent quantum dots/nanocrystals of CdSe, CdSe/ZnS and CdSe/CdS, *Spectrochim. Acta A Mol. Biomol. Spectrosc.*, 179, 2017, 201–210. https://dx. doi.org/10.1016/j.saa.2017.02.028
75. S. Kitagawa, H. Kojima, N. Nakajiama, Studies of mixed-valence states in three-dimensional halogen-bridged gold compounds, Cs2AuIAuIIIX6, (X = Cl, Br or I). Part 2. X-Ray photoelectron spectroscopic study, *Chem. Soc. Dalton Trans.*, 1991, 3121–3125. https://doi.org/10.1039/DT9910003121

76. B. McNeillie, G. Smith, L. Watson, Studies of mixed-valence states in three-dimensional halogen-bridged gold compounds, Cs2AuIAuIIIX6, (X = Cl, Br or I).Part 2. X-Ray photoelectron spectroscopic study, *J. Chem. Soc. Dalton Trans.*, 1980, 767–770. https://doi.org/10.3390/s90503337
77. B. Bona, E. Brusa, S. Carabelli, M. Chiaberge, C. Delprete, G. Genta, The mechatronics laboratory at Politecnico di Torino, *Mechatronics*, 1997,413–427. https://doi.org/10.1016/S0957-4158(97)00007-X
78. J. Wang, B. Koel, RAS studies of NO_2, N_2O_3, and N_2O_4 adsorbed on Au(111) surfaces and reactions with co adsorbed H2O, *J. Phys. Chem A*, 1999, 8573–8579. https://doi.org/10.1016/S0957-4158(97)00007-X
79. D.G. Wu, D. Cahen, Direct detection of low concentration no in physiological solutions by a new gaas-based sensor, *Chem. Eur. J.*, 2001, 1743–1749. https://doi.org/10.1002/1521-3765(20010417)7:8<1743::AID-CHEM17430>3.0.CO;2
80. M. Feelisch, J.S. Stamler, *Methods in Nitric Oxide Research*; Wiley: New York, 1996, 5–115. https://eprints.soton.ac.uk/340425/
81. K. Rovira, B. Hutter, M. Parrinello, Bond characterization of chromium fischer carbene complexes: a combined study of experiment and theory, *J. Phys. Chem. A*, 1997, 8914–8925. https://doi.org/10.1021/jp971848l
82. B. Hamilton, *Sulphate-Reducing Bacteria: Environmental and Engineered Systems*, Cambridge University Press, 2007. https://doi.org/10.1017/CBO9780511541490
83. H. Abdal, Biocorrosion behavior and cell viability of adhesive polymer coated magnesium-based alloys for medical implants, *Appl. Surf. Sci.*, 2012, 536–546. 10.1016/j.apsusc.2012.08.051
84. B. Galicia, Can acid producing bacteria be responsible for very fast MIC pitting, *Corrosion*, 2012. https://doi.org/10.1016/j.crbiot.2022.02.003
85. C. Hegde, J. Su, J. M. Rui Tan, K. He, X. Chen, and S. Magdassi, Sensing in soft robotics, *ACS Nano.*, 2023 17(16), 2023, 15277–15307. https://doi.org/10.1021/acsnano.3c04089
86. D. Festy, Detection and mapping of SRB influenced corrosion of C-steel, In: *Corrosion*, 2002, 200–202. https://doi.org/10.1017/CBO9780511541490
87. R. Kumar, R.K. Ratnesh, J. Singh, R. Chandra, G. Singh, V. Vishnoi, Recent prospects of medical imaging and sensing technologies based on electrical impedance data acquisition system, *J. Electrochem. Soc.*, 170(11) , 2023, 117507. https://doi.org/10.1149/1945-7111/ad050f
88. S. Pathak, N. Kishore, G. Upadhyay, R.K. Ratnesh, R. Mishra, *A Compact Size Planar Microstrip-Fed Patch Antenna with Hexagonal DGS Slot for WLAN Application, Recent Trends in Electronics and Communication*, LNEE, Springer, 2021, 777, 263–271. https://doi.org/10.1007/978-981-16-2761-3_25
89. D. Sepulveda, Bioelectrochemical detection of mycobacterium tuberculosis ESAT-6 in an antibody-based biomicrosystem, *Sensors*, 2017, 2178. https://doi.org/10.3390/s17102178
90. K. Salimiyan rizi, The smartphone biosensors for point-of-care detection of human infectious diseases: Overview and perspectives—A systematic review, *Current Opinion in Electrochemistry*, 32, 2022, 100925. https://doi.org/10.1016/j.coelec.2021.100925
91. S. Chen, Three-dimensional graphene nano sheet doped with gold nanoparticles as electrochemical DNA biosensor for bacterial detection, *Sens. Actuator B: Chem.*, 2018, 860. https://doi.org/10.1016/j.snb.2018.02.093
92. B. Yılmaz, Lab-on-a-Chip technology and its applications, In: *Omics Technologies and Bio-Engineering, Elsevier*, 2018, 145. https://doi.org/10.1016/B978-0-12-804659-3.00008-7
93. G. Zhuang, Environmental application and risks of nanotechnology: a balanced view. In: S. Ripp and T. Henry (eds), *Biotechnology and Nanotechnology Risk Assessment: Minding and Managing the Potential Threats Around Us*, Washington, DC: ACS Publications, 2011, 41–67. https://doi.org/10.1021/bk-2011-1079.ch003

94. S. Elliott, B. Amara, Atomistic modelling of CVD synthesis of carbon nanotubes and graphene, *Nanoscale*, 2013, 6662–6676. https://doi.org/10.1039/C3NR01925J
95. H. Hanwell, C. Richardson, I.M. Ross, Gas and vapour sensing characteristics of Langmuir-Schaeffer thiol encapsulated gold nanoparticle thin films, *Coll. Surf. A*, 2006, 379–383. https://doi.org/10.1016/j.colsurfa.2006.03.019
96. X. Lu, X. Xu, X. Wang, Q. Zhang, Bonding of NO2 to the Au Atom and Au (111) surface: a quantum chemical study, *J. Phys. Chem A*, 1999, 10969. https://doi.org/10.1021/acs.jpcc.2c04775
97. L. Han, David R. Daniel, Mathew M. Maye, and Chuan-Jian Zhong, Core–shell nanostructured nanoparticle films as chemically sensitive interfaces, *Analytical Chemistry*, 73(18), 2001, 4441–4449. https://doi.org/10.1021/ac0104025)
98. S. Baratto, G. Comini, E. Faglia, G. Benussi, V. La Ferrara, L. Quercia, G. Di Francia, V. Guidi, D. Vincenti, D. Boscarino, V. Rigato, Gold-catalysed porous silicon for NOx sensing, *Sens. Actuat.*, 2000, 74–80. https://doi.org/10.1021/acs.jpcc.2c04775
99. H. Steffes, C. Imawan, O. Solzbacher, Enhancement of NO_2 sensing properties of In2O3-based thin films using an Au or Ti surface modification, *Sens. Actuat., B* 2001, 78, 106–112. https://doi.org/10.1016/S0925-4005(01)00799-7
100. G. Cahen, N. Naaman, D. Shvarts, Direct detection of low concentration no in physiological solutions by a new gaas-based sensor, *Chem. Eur. J.*, 2001, 1743–1749. https://doi.org/10.1002/1521-3765(20010417)7:8<1743::aid-chem17430>3.0.co;2–x
101. K. Rovira, B. Hutter, P. Ballone, Bond characterization of chromiumfischer carbene complexes: a combined study of experiment and theory, *J. Phys. Chem. A*, 101, 1997, 8914–8925. https://doi.org/10.1021/jp971848l
102. J. Bell, C.F. McConville, Accumulation layer profiles at InAs polar surfaces, *Appl. Phys. Lett.*, 71, 1997, 3688–3690.
103. X. Qiu, N. Hildebrandt, Rapid and multiplexed microRNA diagnostic assay using quantum dot-based forster resonance energy transfer, *ACS Nano*, 2015, 8449–8457. https://doi.org/10.1021/acsnano.5b03364
104. R. Bilan, I. Nabiev, A. Sukhanova, Quantum dot-based nanotools for bioimaging, diagnostics and drug delivery. *Chem. Biochem.*, 2016, 2103–2114. https://doi.org/10.1002/cbic.201600357
105. M. Ruzycka Ayoush, P. Kowalik, A. Kowalczyk, P. Bujak, A.M. Nowicka K. Wojewodzka, I.P. Grudzinski, Quantum dots as targeted doxorubicin drug delivery nanosystems in human lung cancer cells, *Cancer Nano*, 2021. https://doi.org/10.1186/s12645-021-00077-9
106. W. Li., H. Zhang, Y.G. Assaraf, K. Zhao, Overcoming ABC transporter-mediated multidrug resistance: molecular mechanisms and novel therapeutic drug strategies, *Drug Resist.*, 27, 14–29. https://doi.org/10.1016/j.drup.2016.05.001
107. W. Cheng, Z. Wu, C. Wu, X. Wang, S.S Liow, Overcoming STC2 mediated drug resistance through drug and gene co-delivery by Phb-Pdmaema cationic polyester in liver cancer cells, *Mater. Sci. Eng. C Mater. Biol. Appl.*, 2018, 210–217. https://doi.org/10.1016/j.msec.2017.08.075
108. S. Javanbakht, A. Shaabani, Encapsulation of graphene quantum dot-cross linked Chitosan by carboxymethylcellulose hydrogel beads as a pH-responsive binanocomposite for the oral delivery agent, *Int. J. Biol. Macromol.*, 2019, 389–397. https://doi.org/10.1016/j.ijbiomac.2018.11.118
109. N. Javanbakht, Doxorubicin loaded carboxymethyl cellulose/grapheme quantum dot nanocomposite hydrogel films as a potential anticancer drug delivery system, *Mater. Sci. Eng. C*, 2018, 50–59. https://doi.org/10.1016/j.msec.2018.02.010
110. R.K. Ratnesh, M.K. Singh, V. Kumar, S. Singh, R. Chandra, M. Singh, J. Singh, Mango leaves (mangifera indica) derived highly florescent green graphene quantum dot nanoprobes for enhanced on-off dual detection of cholesterol and Fe^{2+} ions based on molecular logic operation, *ACS Biomater. Sci. Eng.*, 7, 2024, 4417–4426. https://doi.org/10.1021/acsabm.4c00292

10 Investigate Advanced Imaging Algorithms for Nanotechnology Applications

Rohit Kumar Singh, Jagbeer Singh, Vibhor Kumar, and Ratneshwar Kumar Ratnesh

10.1 INTRODUCTION

In this section, we first provide some basic facts on nanoparticles (NPs). Firstly, what is a NP? NPs are materials that exhibit unique behavior not commonly observed in bulk materials. Such behavior is directly correlated to their physical confinement in the range of 1–100 nm, which imparts well-defined characteristics. These tiny particles become exceedingly valuable in addressing challenges related to the fields of electronics, energy, and biomedicine because of their distinct electrical, optical, and magnetic properties. By altering atoms at the smallest scale, scientists can create nanomaterials (NMs), in contrast to classical physics and chemistry, mainly because of their large surface area-to-volume ratio. NMs follow the principles of quantum mechanics. The design and construction of operational systems occur at the molecular or atomic level. In the field of biomedicine, NPs are highly versatile and play various roles that are essential for modern medical diagnostics and treatments. They are used in thermal ablation, which involves the destruction of cancer cells through heat. NPs also enhance the contrast in MRI scans, making it easier to visualize details in the images. They are useful in delivering drugs and genes directly to target cells or tissues, improving the effectiveness of treatments.

Understanding the properties of nanoparticles and how they can be used effectively involves incorporating several important factors. Some of these factors are their average size, which is important for studying their interaction with the body's cells and tissues; their distribution type, for identifying how they are spread throughout the body for consistent and satisfactory results; their shape, which can affect the outcome of the experiment for certain applications; and their chemical composition, which is essential for determining their reactivity with certain substances to avoid medical casualties and ensure safety. These methods provide essential information about the physical and chemical properties of NPs. By measuring these attributes, researchers and scientists can optimize the NPs for their intended use.

DOI: 10.1201/9781003634553-10

In the context of nanotechnology, the author Nick Bostrom [1] quotes, "Nanotechnology has been moving a little faster than I expected, virtual reality a little slower." This technology influences many industries and aspects of society, revolutionizing products, ensuring their safety and durability, and adding additional features. NMs, tiny substances at the nanoscale, are used in many everyday items like sunscreens, cosmetics, sports equipment, tires, and electronics.

When NMs are intermixed with existing products or used to fabricate new advanced devices, they significantly enhance these products. They reduce the manufacturing costs of these products and support eco-friendly manufacturing processes. Nanotechnology has applications in many fields, including medicine, communications, daily life, and agriculture. They help make products better, cleaner, and longer-lasting. This means they provide us with higher-quality end products that are safer for the environment, sustainable, and stable.

Nanotechnologies have particularly revolutionized the medical field, impacting diagnostic methods, imaging, tissue engineering, implants, pharmaceutical therapeutics, and drug delivery. Nanotechnology has made remarkable strides in advancing therapies for a range of ailments, including cardiovascular diseases, cancer, neurological disorders, and bacterial infections. The broad spectrum of benefits provided by NMs is poised to reshape the landscape of numerous industrial sectors, promising a more sustainable and efficient future [2,3].

Cancer is a complex group of diseases characterized by abnormal cell growth. It develops in stages, starting with benign lesions and potentially progressing to malignant tumors. Understanding its genesis and creating personalized diagnoses and treatment approaches is crucial because each type of cancer can be different, and some can be fatal. Every year, over a million people are diagnosed with cancer, making it a major health problem with high death rates. The cancer treatment process becomes difficult since each patient has unique biomarkers and molecular profiles. This means that the genetic makeup and mutations of tumors can vary significantly, making it hard to find treatments that accommodate everyone. One of the significant challenges for effective cancer treatment arises from tumor heterogeneity, or the differences among tumors. Detecting cancer early and diagnosing it accurately are crucial for better clinical outcomes. Early detection often leads to higher survival rates for certain types of cancer because treatments can be more effective when identified in the early stages. Therefore, improving methods for early detection and accurate diagnosis is essential in the fight against cancer.

Medical imaging (MI) plays a key role in early detection and therapeutic assessment, using techniques like MRI, ultrasound, PET, and fluorescence imaging. Often, multiple imaging methods are combined to enhance lesion detection and provide detailed anatomical and functional information. Contrast agents are essential for distinguishing healthy tissues from abnormal lesions, improving image contrast, and aiding in accurate diagnosis and treatment planning.

Current clinical imaging employs small-molecule contrast agents, but these have limited resolution in the spatial domain. This can severely affect the imaging results for analysis and even pose toxicity risks. Thus, it becomes a central task to address these aspects, which hold future prospects for advancing MI and increasing diagnostic accuracy. Researchers are working on new contrast agents and imaging techniques to address these challenges, aiming to revamp cancer diagnosis and treatment.

By overcoming these issues, patient outcomes are likely to improve, ultimately reducing the impact of cancer on individuals and healthcare systems.

Traditional cancer treatment often begins with surgery to remove localized tumors. For more advanced stages, doctors typically use radiation and chemotherapy. However, chemotherapy has a major drawback: it targets not only cancer cells but also affects healthy cells. This can cause significant side effects and harm to the patient. This issue necessitates the use of targeted therapies. The aim behind targeted therapies is to reduce the side effects that patients experience during cancer treatment. These treatments directly target cancer cells without affecting healthy ones, thereby minimizing the need for complicated strategies in cancer treatment.

Nanotechnology has also made strides in developing precision medicine. Nanomedicine, complemented by the expansion of nanodelivery systems, can provide timely treatments at the target tumor site. This technological innovation can now successfully deliver multiple drugs to target sites. This will aid in preparing the patient's specific disease report card, streamlining the assessment process. Researchers are also exploring the use of natural compounds, such as curcumin from turmeric, in cancer therapy. These natural compounds, when combined with NPs like quantum dots—which can also serve as ideal materials for designing photodetectors, as highlighted in the works of [4] —can provide treatments tailored to the needs of the patient. This one-on-one approach aims to improve treatment outcomes.

Doctors can customize therapies suitable for each individual by referring to the patient's profile, which contains necessary information about the suggested treatment approaches. This allows for more accurate and effective treatments. Appropriate therapy with specific attributes for each patient drives intended outcomes. At the same time, it also eliminates any chances of impacting healthy cells, marking a significant step forward in the battle against cancer.

NP applications extend to the domain of tumor imaging for information retrieval to design treatments and prescriptions for a diseased patient. They accumulate and remain in tumors and are more effective than larger particles. Their longer stay ensures easy penetration of tumor tissues. This process ensures clearer visualization and suggests an appropriate diagnosis. NPs with dimensions between 10 and 60 nm allow better absorption by cells, making imaging techniques more precise.

Ligands have specific applications in the study of NPs. Ligands are found in peptides, proteins, etc., and the study of the interaction between proteins and ligands is made possible by surface plasmon resonance [5]. Ligands attach to target protein molecules in tumor cells and have the ability to activate new cellular responses. This successfully creates a localization effect, and a contrasting effect is observed between diseased and healthy cells, which is the core of cellular imaging. Specific areas of the body where there are high chances of tumor cell development can be targeted with physical methods like magnetic, acoustic, and photo-thermal targeting, as well as surface labeling methods like peptide tagging. In the physical targeting method, NPs flow inside the tissue through physical forces that direct NPs to specific areas of the body. This external push aims for the optimum flow of drugs in the body. Stimuli-driven nanoprobe systems, which activate whenever there is a change in pH or temperature, help control the release of drugs to target sites. Tumors can be treated more effectively when both extrinsic and intrinsic factors are considered.

NPs, compared to their traditional counterparts, remain in the bloodstream for a longer duration. This extended stay of NPs improves imaging procedures and provides the best diagnostic test results. Scientists have achieved this by altering some characteristics of NPs. One such application is tumor detection in the early stages. Appropriately sized NPs, under the influence of imaging compounds, can meet clinical requirements and provide new insights in biomedical engineering. Biomedical NMs, when operating at a minuscule level, significantly improve the detection and treatment of diseased cells. Thus, their size and type are crucial in biological systems. Researchers have made this possible by collecting data about the properties of NMs based on experimental conduct. Knowledge about these properties helps them build the requisite materials. This thoughtful approach paves the way for novel medical applications.

Artificial intelligence (AI) has rapidly become a formidable tool in materials science. AI algorithms accelerate data processing, recognize patterns, and are powered by informed decision-making. AI has shown promise in neutralizing the trial-and-error types of errors generated from experimental tests. AI-based simulations can suggest optimal synthesis routes and extract data from material characterization. The optimal routes depend on the design technique, synthesis, and characterization of the material.

Current upgrades in the software and hardware industries are a result of advancements in machine learning (ML), which have sparked significant transformations in object detection, speech recognition, self-driving vehicles, robotics, and healthcare systems. ML techniques have reshaped the way we address complex problems. This influence also extends to materials science, where ML has shown promising results in the study of chiral NMs. Accurate property analysis of chiral NMs has been made possible largely due to ML algorithms. They examine finer interconnections among different parameters. This study aims to enhance research competence and drive innovation in this field.

The use of ML has grown rapidly of late, and scientists have successfully exploited its advantages, especially in nanotechnologies and NMs. Although ML in interdisciplinary fields has been a revelation, everything has a cost. The challenge of handling large datasets is one such drawback. Despite these adversities, ML can boost manpower and create jobs in various materials science companies. It has bright prospects for increasing turnover and profits for those companies and ultimately contributing to the industrial revolution.

The amalgamation of interdisciplinary domains like nanoscience, bioinformatics, and ML opens up new possibilities for automation features in NMs. One of the recent advancements made in this direction is the proposal for the development of self-contained experimental devices. ML can also ensure the maintenance of ecological balance and that no harm is done to society with the design of sustainable NMs.

10.2 LITERATURE REVIEW AND METHODS

The study of NPs has been initiated with the help of electron microscopy. It examines the overall structure of these tiny particles. It interacts with the NPs using a beam of electrons, which is used to produce detailed images more effectively from a region

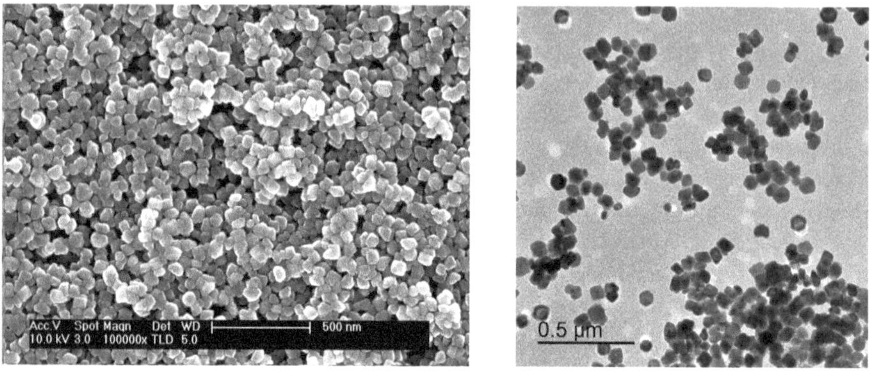

(a) (b)

Fig 1 (a) SEM Image

Fig 1 (b) TEM Image

FIGURE 10.1 (a) SEM image (b) TEM image [6].

under observation. The two known techniques for observing finer details about the region with NP concentrations are scanning electron microscopy (SEM) and transmission electron microscopy (TEM). Each technique has its own advantages and disadvantages. SEM can prepare 3D representations of NP surfaces. It examines the surface features of NPs with precision, easing the analysis of the region and revealing tiny bulges on the surface of the particles. TEM manages the internal structure of NPs exhaustively, disclosing what lies inside these particles. It focuses on the internal structure, particle size, and distribution statistics. These techniques help scientists understand and explore the scope of new modifications made in the particle's properties [6,7]. Figure 10.1 shows a pictorial representation of the observed images through SEM and TEM.

Researchers examine NP behavior by collecting samples of a thin film onto which NPs are deposited. The wet sample is dried to remove the liquid suspension [8]. Image acquisition and measurement are done manually with the help of programs like ImageJ [9].

While automated or semi-automated detection methods exist, their adaptation to images makes them less commonly employed. The progressive evolution of techniques for automating the detection and segmentation of particles in microscope images, evolving from classical methodologies to advanced approaches employing convolutional neural networks (CNNs), is also explored[6]. Figure 10.2 represents the methodology adopted by the authors of [6] by partitioning their work into dataset generation by artificial means, which is eventually used for training two different models involving segmentation, leading to image reconstruction of the input data.

Initially, classical methods relied on image pre-processing, feature extraction, and classification algorithms, including template matching, edge detection, and classification models like support vector machines [10–12] and CNN [13]. However, these methods faced challenges in generalization, computational costs, and expert parameter tuning. Although the Hough transform addressed some limitations, particularly for round-shaped particles [14–16], its adaptability to diverse particle

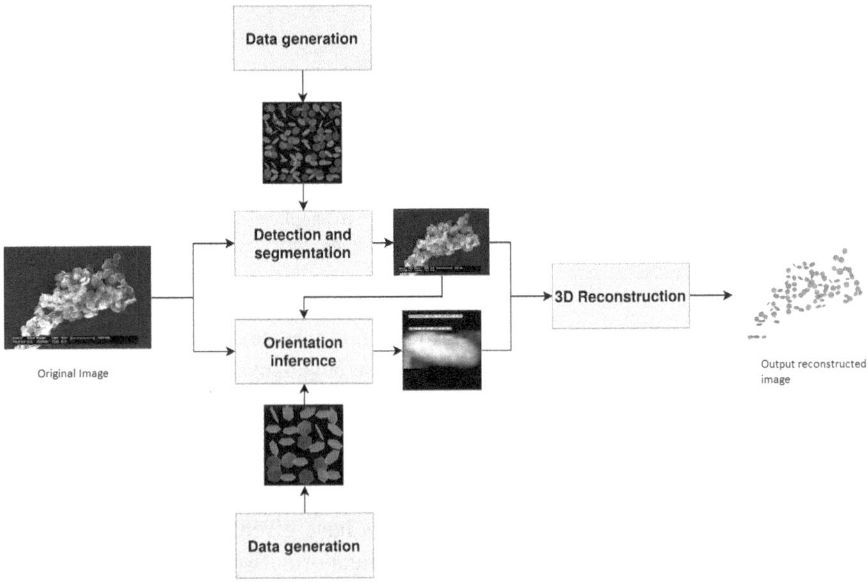

FIGURE 10.2 Nanoparticle image reconstruction using artificial dataset [6].

shapes remained a concern. The rise of deep learning algorithms like CNN has revolutionized the field by learning hierarchical representations and overcoming manual tuning issues.

CNNs excel in tasks like detection, segmentation, and classification in microscopy imaging, with applications ranging from nucleus/cell detection in tissues [16] to image enhancement [17,18]. While SEM and TEM provide exceptional accuracy and speed in analyzing NPs, they often depend on annotated datasets, which can be challenging to obtain for certain types of particles. In the field of bioinformatics, deep learning has produced groundbreaking results, particularly in biomedical imaging [19]. CNNs are especially effective for tasks such as segmenting MRI and microscope images. An exemplary case is Shahin et al.'s work [20], which utilized CNNs to identify white blood cells, demonstrating the potential of deep learning to transform such analyses. To address the difficulty of acquiring annotated datasets, some researchers have developed automated methods for training deep CNNs in cell segmentation [21]. This innovation simplifies the process by reducing the need for datasets curated by experts, making it more efficient and accessible for a wider range of applications. Importantly, the method described in a previous study [22] offers a way to evaluate and compare the accuracy of tumor reconstruction from individual images. This approach involves assessing how well the images represent the actual tumor. By comparing different images, scientists can determine which methods provide the most accurate representations. This is crucial for improving diagnosis and treatment planning. The more straightforward the approach, the more likely it is for researchers and medical professionals to use advanced imaging techniques, leading to better patient outcomes. In a nutshell, the pivotal role of deep learning is underscored in advancing biomedical imaging techniques. Data collection hurdles are also addressed, and the precision of crucial medical analyses is highlighted.

TABLE 10.1
Few Nanoparticles Used as a Contrast Agent [23]

Nanoparticle Name	Area of Specialization
Perfluorocarbon	Angiogenesis imaging
Quantum dots	Cellular imaging
Silicon particles	MRI applications
Gold particles	X-Ray/CT

The process of disease diagnosis in healthcare requires three qualities: speed, precision, and specificity to avoid the pitfalls of "false negatives," as discussed in a previous study [23]. In this pursuit, in-vivo diagnostic imaging serves as a cornerstone offering non-invasive insights into live tissues, thereby revolutionizing disease detection [24]. A pivotal stride in diagnostic imaging involves the incorporation of nanotechnologies, ushering in the era of biological markers for early disease detection at the cellular level [25]. Thus, nanotechnologies have given rise to high-precision molecular imaging agents, a testament to their transformative impact on diagnostic capabilities, as shown in Table 10.1.

Moving beyond diagnosis, this work highlights different MI techniques such as X-ray, ultrasound, computed tomography (CT), nuclear medicine, and magnetic resonance imaging, which have long been stalwarts in research but are limited by their effectiveness in detecting changes at later stages of disease progression

In a previous study [26], MRI, a widely accepted clinical imaging technique, is examined; it operates by maneuvering the spin of the protons within the human body under the influence of an external magnetic field. This technique offers high spatial and temporal resolution, along with exceptional intrinsic soft-tissue contrast, presenting anatomical tomographic information in three dimensions. Despite its strengths, MRI comes with limitations such as cost, extended imaging times, susceptibility to motion artifacts, and potential artifacts from foreign bodies or implants. To tackle these obstacles, the authors assert that MRI contrast agents are pivotal in improving lesion identification and differentiating between abnormalities and normal tissues. Active targeting involves surface labeling to influence the biodistribution of NPs, while passive targeting relies on the size of the NPs. The enhanced permeability and retention (EPR) effect is a key outcome of this strategy. In the current study, gold nanoshells, an excellent example, encapsulate a dielectric core in gold shells, exhibiting low toxicity, cost-effectiveness, and exceptional resolution in optical coherence tomography, particularly in near-infrared imaging of cancer cells. Future developments anticipate the integration of nanosized sensors for detecting chemicals, viruses, bacteria, and pH, thereby expanding the range of applications.

In the study [26], the authors also explore another MI method known as CT, which utilizes variations in X-ray absorption in tissue to produce 3D images. This technique offers several benefits over traditional radiography, including quicker examination times, lower expenses, and enhanced efficacy. Despite the increased ionizing radiation exposure, CT has become the preferred choice for clinical imaging. Contrast agents in CT are essential for distinguishing tissues with similar attenuation coefficients. Current CT agents for providing contrast are mostly derived from iodine;

however, iodine-based agents suffer from drawbacks such as rapid release, which can harm the kidneys in some cases. To overcome these limitations and expand CT imaging applications, nanoscale contrast agents have been developed. These agents generate robust CT attenuation, highlighting the versatility of nanosized CT contrast agents in various biomedical contexts. Figure 10.3 shows NPs of different structures for MRI and CT applications [26].

Other MI techniques covered in [26] include positron emission tomography (PET), a powerful nuclear medicine technology widely utilized for its tissue penetration capabilities, and real-time imaging. Single photon emission CT (SPECT), similar to PET, detects abnormal biochemical functions before anatomical changes manifest. However, the high cost and radioactive exposure pose limitations to PET/SPECT. In this context, NPs play a significant role in PET/SPECT tumor detection. Tumor imaging encompasses targeting specific receptors or utilizing the EPR effect. This underscores the dynamic use of PET/SPECT-based imaging, with nanotechnology at the forefront for various biomedical purposes [27].

To develop a thorough understanding among the intended audience, the authors of the study [26] have drawn a comparison between conventional and fluorescence-based imaging. While conventional imaging methods provide effective lesion detection with macroscopic resolution, fluorescence imaging, particularly near-infrared fluorescence imaging, offers superior microscopic spatial resolution. Near-infrared fluorescence addresses challenges such as limited penetration depth, tissue auto-fluorescence, and scattering properties. To overcome these limitations, NPs play a crucial role in fluorescence imaging. They enable a higher load of fluorescent dye molecules, preventing potential quenching of NIR fluorescence through structural modifications. Active and passive strategies increase NP concentrations in lesions, enhancing local concentrations of fluorescent dye. Prolonged circulation times contribute to increased uptake in target lesions. The development of fluorescent NP platforms reflects ongoing efforts to leverage these advantages. Fluorescence imaging with NPs is utilized across diverse fields, including tumor detection and imaging, protein assessment, and enzymatic analysis in real-time, as summarized in Table 10.2.

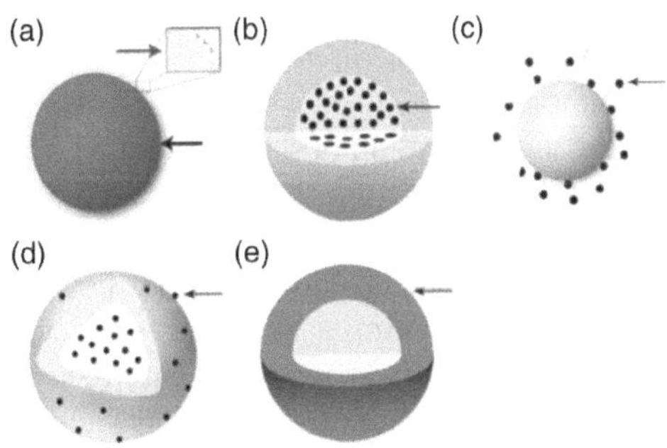

FIGURE 10.3 Nanoparticles of different structures where; (a) basic structure; (b) core–shell structure; (c) vector structure; (d) mixed structure; (e) core–shell structure; for imaging in MRI/CT applications [26].

TABLE 10.2
Fluorescence Applications of Selected Nanoparticles [26]

Imaging Agent	Measurement	Application
Persistent luminescence nanoparticles (PLNP)	141	Tumor detection
Quantum dots	12	Tumor imaging
Up-conversion nanoparticles (UCNP)	120	Breast cancer detection
Cy 5.5	20	Detecting protease activity

This compilation offers insights into the diverse potential of MRI contrast NPs in biomedical research and clinical settings.

Distinct imaging modalities possess individual strengths and limitations. While fluorescence imaging excels in spatial resolution, its tissue penetration remains limited. PET offers exceptional sensitivity and molecular-level functional insights with sufficient tissue penetration but is constrained by spatial resolution. CT provides remarkable spatial precision at the cost of ionizing radiation. To address these constraints, there has been a growing interest in multimodality imaging methods.

The authors of the study [28] have discussed the research scope of biomedical NMs, encompassing medical detection and diagnosis, therapeutic drugs, and functional NMs serving roles in imaging, antibacterial functions, and drug delivery. These are broadly identified as carbon-based NMs, inorganic NMs, organic polymer NMs, and nanocomposites. NMs act as carriers for sustained drug delivery, with gold NPs being effective carriers for various drugs. Additionally, NMs play a crucial role in high-resolution, high-contrast bioimaging applications, especially iron-based magnetic NMs in MRI. The distinct physical, chemical, and biological properties of biomedical NMs significantly contribute to advancements in medical science, driving innovative breakthroughs in disease treatment and biological research. To comprehensively assess NMs, various characterization methods are available. Microscopic techniques like SEM and TEM, as mentioned in the study [6], are used for surface and internal structure observation. X-Ray diffraction analyzes crystal structures, while Raman and Fourier transform infrared spectroscopy provide chemical information. These methods, among others, form a comprehensive toolbox for NM characterization, enhancing the understanding of their structure and properties.

The role of AI is consolidated in [28] in the synthesis phase of biomedical NMs by optimizing and designing parameters, thereby significantly enhancing efficiency. This is done with the assistance of ML algorithms, as shown in Figure 10.4 [28], which are employed to analyze and optimize synthetic routes for various NMs, contributing to innovation and efficiency in chemical research and development [29]. In this work, microbial synthesis, an eco-friendly method, is studied with AI techniques for efficient control, enabling applications in environmental monitoring, medical diagnosis, and biotechnology. These innovations, guided by AI and ML, offer precise control over synthesis processes, contributing to the development of high-quality and reproducible NPs with desired properties.

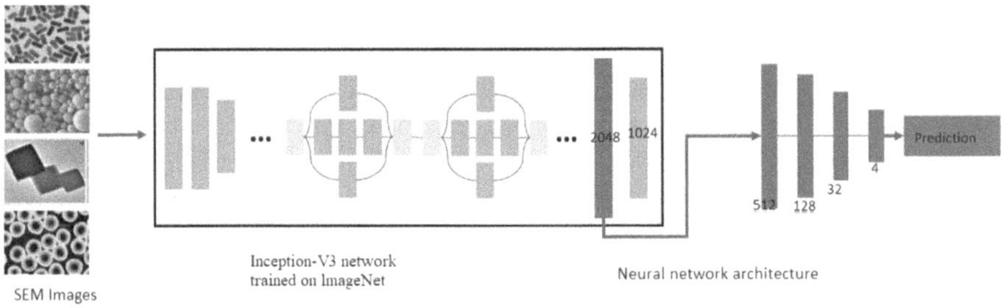

FIGURE 10.4 Deep neural network-based SEM image classification [28].

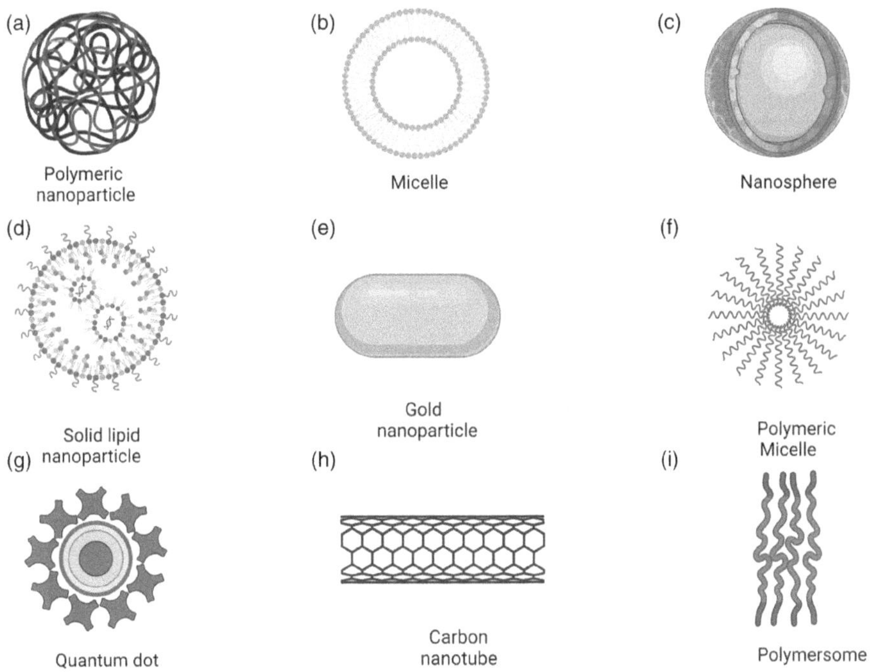

FIGURE 10.5 (a–i) Different types of nanoparticles used in drug delivery [30].

In the review article [30], the authors have highlighted the essence of nanotechnology for drug delivery and precision medicine. NPs (NPs) have gained prominence across various domains, ranging from drug delivery to tissue engineering. NPs, synthesized from diverse materials, show promise in delivering recombinant proteins, nucleotides, and vaccines. Figure 10.5 shows different types of NPs for drug delivery scenarios [30].

Despite their potential, challenges exist in translating NP-mediated drug delivery from extensive research to clinical application. In recent decades, substantial progress has been made in cancer treatments, driven by advances in understanding carcinogenesis, cell biology, and the tumor microenvironment. Despite these successes, cancer remains a complex disease with high fatality rates for many types. Targeted drug delivery is crucial for enhancing the survival rates of cancer patients and ensuring the precise delivery of anticancer drugs. In reference to the above criticality, NPs offer two targeting methods: passive and active. Passive targeting achieves NP localization within the tumor, while active targeting involves the recognition of NPs taken up by tumor cells. The EPR effect, a key consideration for effective drug delivery, has played a crucial role during the chemotherapy process. Nanotechnology faces challenges such as limited knowledge about NP components, uniformity of toxicity, standardized assays, synthesis protocols, and advanced analytical tools in drug administration. AI holds promise in this direction, as shown in [30] with the help of Figure 10.6, addressing these challenges, and its utility is also found in different applications in the study [31].

The authors of the study [30] have discussed the active engagement of researchers in the development and application of microrobotics in the field of medicine, particularly focusing on nanomachines. These nanomachines serve various medical purposes, including in situ drug delivery, targeting disease cell membranes, and conducting microsurgeries. Despite being in their nascent stages, nanorobots hold considerable promise. Recent research showcases an application of in vivo imaging using mesoporous silica and gold NPs. The integration of PET technology with this combination allows for precise monitoring of nanomotor movement, thereby enhancing real-time imaging. Thus, developing powerful sensor-based nanorobotic systems

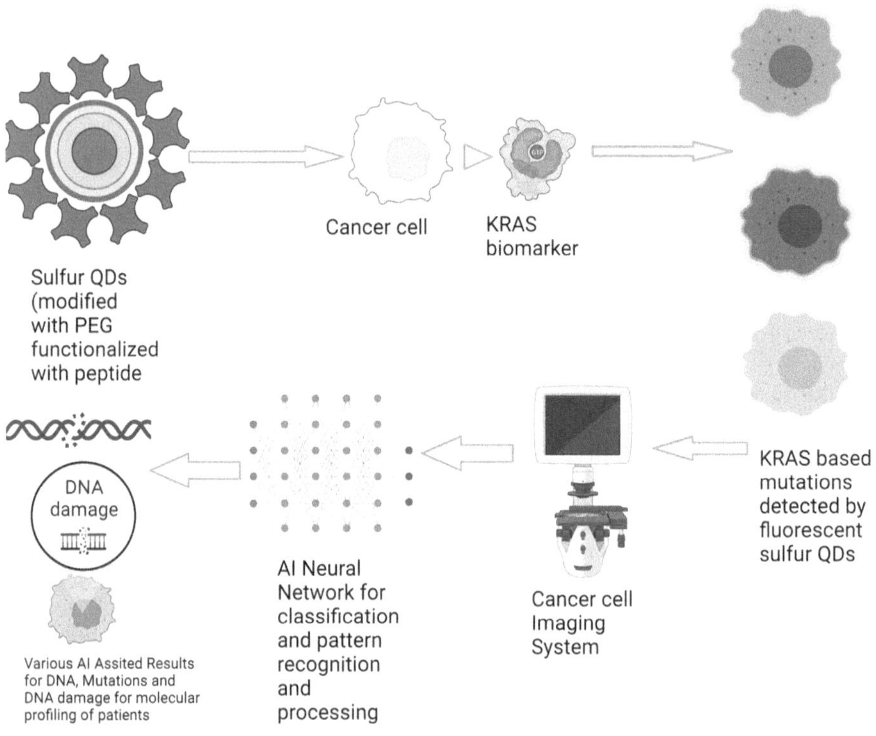

FIGURE 10.6 Biomarker sensing using artificial intelligence [30].

using AI is essential, and its future role and scope are further investigated in molecular production to advance nanorobot development.

The study [32] continues the exploration of chirality in NPs and NMs. Chirality refers to the absence of mirror symmetry in an object, characterized by distinct left and right forms, such as our hands, known as enantiomers. In biological organisms, organic and biological molecules interact with each other, known as the "lock–key principle," resulting in different biological activities, including toxicity. Currently, the chirality of drugs and synthetic biologically active substances is approximately 80%. Chiral substances also interact uniquely with circularly polarized light, exhibiting chiroptical activity. This study formalizes the chiral nanostructures demonstrating significant CD values and circularly polarized emission intensities across the visible and NIR spectrum, offering opportunities in optoelectronics. The field of chiral NMs has gained attention due to their unique properties and applications, yet there is a need to harness the power of ML, as demonstrated in general NM development, to advance research in this specific area.

The review of [32] emphasizes the potential of ML in nanotechnology, with a particular focus on its applicability to chiral NMs. Advances in ML methods for chiral NMs are discussed, including electron microscopy, optical microscopy, and scanning probe microscopy. Notable examples presented in this work include the use of DL models for identifying chiral morphology from SEM images. Additionally, CNN models are applied to analyze HR-TEM images to determine the chirality of carbon nanotubes, which finds application in the development of Microstrip Patch antennas discussed in [33]. Optical microscopy, particularly fluorescent microscopy, combined with ML simplifies the analysis of fluorescent images, aiding in addressing questions related to NP transport through tumor vessels. Moreover, the higher resolution of TEM, favored for NM characterization, provides in-depth information on defects and space groups [34]. The potential of scanning probe microscopy in investigating properties and constructing new chiral species on an atomic level is highlighted, with ongoing efforts in developing fully automated SPM processes. Overall, the integration of ML techniques holds great promise for advancing research in chiral NMs, with applications across various microscopy techniques.

10.2.1 ML Approaches in Nanotechnology

In the purview of nanotechnology, as presented in [32], the intersection of massive data generation and materials science has spurred significant growth. The application of ML plays a pivotal role in expediting the discovery and optimization of novel NMs by unraveling the intricate structural and parametric relationships between them. Figure 10.7 discusses some of the common ML techniques used for NMs [32], while Figure 10.8 points out some of the chiral NM's properties and the study of complex relationships [32]. This approach reduces reliance on traditional and time-bound experiments and simulations, which helps in preparing quality data suited to accurate ML-based predictions.

In the expansive landscape of NMs, the continuous increase in experimental data, through techniques such as wet chemistry approaches, yields NMs with fine structural and compositional differences. To comprehend this intricate interrelation, a

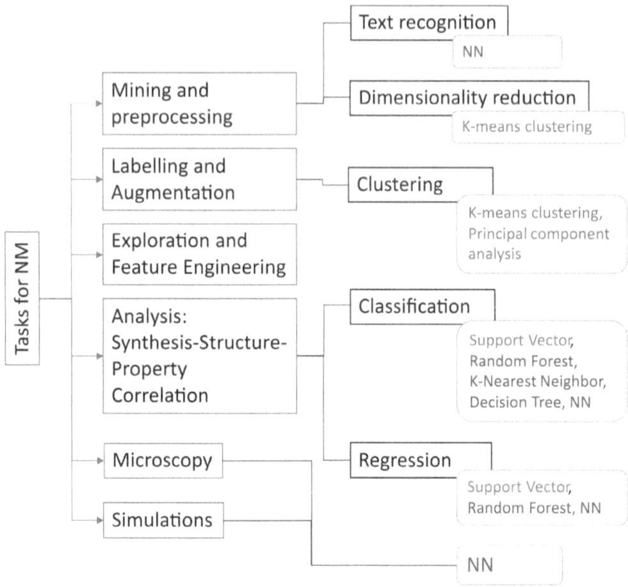

FIGURE 10.7 Machine learning methodologies for different applications on nanomaterials [32].

parameter space is constructed encompassing synthesis-related and property-related data. Synthesis-related data involves systematically varying one parameter while keeping others constant, which demands significant time and resources. Property-related data includes diverse characteristics like morphology, optical, and electrical properties, often assessed through various characterization methods.

However, challenges arise due to inconsistent reporting of parameters and uncertainties in inferred relationships, highlighting the complexity of navigating the parameter space in NM research. Numerous databases have been proposed such as NanoE-Tox, which focuses on nanotoxicity interactions, providing crucial information for predicting and exploring materials with desired properties. However, existing datasets may suffer from disorganization, incompleteness, and inconsistency, influencing data quality. ML, particularly in a data-centric approach, prioritizes improving data quality over exclusive model refinement, recognizing the critical criteria of accuracy, completeness, reliability, relevance, consistency, and timeliness. Challenges in data quality assessment include heterogeneity, incompleteness, inaccuracy, inconsistency, insufficiency, and an unstructured nature. To enhance data quality, researchers must adhere to the principles of findability, accessibility, interoperability, and reusability.

Data labeling, involving the assignment of labels to objects, is a crucial step in data preparation for supervised and semi-supervised ML methods. Relying solely on a single expert for labeling can lead to errors and discrepancies. Cross-validation and expert agreement, although effective, can be expensive and time-consuming. To address this, annotation services and crowdsourcing are employed, as demonstrated in nanomedicine databases. Automatic labeling methods are actively being

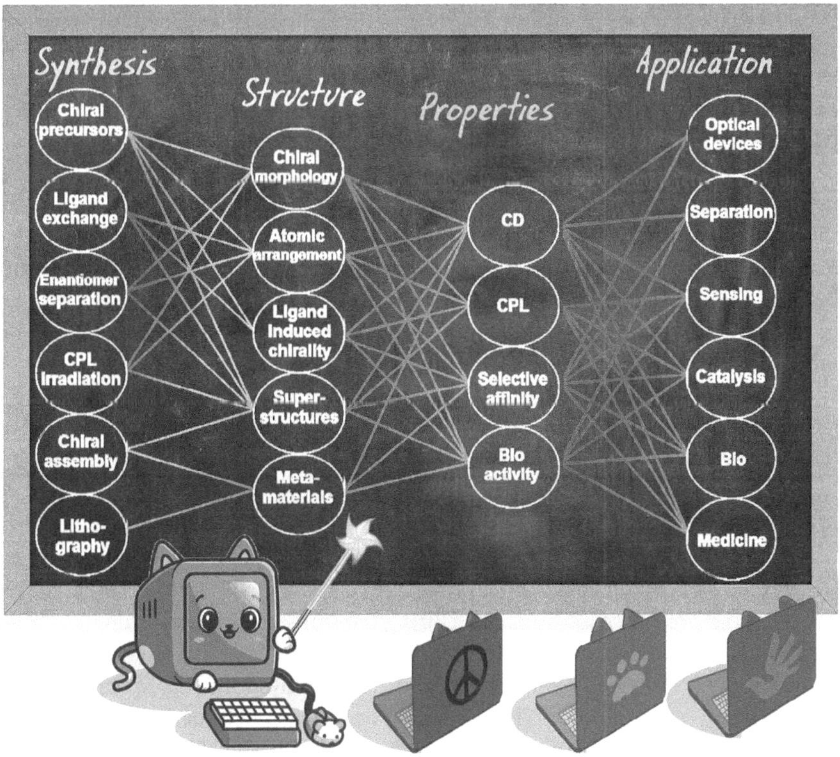

FIGURE 10.8 Study of different properties and applications of chiral nanomaterials [32].

developed, such as the toolkit by Schwenker et al. [35], which creates a self-labeled database of electron microscopy images. This involves using ML-generated figure captions derived from open-source articles, providing an efficient means of labeling large datasets. This is followed by data augmentation techniques, like those employed by Luo et al. [27], which contribute to training neural networks (NNs) by generating modified copies of existing datasets. In their study, data augmentation and transfer learning enhanced the deep learning classification of TEM images of carbon NMs. These approaches collectively streamline data labeling processes and contribute to the effective training of ML models for materials science applications.

In the exploratory data analysis phase, understanding the collected data is crucial. Various statistical measures, such as mean, standard deviation, minimum and maximum values, quartiles, medians, and outliers, are assessed for each parameter. However, estimating target variables, especially in large and sparse datasets, can be challenging. In such cases, ML proves powerful enough to identify correlations between NM properties, such as electron affinity, ionization potential values, and electret performance, as demonstrated by Mao et al. [36]. Further data exploration involves splitting parameters into smaller groups to enhance analysis accuracy. Classification tasks, solvable through manual methods or supervised ML approaches such as support vector machine, RF, KNN, decision trees, and NNs, aid in this process. Following data exploration, feature engineering is used to uncover correlations between parameters and target variables, utilizing techniques like correlation matrices, SHAP, and outlier detection. The Matminer open-source library offers a "featurizer" tool to identify and derive material features from available data, contributing to the overall understanding of complex relationships in materials science datasets.

10.2.2 ML-Assisted Microscopy

The research work in [32] also provides comprehensive insights into the structure of NMs, which are frequently obtained through microscopy imaging, and ML finds widespread application in this domain. Regardless of the microscopy technique employed, be it electronic, optical, or scanning probe microscopy, ML is commonly utilized for two primary objectives: (i) enhancing image quality through image restoration and (ii) recognizing and classifying images.

Numerous works already exist in image reconstruction and enhancement using ML and DL. The most preferred technique in the domain of deep learning for image analysis is CNNs. However, to attain optimal output, experimentation with extensive datasets like the CNR-IOM dataset and ImageNet, which work well with SEM and TEM, respectively, is necessary. Building precise deep NN architectures is critical for training pre-processed images. During the transmission stage, images or any other signals encounter noise in different forms. To remove noise from images, several strategies and filtering techniques have been employed that are specific to dealing with certain types of noise. The two most prominent and effective functions used for noise minimization are Noise2Noise and Noise2Void. The Noise2Noise approach employs noisy data for training purposes, while Noise2Void directly uses an NN with input data to segregate useful information from noise. These developments have become a preliminary need for conducting microscopic analysis on input samples to generate larger volumes of data and make our results more reliable.

One of the key tasks in image analysis is segmentation. Segmentation labels specific parts of an image and converts them into a machine-understandable format that can be analyzed later. Some of the most accepted architectures used for performing image segmentation are SegNet, U-Net, and Mask R-CNN. ResNet18, which was designed to work with natural images, has the best computational efficiency among its known counterparts.

These computational simulations are required for discovering new compounds with unconventional properties. To perform complex simulations on NMs that were otherwise not possible with standard techniques like density functional theory (DFT) has been made possible with ML techniques like AtomAI. They assist in the analysis of spectroscopic data. The DeepImageJ plugin, which works with software like ImageJ, allows for the selection of appropriate NNs for image analysis and accurate results. The development of new data processing tools guarantees a new dawn in the areas of biomedical imaging and materials science.

The progress made in the form of novel NMs or the modernization of old tools and software through the assimilation of AI is increasing the popularity of nanoscience. The ability to perform complex simulations is assisting scientists in their study and development of new materials. These innovations make possible the discovery of new and improved compounds.

10.3 RESULTS AND DISCUSSIONS

This section sheds light on the cumulative findings of the existing research work in the field of NMs with and without AI, as proposed by several authors in the preceding sections. Firstly, the author's innovative approach transcends the limitations of conventional datasets for NP analysis. By neatly crafting an artificial dataset that simulates electron microscope images, this method operates without the need for a real annotated dataset of NPs. Figure 10.9 shows the images obtained in [6] using the methodology adopted in Figure 10.2 [6]. Table 10.3 shows the artificial dataset prepared for NPs for the purpose of validation [6].

The groundbreaking aspect lies in the meticulous creation of 3D models, mirroring the appearance of particles in electron microscopy images and integrating them with corresponding annotations. Two distinct CNNs are employed to meet our above requirements; one excels in detecting and segmenting NPs of diverse shapes and sizes, while the other predicts their orientations. Notably, this addresses the challenge of obtaining annotated datasets specific to NP shapes, offering a versatile solution.

The significance of the work extends beyond mere detection, as shown in Table 10.3, where the detection of different shapes of NPs is performed.

For the purpose of detection and segmentation, the networks employed here are trained for a duration of 100 epochs, each consisting of 50 iterations and a batch size of 1. The optimizer used is stochastic gradient descent. These trained datasets are applied to artificial and real NPs. Table 10.3 indicates that apart from needle NPs, all other particles are detected correctly, denoted by their mean average precision. The dataset used for segmentation consists of 1506 images. Nanotechnologists often

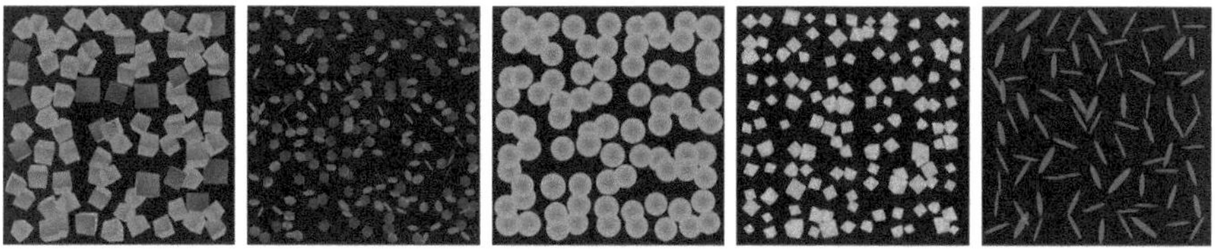

(a) Cube (b) Hexagon (c) Octahedron (d) Sphere (e) Needle

FIGURE 10.9 (a–e) Dataset generation involving masked R-CNN model [6].

TABLE 10.3
Artificial Dataset for Different Nanoparticles With Mean Average Precision (mAP) [6]

Nanoparticle Shape	mAP	mAP_{50}	mAP_{75}
Cube	0.7886	0.9965	0.9808
Octahedron	0.8173	0.9836	0.9340
Needle	0.5568	0.9606	0.5848
Sphere	0.8573	0.9871	0.9847
Hexagonal disc	0.7367	0.9786	0.9115

grapple with the problem of manually characterizing NP structures, complicated further by the absence of standardization in measurements. The above method revolutionizes this process by automating the extraction of measurements from segmented NPs. It accelerates the characterization process with a known technique, thereby minimizing inconsistencies in measurements arising from the different approaches of researchers. This work is positioned at the forefront of NP analysis, offering efficiency, versatility, and standardization in a rapidly evolving field.

This study, which also focuses on the 2D and 3D representations analyzed by nanotechnologists, delves into the imaging intricacies encountered with various electron microscopy techniques. While analyzing materials in 2D generally yields shapes in the form of polygons, circles, and ellipses, 3D views reveal much more detailed structures in the form of polyhedrons, cones, and ellipsoids. This investigation exclusively addresses 3D representations, which are created from 2D projections of objects. The process involves three key steps: firstly, dataset preparation; secondly, supervised algorithms for training the model; and finally, output scene reconstruction (Figure 10.2) [6]. This work opens up promising avenues in the domains of NP detection, segmentation, and 3D reconstruction.

Additionally, NPs with multiple functionalities have been engineered to confer multimodality imaging capabilities, presenting potential advantages in the form of reduced contrast agent dose and enhanced targeting (Table 10.1 [23] citing some applications of NPs). The authors have touched upon, for instance, a multimodal platform like 64Cu-Fe-RGD-PEG-MNP, serving as a photoacoustic, PET, and MRI imaging contrast agent. Despite the promise of multimodality imaging with NPs, challenges persist. The diverse designs of these NPs make establishing standard criteria challenging. Multimodality imaging is a relatively new modality, and its transition to clinical practice requires further exploration. Specialized equipment with multiple imaging modules must be developed for broader application. The NPs capacity for multimodal imaging has particularly fueled interest in early tumor detection. Beyond oncology, the study also explores additional applications in the cardiovascular, renal, neural, and lymphatic systems.

The findings of the work highlighted in [28] are achieved with the integration of AI into NMs' characterization. This has proven to be instrumental in extracting valuable information and advancing the understanding of materials. Characterization data, encompassing images from TEM and curves from UV-Vis absorption spectroscopy, presents a complex landscape that AI adeptly navigates. By automatically identifying and classifying nanostructures, crystal structures, and morphology in samples, AI-driven intelligent characterization aids in interpreting experimental results, facilitating researchers in analyzing intricate data and extracting meaningful insights. The critical information pertaining to NP size, shape, surface, and other characteristics, typically acquired through microscopy, benefits from recent AI advancements. While optical microscopes offer size and shape details, high-resolution TEM (HRTEM) excels in capturing crystallographic intricacies [29].

Noteworthy progress in AI methodologies has streamlined the analysis of microscopic images that are subsequently presented to the targeted audience [37]. For instance, Xu et al. [31] implemented an ML model utilizing images from a defocused scanning optical microscope to precisely determine the size of silver NPs, achieving an estimation error of less than 5% for individual particles [31]. This method ensures accurate visualization of particle size distributions, even in samples with broad size variations. Groschner et al. [38] applied automatic analysis methods, combining U-net and random forest models, to segment NPs in HRTEM images and classify different defect types predictively [38]. Additionally, systems for frequency domain information analysis based on the U-net model have been developed to effectively analyze lattice fringe information in HRTEM images through Fourier transform calculations [39]. This innovative approach has been successfully applied to the study of iron oxide NMs.

The use of AI in NM characterization has led to a major boost, especially in analyzing SEM images. The automation of near-accurate analysis of NM images is largely possible due to deep learning models. This automation has transformed the study of the surface composition and specifications of NMs. Running simulations on large image datasets with shape and type as inputs for performing classification of SEM images is found in the studies of Aversa et al. [40] and Dahy et al. [41]. The accuracy of these models reaches up to 97%, a remarkable achievement.

To extract detailed information from SEM images, an algorithm was developed by Kim et al. [42]. The algorithm is specifically designed to handle parameters that alter the behavior of NPs and define their role in different applications, such as their size, arrangement of particles, and shape. It has overcome challenges like overlapping and complex structures, which significantly hinder their analysis. The algorithm is organized into systematic steps that begin with dataset preparation, followed by training on the calibrated data, and finally performance evaluation through metrics. This ensures more accurate analytical outcomes.

The adoption of AI-based algorithms in NM characterization ensures efficient and reliable data analysis. The use of AI streamlines the analytical study of NMs, which are often quite durable, making the material responsive and versatile for use in various applications. The benefits of AI are realized when it helps minimize the duration of temporal analysis and reduce the potential for human error, leading to more comprehensive studies. This enables researchers to make timely decisions based on properties and future implementations.

Overall, AI in NMs is enhancing our understanding of materials at the nanoscale level. This forward leap is creating favorable conditions for new inventions and modernization in nanotechnology that were previously not feasible with conventional strategies.

The chiral NM properties have been effectively studied using the DFT algorithm. DFT-based simulations are able to deliver results on important tasks such as hypothesis testing, ligand–NP interaction, and chiral defect identification in crystals. DFT can locate asymmetry in nanorods and reveal the underlying reasons behind certain patterns in gold nanooctopods. This gives it an added advantage in studying CdTe and gold NPs. Another approach that is entirely different from the DFT approach is GFN-xTB, which is used for inspecting chiral silver NMs with great success. Compared to these two methods, AI-enabled measures in chiral NMs can improve simulation productivity for accurate predictions with a significant reduction in thermal fluctuations.

ML has become a potent tool in reducing the computational costs of software used for simulating NPs and NMs and in effective resource management. NNs can predict the mechanical properties of chiral substances based on the trained data from molecular dynamics simulations. This approach has been used in designing structures like nanocomposites and single-walled carbon nanotubes.

Advanced imaging techniques with sophisticated processing platforms have emerged as essential for a complete understanding of NMs. Characterization of SEM images and analysis of chirality in nanostructures are gradually helping us recognize their importance in nanotechnology, which cannot be overlooked while pursuing innovation in the preparation of customized NMs.

AI technologies have also reshaped the biomedical sector. New drug discovery requires compound identification and the fulfillment of certain regulatory standards to secure compliance approval, where AI finds extensive use.

The authors of the study [30] have examined the use of AI in scanning probe microscopy, code consistency, and nanocomputing. They have employed artificial NNs (ANNs) to detect changes from spectroscopic reactions. This method considers only key variables that contribute to the detection process, making it effective.

One of the major challenges faced during research on nanosized materials is producing accurate results. AI employing NNs can predict, classify, and customize outcomes in pharmaceutical research. These networks analyze molecular structures through their extensive training data. The authors, while conducting AI-based simulations of NPs, aim to reduce computational time and perform efficient parameter estimation. Developing an appropriate drug formula for manufacturing, storage, and transportation is crucial for circulating that drug in the pharmaceutical ecosystem and enhancing treatment outcomes. It can address issues like drug overdosage, drug instability, and inconsistent response times at the target sites of different patients. They analyze genetic information to suggest therapies, accelerate novel drug development, and predict molecular behavior. It has also shown potential in creating combinational drugs to anticipate situations where a single drug cannot fulfill its designated task. This will improve medical treatments and personalized patient care.

The treatment of cancer, whether at an early or advanced stage, requires upgrades in either processing techniques, devices, or sometimes both. This will assist doctors in preparing dedicated treatment procedures based on the patient's profile and their response to drugs. The generation of vast amounts of data makes the test results authentic, and the therapies designed from it meet individual patient needs. This makes the entire procedure effective, with customized outcomes and minimal side effects.

Due to the complex structure of the human body, comprehending membranes and nerves makes drug transportation to the affected cells a tedious task. Drug delivery requires timely dispatch, distribution, and proper absorption of drugs into specific cells in the targeted area. AI will be helpful in creating optimum paths for drug flow as it uses critical information like drug interactions and genetic knowledge. It ensures that medications effectively reach their destination once administered to patients, reducing the chances of cancerous cell growth. The optimization through AI for drug delivery is summarized in Table 10.1 [23].

Biomarkers have emerged as effective tools for cancer treatment through risk estimation and progress surveillance. Biomarkers are important for assessing the true condition of any emerging symptoms in specific parts of the body and eventually finding the correct treatment strategies in cases of severe growth. Biomarker tests, such as genetic markers and proteins, create timelines on which the health standards of specific patients are evaluated. AI-driven solutions can automate the identification of biomarkers and the creation of visually interactive dashboards to improve treatment efficiency.

Nanosized crystals, such as quantum dots, have also opened up promising prospects for cancer detection and treatment. Due to their 3D confinement, they are sometimes referred to as zero-dimensional particles. They are extremely sensitive, highly fluorescent, and resistant to photobleaching, making them ideal candidates for identifying cancer markers, performing diagnosis, and improving outcomes as seen in the works of [43]. Recent systems that employ NPs in cancer diagnosis ensure timely interactions with tumor cells, surpassing traditional methods by a wide margin. Though advanced medical care facilitates this, researchers are transforming the lives of cancer-stricken patients and continuously fighting against this global adversary.

Gold and silver NPs are the two most effective materials for the accurate imaging of tumor sites. Due to their binding properties with amine groups and superior light scattering properties, gold NPs are well suited for clinical applications. On the other hand, silver NPs possess antimicrobial properties that enable them to tackle vital tasks such as treating skin burns and addressing viral symptoms in patients with higher precision and accuracy.

Cancer imaging currently requires detailed information, multiple examinations, post-processing of acquired images, and clinical investigation for effective description and treatment formulation. Recent developments focus on combining molecular imaging and drug delivery to visually assess real-time drug delivery processes and understand in vivo efficacy. Researchers aim to create multifunctional theranostic probes for targeted drug delivery. Traditional methods encounter challenges in evaluating pharmacokinetics and pharmacodynamics, but molecular imaging reduces the workload, yielding more precise data for developing candidate drugs with optimal target specificity and pharmacodynamic efficacy. The integration of imaging in drug delivery seeks to provide valuable insights into pharmacokinetics, biodistribution, and drug accumulation at the targeted tumor site. These imaging outcomes offer evidence of nanomedicine uptake, confirming the EPR effect, which exhibits high variability among cancer patients.

AI-based simulations and modeling contribute to refining nanorobot design for controlled drug delivery and effective nanocommunication. Notably, novel technologies like DNA nanorobots are emerging for drug delivery and biosensing, with ANNs playing a crucial role in enhancing prediction capabilities and optimizing performance for detecting tumor cells in targeted drug delivery. As AI has gained prominence in designing nanorobots, the concept of fuzzy logic enhances the estimation of drug dosage at the intracellular level following tumor diagnosis.

10.4 CHALLENGES AND OPPORTUNITIES

The advancements made in the nanosector have captured many industries by equipping them with novel problem-solving strategies and symbolizing their true worth to society as a whole. However, society needs to incur some costs in order to ensure safety while availing of these benefits. Thus, merely maximizing profits at high risk should not be the preferred path of any industry; understanding and assessing risk should be the end goal.

The impact of nanosized materials compared to larger-dimension particles on biological systems is one of the main issues that requires thorough examination with sufficient proof, which at times can be difficult to attain. This problem is generally encountered with low-solubility NPs, as they effortlessly pass through membranes and interact with cells. It becomes crucial to safeguard human health from such groundbreaking scientific discoveries.

This is only possible by imposing strict regulations and fostering scientific awareness before their dissemination to the intended patients to manage the risks effectively.

Apart from affecting individual health, the physical properties of NMs can have a detrimental effect on our surrounding environment, especially when they are present in large amounts. Some NMs can cause explosions or unwanted reactions due to their reactive nature. They need to be handled cautiously, stored under optimal conditions, and tested several times before being exposed to industrial sites.

The industrial integration of nanotechnology can only be implemented while adhering to safety norms and possessing detailed knowledge about NMs. This also involves bringing scientists and industrial experts together on a single platform where they can exchange valuable information on the design and development of NMs for public use, as well as being informed of their merits and threats.

The widespread use of nanotechnology in the fields of environment, health, and sports incorporates various elements of risk. Responsible and ethical considerations in materials design are driven by the development of sound strategies that seek to mitigate the negative impacts of technological advancements. Being proactive in designing safety protocols for the development of nanostructures for varied applications can serve as a key measure for encouraging steady growth. Thus, the role of policymakers becomes paramount, along with collaboration with industrial and scientific experts for product authenticity.

It is crucial to recognize that the risk factors linked to different NMs vary, and a comprehensive review is required to identify the potential hazards associated with distinct NPs. In a laboratory setting, the mere presence of NMs does not inherently pose a threat to humanity or the environment. However, ongoing research aims to unravel the intricate dimensions of nanotechnology risks.

In order to diminish the adversarial effect of AI in nanotechnology, the significance of quality data in ML tools is emphasized in nanobiotechnology and nanomedicine, where the absence of standardized reporting practices undermines reproducibility and comparative studies. The current reporting heterogeneity, particularly regarding physicochemical properties, administered doses, and composition of nanodelivery systems, hampers progress in nanomedical research. The lack of explicit information on parameters such as composition, injected volume, concentration, and route of administration further hinders meaningful analysis. Shortcomings extend to the characterization of nanodelivery materials and assay endpoints, with insufficient reporting of crucial details like the initial dose and tumor mass. The complexity of designing NMs underscores the inadequacy of current experimental characterization practices, necessitating a shift toward robust reporting standards to support the automated development of nanodelivery materials and avoid inhibiting innovation and research throughput in healthcare.

For complete datasets, comprehensive reporting in nanotechnology is highlighted by the potential construction of nanotechnology databases akin to ChEMBL. Emphasizing the importance of quality and complete data, an analysis of 315 research manuscripts is presented in the study [44] on iron oxide NPs, revealing deficiencies, with 68% lacking information on size, shape, or zeta potential. Similarly, 51% and 45% of 322 gold and 257 silica NP manuscripts, respectively, lacked full physicochemical characterization. Pharmacokinetics and dosing data were also frequently absent. This pervasive limitation in the nanotechnology field necessitates urgent attention.

In the domain of biomedical NMs, the versatility and advantages of AI are evident, extending to material design, intelligent synthesis, and characterization analysis. However, the practical application of these algorithms and the reduction of experimental costs and time pose significant challenges [45].

10.5 FUTURE PROSPECTS

AI combined with nanotechnology will certainly show promise in elevating various sectors, especially healthcare and manufacturing, to new heights. Scientists are adopting natural language processing and deep learning to tackle challenges in nanotechnology. By using AI to gather valuable information from scientific papers, researchers hope to identify patterns that can expedite the use of ML tools. This approach could simplify the prioritization of experiments and enhance our understanding and application of nanotechnology.

In discussing this convergence, it is important to consider the broader implications and potential benefits of integrating AI with nanotechnology. This includes not only the immediate developments in specific fields but also the long-term impact on research methodologies, interdisciplinary collaboration, and technological innovation. Addressing ethical considerations and establishing best practices for data management and AI applications will also be crucial in ensuring that these advancements are both effective and responsible.

However, nanotechnology driven by AI presents challenges, mainly due to the high costs and time required for experimental work. To address these issues, researchers stress the need for large databases and specific descriptors during the material design stage. These databases can improve the performance of algorithms, making AI tools more efficient and adaptable for creating new materials. Developing algorithms that can thoroughly analyze experimental data and provide accessible visual interfaces will also make AI more valuable and useful in biomedical research.

In drug delivery, AI provides solutions to current problems, enhancing healthcare practices. AI techniques can handle complex data, making drug design and optimization more efficient. This integration helps address challenges such as data noise, unknown parameters, and differentiating between healthy and cancerous cells during medication delivery. Methods like genetic programming, along with tools like NNs, fuzzy logic, and decision trees, help interpret complex patterns, speeding up the drug discovery process. Techniques like supervised and reinforcement learning help explore relationships between variables, improving drug delivery scheduling and automation.

When discussing these advancements, it is essential to cover the entire spectrum. Integrating AI with nanotechnology and drug delivery not only solves specific technical issues but also has the potential to transform research approaches, promote interdisciplinary collaboration, and drive innovation in various fields. Addressing ethical considerations and ensuring transparent data management will be key to maximizing the benefits of these technologies while minimizing potential risks.

Moreover, creating composite materials requires new tools and methods that are similar to those used in discovering small molecules. These tools allow for the development and customization of composite materials, widening the possibilities and accuracy of material design. As AI becomes more intertwined with nanotechnology and pharmaceutical research, developing these new tools enhances AI's competence, promoting innovation and effectiveness in creating advanced materials and drug delivery systems.

As nanotechnology research grows to meet market demands, these solutions aim to address significant challenges, paving the way for a new digital era in nanotechnology. This era will focus on quickly testing out new treatments using advanced technology. By combining computers with robots, scientists hope to speed up their work in nanotechnology, discovering new things and finding more uses for this technology.

10.6 CONCLUSION

In this section, the findings of the results in the preceding sections are outlined, imparting substance to AI-based nanotechnological advancements in the design and study of NMs for various medical and biomedical applications.

A methodology is devised for detecting and segmenting NPs in microscope images, along with approximating their 3D orientation and reconstructing them in three dimensions with the help of a simulated dataset rather than annotated images. A scope for improvement observed in this case is the use of masked R-CNN for optimal outputs. Extracted segmentations to streamline dataset creation would enable human experts to focus on data validation. Nanotechnology has empowered single imaging modalities and shown promising results in multimodality imaging. The structural properties and biomedical imaging applications of NPs span various modalities, such as fluorescence imaging, MRI, CT, ultrasound, PET, and SPECT. However, challenges persist in clinical practice, mainly related to biodistribution and safety concerns; NP size, shape, surface charge, and labeling affect biodistribution and toxicity, necessitating further understanding. Theranostic nanoplatforms, integrating diagnostics and therapeutics, hold future prospects in clinical imaging. The design of NPs, nanocages, nanodiamonds, nanoshells, and nanotubes is another development offering promising options for modern cancer therapy. Some issues, such as cytotoxicity, nonbiodegradability, and biocompatibility, require targeted efforts in the direction of cellular interactions and enhanced imaging.

The amalgamation of AI with biomedical NMs has ushered in significant advancements, revolutionizing the optimization and exploration processes. By exploiting AI-driven methods, researchers can achieve superior efficiency compared to traditional manual analysis, accelerating the attainment of valuable knowledge and information. This fusion rationalizes the synthesis stage, enabling the intelligent design of synthesis reaction parameters and augmenting NM characterization analysis. Moreover, the underutilization of ML in chiral NMs presents an opportunity for innovative scrutiny. ML can optimize synthesis–structure–property–application relations, leading to the design of NMs with high optical activity and chiroptical responses. By customizing treatments based on individual patient molecular profiles and leveraging AI's capabilities in real-time monitoring and clinical imaging, nanotechnology-based therapeutics aim to improve outcomes and address treatment limitations. Ultimately, with the advancements made in AI, the development of biomarker detection systems and NP analyzers is expected to transpire in the future of biomedicine. In conclusion, ML's pivotal role in the above scenario underscores the importance of interdisciplinary collaboration and innovation in driving materials science forward.

REFERENCES

1. Nick Bostrom Quotes. BrainyQuote.com, BrainyMedia Inc. 2024. https://www.brainyquote.com/quotes/nick_bostrom_554505
2. Mashaghi, S., T. Jadidi, G. Koenderink, and A. Mashaghi. 2013. "Lipid nanotechnology". *Int. J. Mol. Sci.*, 14, p. 4242–4282.
3. Yadav, M. K., R. Kumar, R. K. Ratnesh, J. Singh, R. Chandra, A. Kumar, V. Vishnoi, G. Singh, and A. K. Singh. 2024. "Revolutionizing technology with spintronics: devices and their transformative applications." *Mat. Sci. Eng. B*, 303, p. 117293. https://doi.org/10.1016/j.mseb.2024.117293
4. Ratnesh, R. K., M. K. Singh, and J. Singh. 2024. "Enhancing ZnO/si heterojunction photodetector performance for ultra high responsivity across wide spectral range." *J. Mat. Sci.: Mater. Electron.*, 35, p. 756. https://doi.org/10.1007/s10854-024-12516-5
5. Kothiyal, S. R., R. K. Ratnesh, and A. K. Shukla. 2023. *Enhancing Sensing with Surface Plasmons Resonance: Theoretical Insights and Design Strategies in Photonic Crystal Fiber, International Conference Artificial Intelligence and Information Technologies, (ICAIIT)-2024*, Taylor & Francis, pp. 279–285. https://doi.org/10.1201/9781003510833-46
6. A. Cid-Mejias, Alonso-Calvo, H. Gavilán, J. Crespo, and V. Maojo. 2021. "A deep learning approach using synthetic images for segmenting and estimating 3D orientation of nanoparticles in EM images." *Comput. Methods Program. Biomed.*, 202, p. 105958. Epub 2021 Feb 2. PMID: 33588253. https://doi.org/10.1016/j.cmpb.2021.105958
7. Su, D., 2017. "Advanced electron microscopy characterization of nanomaterials for catalysis". *Green Energy Environ.,* 2(2), pp. 70–83. https://doi.org/10.1016/j.gee.2017.02.001
8. Benjamin, M., C. Geers, D. Vanhecke, C. Endes, B. Rothen-Rutishauser, S. Balog, and A. Petri-Fink. 2015. "Avoiding drying-artifacts in transmission electron microscopy: Characterizing the size and colloidal state of nanoparticles." *Sci. Rep.*, 5(1), p. 9793. https://doi.org/10.1038/srep09793
9. C. T. Rueden, J. Schindelin, M. C. Hiner, B. E. DeZonia, A. E. Walter, E. T. Arena, and K.W. Eliceiri. 2017. "ImageJ2: ImageJ for the next generation of scientific image data." *BMC Bioinform.*, 18(1), p. 529, PMID: 29187165; PMCID: PMC5708080. https://doi.org/10.1186/s12859-017-1934-z
10. Nicholson, W. V., and R. M. Glaeser. 2001. "Review: automatic particle detection in electron microscopy." *J. Struct. Biol.*, 133(2), pp. 90–101. https://doi.org/10.10 06/jsbi.20 01.4348

11. Ranzato, M., P. Taylor, J. House, R. Flagan, Y. LeCun, and P. Perona. 2007. "Automatic recognition of biological particles in microscopic images." *Pattern Recognit. Lett.*, 28(1), pp. 31–39. https://doi.org/10.1016/j.patrec.2006.06.010
12. Kuba, H., K. Hotta, and H. Takahashi. 2009. "Automatic particle detection and counting by one-class SVM from microscope image." In: M. Köppen, N. Kasabov, and G. Coghill (Eds.), *Advances in Neuro-Information Processing, Lecture Notes in Computer Science*, Springer, Berlin, Heidelberg, pp. 361–368. https://doi.org/10.1007/978-3-642-03040-6-44
13. Garg, P., T. Chandra, R. Ahlawat, N. Mittal, R. K. Ratnesh, and S. K. Tripathi. 2022. *Star Galaxy Image Classification Via Convolutional Neural Network, ICOSEC-2022*, IEEE, pp. 1156–1168. https://doi.org/10.1109/ICOSEC54921.2022.9952065
14. Kook, S., R. Zhang, Q. N. Chan, T. Aizawa, K. Kondo, L. M. Pickett, E. Cenker, G. Bruneaux, O. Andersson, J. Pagels, and E. Z. Nordin. 2015. "Automated detection of primary particles from transmission electron microscope (TEM) images of soot aggregates in diesel engine environments". *SAE Int. J. Eng.*, 9(1), pp. 279–296, https://doi.org/10.4271/2015-01-1991
15. Mirzaei, M., and H. K. Rafsanjani. 2017. "An automatic algorithm for determination of the nanoparticles from TEM images using circular Hough transform". *Micron*, 96, pp. 86–95. https://doi.org/10.1016/j.micron.2017.02.008
16. Al-Kofahi, Y., A. Zaltsman, R. Graves, W. Marshall, and M. Rusu. 2018. "A deep learning- based algorithm for 2-D cell segmentation in microscopy images". *BMC Bioinform.,* 19(1), p. 365. https://doi.org/10.1186/s12859-018-2375-z
17. Rivenson, Y., Z. Göröcs, H. Günaydin, Y. Zhang, H. Wang, and A. Ozcan. 2017. "Deep learning microscopy". *Optica*, 4(11), pp. 1437–1443. https://doi.org/10.1364/OPTICA.4.001437
18. Nehme, E., L. E. Weiss, T. Michaeli, and Y. Shechtman. 2018. "Deep-STORM: super-resolution single-molecule microscopy by deep learning". *Optica*, 5(4), pp. 458–464, https://doi.org/10.1364/OPTICA.5.000458
19. Min, S., B. Lee, and S. Yoon. 2017. "Deep learning in bioinformatics, brief". *Bioinform.*, 18(5), pp. 851–869. https://doi.org/10.1093/bib/bbw068
20. Shahin, A. I., Y. Guo, K. M. Amin, and A. A. Sharawi. 2019. "White blood cells identification system based on convolutional deep neural learning networks". *Comput. Methods Prog. Biomed.*, 168, pp. 69–80. https://doi.org/10.1016/j.cmpb.2017.11.015
21. Sadanandan, S.K., P. Ranefall, S. Le Guyader, and C. Wählby. 2017. "Automated training of deep convolutional neural networks for cell segmentation". *Sci. Rep.*, 7(1), p. 7860, https://doi.org/10.1038/s41598-017-07599-6
22. Piccinini, F., A. Tesei, and A. Bevilacqua. 2016. "Single-image based methods used for non- invasive volume estimation of cancer spheroids: a practical assessing approach based on entry-level equipment". *Comput. Methods Prog. Biomed.*, 135, pp. 51–60. https://doi.org/10.1016/j.cmpb.2016.07.024
23. Sim, S., and N. K. Wong. 2021. "Nanotechnology and its use in imaging and drug delivery (Review)". *Biomed. Rep.*, 14(5), p, 42. https://doi.org/10.3892/br.2021.1418
24. Medina, C., M. J. S. Martinez, A. Radomski, O. I. Corrigan, and M.W. Radomski. 2007. "Nanoparticles: pharmacological and toxicological significance". *Br. J. Pharmacol.*, 150(5), pp. 552–558. https://doi.org/10.1038/sj.bjp.0707130. PMID: 17245366; PMCID: PMC2189773
25. Ratner, B. D., and S. J. Bryant. 2004. "Biomaterials: where we have been and where we are going". *Annu. Rev. Biomed. Eng.*, 6, pp. 41–75. https://doi.org/10.1146/annurev.bioeng.6.040803.140027. PMID: 15255762
26. Han, X., K. Xu, O. Taratula, and K. Farsad. 2019. "Applications of nanoparticles in biomedical imaging". *Nanoscale*, 11(3), pp. 799–819. https://doi.org/10.1039/c8nr07769j. PMID: 30603750; PMCID: PMC8112886

27. Luo, Q., E. A. Holm, and C. Wang. 2021. "A transfer learning approach for improved classification of carbon nanomaterials from TEM images". *Nanoscale Adv.*, 3, pp. 206–213. https://doi.org/10.1039/D0NA00634C
28. Zhu, X., Y. Li, and N. Gu. 2023. "Application of artificial intelligence in the exploration and optimization of biomedical nanomaterials". *Nano Biomed. Eng.*, 15(3), pp. 342–353. https://doi.org/10.26599/NBE.2023.9290035
29. Ramkumar, V. S., A. Pugazhendhi, K. Gopalakrishnan, et.al. 2017. "Biofabrication and characterization of silver nanoparticles using aqueous extract of seaweed Enteromorpha compressa and its biomedical properties". *Biotechnol. Rep.*, 14, pp. 1–7. https://doi.org/10.1016/j.btre.2017.02.001
30. Das, K. P., and C. J. 2023. "Nanoparticles and convergence of artificial intelligence for targeted drug delivery for cancer therapy: Current progress and challenges". *Front. Med. Technol.*, 4, p. 1067144. https://doi.org/10.3389/fmedt.2022.1067144. PMID: 36688144; PMCID: PMC9853978
31. Xu, Y., D. Xu, N. Yu, et al. 2023. "Machine learning enhanced optical microscopy for the rapid morphology characterization of silver nanoparticles. *ACS Appl. Mat. Interfaces*, 15(14), pp. 18244–18251. https://doi.org/10.1021/acsami.3c02448
32. Kuznetsova. V., Á. Coogan, D. Botov, Y. Gromova, E. V. Ushakova, and Y. K. Gun'ko. 2024. "Expanding the horizons of machine learning in nanomaterials to chiral nanostructures". *Adv Mater.*, p. e2308912. https://doi.org/10.1002/adma.202308912. Epub ahead of print. PMID: 38241607
33. S. Pathak, N. Kishore, G. Upadhyay, R. K. Ratnesh and R. Mishra. 2021. "A compact size planar microstrip-fed patch antenna with hexagonal DGS slot for WLAN application". *Recent Trends Electron. Commun.*, 777, pp. 263-271. https://doi.org/10.1007/978-981-16-2761-3_25
34. Seo, J.W., and D. Schryvers. 1998. "TEM investigation of the microstructure and defects of CuZr martensite. Part II: Planar defects". *Acta Mater.,* 46(4), pp. 1177–1183. https://doi.org/10.1016/S1359-6454(97)00334-0
35. Schwenker, E., W. Jiang, T. Spreadbury, N. Ferrier, O. Cossairt, and M. K. Y. Chan. 2023. "EXSCLAIM!-- An automated pipeline for the construction of labeled materials imaging datasets from literature". *Patterns*, 4(11), Article100843. https://doi.org/10.1016/j.patter.2023.100843
36. Mao, Z., C. Chen, Y. Zhang, K. Suzuki, and Y. Suzuki. 2023. "AI-driven discovery of amorphous fluorinated polymer electret with improved charge stability for energy harvesting". *Adv. Mater.*, 2303827. https://doi.org/10.1002/adma.202303827
37. Ioku, K., and Kamitakahara, M. 2008. "Hydrothermal synthesis of hydroxyapatite ceramics for medical application". In: *Biomaterials in Asia: In Commemoration of the 1st Asian Biomaterials Congress, Tsukuba, Japan, 6–8 December 2007,* World Scientific Publishing Co., pp. 317–326. https://doi.org/10.1142/9789812835758_0020.
38. Groschner, C. K., C. Choi, and M. C. Scott. "Machine learning pipeline for segmentation and defect identification from high-resolution transmission electron microscopy data". *Microsc. Microanal.*, 27(3), pp. 549–556. https://doi.org/10.1017/S1431927621000386
39. Zhu, X., Z. Zhang, Y. Mao, et al. 2020. "Applying deep learning in automatic and rapid measurement of lattice spacings in HRTEM images". *Sci. China Mater.*, 63(11), pp. 2365–2370. https://doi.org/10.1007/s40843-020-1368-7
40. Aversa, R., M.H. Modarres, S. Cozzini, et al. 2019. "The first annotated set of scanning electron microscopy images for nanoscience". *Sci. Data*, 6(1), p. 3. https://doi.org/10.1038/s41597-019-0007-8
41. Dahy, G., M. Soliman, H. Alshater, et al. 2023. "Optimized deep networks for the classification of nanoparticles in scanning electron microscopy imaging". *Comput. Mater. Sci.*, 223, p. 112135. https://doi.org/10.1016/j.commatsci.2023.112135

42. Kim, H., J. Han, and T.Y.J. Han. 2020. "Machine vision-driven automatic recognition of particle size and morphology in SEM images". *Nanoscale*, 12(37), pp. 19461–19469. https://doi.org/10.1039/D0NR04140H
43. Ratnesh, R. K., M. K. Singh, V. Kumar, S. Singh, R. Chandra, M. Singh, and J. Singh. 2024. "Mango leaves (Mangifera Indica) derived highly florescent green graphene quantum dot nanoprobes for enhanced on-off dual detection of cholesterol and Fe^{2+} ions based on molecular logic operation". *ACS Biomater. Sci. Eng.*, 7, pp. 4417–4426. https://doi.org/10.1021/acsabm.4c00292
44. Lorenc, A., B. B. Mendes, J. Conniot, D. P. Sousa, J. Conde, and T. Rodrigues. 2021. "Machine learning for next-generation nanotechnology in healthcare". *Matter*, 4(10), p. 3078–3080. https://doi.org/10.1016/j.matt.2021.09.014
45. Kumar, R., R. K. Ratnesh, J. Singh, R. Chandra, G. Singh , and V. Vishnoi. 2023. "Recent prospects of medical imaging and sensing technologies based on electrical impedance data acquisition system. *J. Electrochem. Soc.,* 170(11), p. 117507. https://doi.org/10.1149/1945-7111/ad050f

11 Biocompatible Nanoantennas for Enhanced Implantable Medical Devices

Anupma Gupta, Vipan Kumar, Shonak Bansal, Ramesh Kumar, Manish Kumar Singla, and Mohammed H. Alsharif

11.1 INTRODUCTION

Breast cancer is a severe health condition characterized by the uncontrollable growth and division of breast cells, often leading to the invasion of surrounding healthy tissue or metastasis to other parts of the body. It ranks as the second leading cause of death for females globally, with an estimated 685,000 deaths in 2020 [1–3]. In 2020, there were over 2.26 million new cases of breast cancer worldwide, making it the most frequently diagnosed cancer, accounting for 11.7% of all new cancer patients. In the United States alone, the American Cancer Society predicts around 297,790 new cases and 43,700 deaths from breast cancer in 2023. Early detection is crucial, as it can significantly increase the survival rate by up to 97% [1–6]. Imaging modalities play a crucial role in breast cancer diagnosis and treatment, offering a wide range of internal body images and information. Common imaging methods include ultrasound, mammography, magnetic resonance imaging, and positron emission tomography. However, these approaches have limitations, prompting researchers to explore novel microwave-based procedures for breast cancer detection [7–10]. Microwave techniques are attractive due to their non-ionizing radiation and non-invasive nature, making them promising options in this field. Microwave tomography and ultra-wideband (UWB) radar imaging are two commonly used methods for analyzing transmitted and received signals when exposing the body to microwaves. Active microwave imaging (MWI) relies on electromagnetic scattering caused by the variation of dielectric constants among different materials, such as tissues in the body [8–11]. In breast imaging, this contrast is utilized to create 2D or 3D images. Numerous active microwave breast imaging structures have been proposed recently, aiming to reconstruct images of the breast. UWB radar-based imaging, on the other hand, focuses on locating large scattering obstacles like tumors rather than on image reconstruction. Antennas play a crucial role in UWB imaging, serving as the interface between the objects being imaged and electromagnetic waves [12–14].

DOI: 10.1201/9781003634553-11

Sensors operating in the UWB frequency range are used to irradiate the breast with short pulses. Cancerous tumor cells react differently to the radiated wave and affect the backscattered signal. The broad bandwidth of the backscattered signal penetrates different layers of the tissue, differentiating extremely small abnormalities and irregularities within the tissue and providing high-resolution images [11–19].

Antennas play a crucial role in UWB imaging as they serve as the interface between electromagnetic waves and the breast tissue being diagnosed, particularly in breast cancer detection. Microwave antennas operating in the frequency range of 300 MHz–20 GHz are used to irradiate the breast with short pulses, providing valuable information about the object under investigation [20]. In breast imaging, it is essential for antennas to be comfortable and compact, conforming to the breast's shape while maintaining performance. Patient comfort and compliance are critical, and antennas that can adapt to the breast's shape are essential for practical and effective imaging. Decreasing the air-gap between the breast and antenna increases detection sensitivity. The COVID-19 pandemic has highlighted the need for alternative breast imaging devices, driving attention to textile-based wearable antennas. These structures are smaller in overall size. Academicians have presented various textile-based antennas for early breast cancer detection, addressing challenges such as compactness, efficiency, integration with devices, biocompatibility, flexibility, radiation safety, and durability [21–25]. Antennas have the potential to transform breast cancer detection by providing portable, real-time, and patient-friendly solutions, changing the way breast cancer is diagnosed and monitored [26,27].

To meet the requirements of MWI, the antenna used should have certain specific characteristics. These include a directional radiation pattern, high radiation efficiency, compact size, gain, the ability to resonate over a wide range of frequencies, and the maintenance of fidelity throughout this range. This allows for improved penetration into heterogeneous tissue, resulting in high-resolution images with greater accuracy [28–35]. Printed antennas, and specifically microstrip antennas (MSAs), are often adopted in MWI systems due to their planar topology, lower fabrication costs, and thin profile. However, MSAs often have low gain and poor radiation characteristics. To improve these characteristics and make the antenna more compatible with MWI systems, researchers have employed various techniques [36–44]. These include the use of split ring resonator structures [44], defected ground [45], aperture coupled [46], metamaterial-based unit cells [47], cross and cavity [48] backed Vivaldi antennas, slot loaded antennas, and AMC (artificial magnetic conductor) backed uniplanar structures [49–53]. These performance enhancement methods have been incorporated into various types of antennas, including Vivaldi, fractal patch, slotted patch, monopole, biconical, horn, and bow-tie antennas [54–75]. For example, 12 planar antenna array with corrugated-tapered slotting was designed in [54], which enhanced the gain of the antenna while also resulting in a larger antenna structure, higher permittivity of the substrate, and increased lower operating frequency of 5 GHz. In [59], a dark-eyed shaped radiator was used to enhance the 10-dB bandwidth, but the gain obtained was less than 5 dBi, which is not sufficient for medical applications. In [65], a parasitic radiator of eclipse shape was designed to improve coupling with body tissue, but this resulted in larger antenna dimensions and a missed resonance at lower frequencies. A slotted patch antenna with a partial ground plane was used in [29] to achieve wider impedance bandwidth and to accumulate information for calculating the depth of a tumor. However, important characteristics such as gain and radiation characteristics were not provided. In [64], a directional radiation pattern was attained by optimizing antenna dimensional parameters, but the upper cut-off frequency was detuned, which decreased the bandwidth. Additionally, the use of an array of radiators has been employed to improve directivity and gain for high-speed calculations. However, this can result in increased system complexity. A brief description of some of the antennas mentioned above is provided in Table 11.1 for ease of readability and understanding.

TABLE 11.1
Highlights of Some Antennas Used for Microwave Imaging

Ref.	Type of Antenna	Performance Enhancement Factor	Technique	Limitation
[55]	Vivaldi	Gain enhancement	Tapered slotting in corrugated radiator	Larger structure and higher lower resonance frequency
[56]	Vivaldi	Broadside radiation characteristics with improved gain	Adding of dielectric lens with balanced antipodal	Larger dimensions
[57]	Vivaldi	Enhancement of front-to-back ratio (FBR) and directivity	Cavity-backed reflector	Smaller resonating frequency range
[58]	Antipodal Vivaldi	Wide resonating frequency range	Gaussian modulated slot with bent feed line	Lower gain
[59]	Patch	Wide resonating frequency range	fractal geometry of dark-eyed shaped radiator	Lower gain
[60]	Patch	Wide resonating frequency range and stability of radiation characteristics	Fractal geometry of hibiscus petal-shaped radiator	Not suitable for array system due to complex geometry
[61]	Patch	Ultra-wideband performance (1.9-35 GHz)	Traveling wave tapered slot with fractal geometry	Larger geometry and unstable radiation behavior
[62]	Patch	Wide resonating frequency range	Horizontal slots in radiator and partial ground plane	No evaluation for radiation characteristics
[63]	Patch	To attain directional radiation characteristics	Optimization of antenna dimensions	Less bandwidth and supporting only lower frequencies
[64]	Monopole	High directivity and better coupling with body tissue	Adding of eclipse-shaped parasitic radiator	Larger size and poor impedance matching toward lower cutoff frequency

(*Continued*)

TABLE 11.1 (*Continued*)
Highlights of Some Antennas Used for Microwave Imaging

Ref.	Type of Antenna	Performance Enhancement Factor	Technique	Limitation
[65]	Monopole	Radiation efficiency and bandwidth enhancement	T- and E-shaped hybrid slots in ground plane	Low gain and directivity
[66]	Monopole	Wide band performance	Etching of a circular and "+"-shaped slot in octagonal-shaped patch	Unstable radiation performance
[67]	Biconical	To achieve stability of impedance matching	Adjustment of cone angle	Less power coupling with tissue due to null toward body
[68]	Horn	Reducing interference from other EM waves and improvement of efficiency	Dielectric shielding	High cost, large structure difficult to integrate in array configuration
[69]	Horn	Wider bandwidth toward lower frequency	Addition of double-ridges	Complex structure
[70]	Horn	Improvement of fidelity factor over broad frequency spectrum	Pyramidal horn with ridge	Large volume and complex structure
[71]	Bow-tie	Size miniaturization	Meandering with double-layer structure	Narrow band and unsuitable for array system
[72]	Bow-tie	Radiation characteristics improvement	Adding of octagonal cavity with a crossed bow-tie	Narrow bandwidth and large antenna geometry
[73]	Bow-tie	Consistency in directivity for wide spectrum	Adding of lumped six resistors	Large dimensions
[74]	Bow-tie	Improving coupling effect of antenna with body tissue	Adding of lumped resistance and capacitance at load	Narrow bandwidth and complex geometrical structure
[75]	Bow-tie	Enhancing sensitivity with tissue	Use of multiple slot lines	Complex configuration and not appropriate for array

Antenna performance has a major impact on the accuracy and efficiency of the MWI system. Near-field radiations are particularly sensitive to coupling with the surrounding dielectric medium. To achieve this, researchers have explored various approaches and techniques. These methods aim to improve the antenna's gain, bandwidth, radiation pattern, and directivity. In addition to this, other limitations such as size, complexity, and cost are also addressed. By incorporating these advancements into various antenna types, such as Vivaldi, fractal patch, slotted patch, monopole, biconical, horn, and bow-tie antennas, the overall effectiveness of MWI systems can be significantly improved.

The rise of technology has brought the unique properties of graphene into consideration. Graphene is arranged in a honeycomb lattice structure and consists of a 2D layer of carbon atoms; it is also known for its exceptional carrier mobility of 200,000 $cm^2/(V\ s)$, which is much greater than that of other materials. Its outstanding tunable and switching properties make it an efficient material for manufacturing ultra-fast electronic components and devices. The conduction and valence bands of graphene join at the K-point of the Dirac point or Brillouin zone, which makes it a semiconductor with a zero bandgap. This phenomenon allows it to maintain conductivity while reducing transition time. Its flexibility and unprecedented electrostatic confinement are also noteworthy. Due to these properties, graphene has been considered for use in circuits for sensors, radio frequency antennas, switches, and transparent devices. In Table 11.2, a brief summary of different antenna structures designed with graphene-based nanomaterials is presented. Almost all the graphene-based

TABLE 11.2
List of Graphene-Based Antenna Structures

Ref.	Antenna Design	Bandwidth (%)	Operating Frequency (GHz)	Total Efficiency (%)	Gain (dBi)	Applications
[82]	CPW-fed wideband quasi-dipole	98.51%	1.7–5	60	0.6–2.3	Low cost wireless
[82]	CPW fed slot antenna	74.18%	1.73–3.77	74.18%	−1 to 0.2	On-body wearable communication
[84]	Meandered line antenna	6.67%	0.984–1.052	6.67%	−4	Low cost RFID and sensing
[85]	Rectangular microstrip patch antenna	–	1.63	–	–	Wearable sensor
[86]	Dipole antenna	18.70%	0.8–0.965	40	−2.18	Low-cost and environmentally friendly RFID tags
[87]	CPW-fed rectangular slot with chamfer	48%	15	72.98	8.41	Mobile terminal for Fifth Generation (5G)
[88]	Microstrip antenna array	–	29–31	–	15.07	Electromagnetic shielding and radiation

radiators are designed for GHz frequencies and used for wearable devices and RFID tags. Different types of antenna structures, such as reconfigurable, reflect arrays, beam scanning, and tunable antennas, are designed for THz frequencies. Thus, there is a possibility to explore the potential of nanomaterial-based antenna structures in UWB technology for MWI [76–81].

This study proposes and examines a novel UWB patch antenna designed with graphene nanomaterials and polyamide for medical imaging of breast tissue cancer detection using microwave technology. According to the literature, MWI relies on the principle that changes in the dielectric and conductive properties of tissue result in variations in the backscattered signal. The antenna proposed here is intended to detect cancerous tissue, which has a larger permittivity due to higher water content than healthy breast tissue. The deviation of the scattering parameter S11 is examined to detect the presence of unhealthy tissue.

The contributions of the proposed structure are as follows:

1. The antenna has a planar structure with graphene as the conducting part.
2. The broad operating frequency spectrum of the antenna enables the penetration of radiation at varying depths.
3. The antenna provides higher gain with a directional power pattern.
4. The antenna is highly sensitive to variations in tissue properties.

The article is organized as follows: In Section 11.1, cancer detection techniques are introduced, and a literature review of antennas for imaging and graphene technology is presented, while Section 11.2 presents the design technique of the proposed graphene nanomaterial-based defected ground patch antenna. It also describes the results of the antenna, the analysis of S11, and the electric field variation for the graphene nanomaterial-based defected ground patch antenna in healthy and unhealthy tissue surroundings. Section 11.3 explains the conductivity properties of graphene, and Section 11.4 elaborates on the antenna characteristics for MWI. Both frequency and time domain analyses are performed to examine the suitability of the antenna for MWI. SAR is evaluated to check the safety of healthy cells from electromagnetic radiation.

11.2 GRAPHENE NANOMATERIAL BASED DEFECTED GROUND PATCH ANTENNA

The geometrical dimensions of the UWB antenna designed with nanomaterial graphene in the radiator and ground plane are presented in Figure 11.1. The antenna is designed on a Kapton polyimide film substrate with a permittivity (εr = 3.5) and $\tan\delta = 0.002$. The sheet resistivity of graphene varies from 0. 003 to 0.008 Ω.cm and has a nanothickness of about 100 nm. The graphene patch is coated on the upper surface of the patch, and the ground is coated on the bottom layer. The patch is excited with a microstrip line. A partial ground plane leads to a wider impedance bandwidth. The printed graphene-based antenna has higher conductivity and a thinner size, which also allows bending of the antenna on non-planar tissue structures without internal cracks or stretching in the material. It features excellent control over electrical conductivity and plasmonic properties. Besides this, graphene-integrated structures have attained better radiation characteristics than metallic structures [89–91].

The initial design of the antenna aims to obtain a resonance bandwidth greater than 6 GHz in free space. The antenna's radiating patch has slanted edges along its height (vertical axis). This design creates multiple points where the signal resonates, allowing the antenna to function over a wide range of frequencies. To ensure the antenna efficiently maintains the ultra-wide bandwidth and impedance matching, a narrow slot measuring 9.68 mm is etched into a portion of the ground plane. The exact dimensions of the antenna's various parts are detailed in Table 11.3, and its overall layout is illustrated in Figure 11.1a and b.

This antenna is designed to detect unhealthy cells in the breast. It does this by looking for changes in the signal that scatters when cancerous and non-cancerous cells are present. To test this, we created a model of breast tissue shaped like a sphere with a 30 mm radius. This model mimics the electrical properties of real breast tissue, as detailed in a different study [21]. The model is made of three layers: a thin skin layer (2 mm thick), a thicker layer of fat (28 mm thick), and a small spherical tumor (4 mm in radius) placed 5 mm inside the model. You can see a picture of this model with the tumor in Figure 11.1c.

TABLE 11.3
Geometrical Parameters of the Proposed Antenna

Parameters	Value (mm)	Parameters	Value (mm)
Ls	40	Lp	33.19
Ws	40	Lg	19.65
L_F	10	Wg1	13.6
Wp	11.6	Wg2	11.92

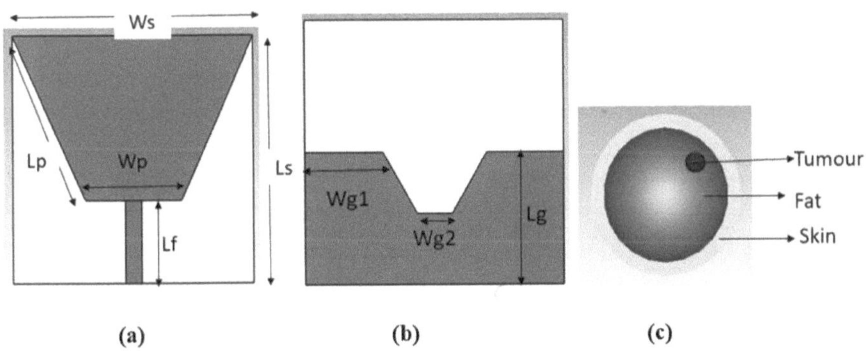

FIGURE 11.1 Antenna geometry (a) upper surface (b) lower surface view (c) breast tissue with tumor.

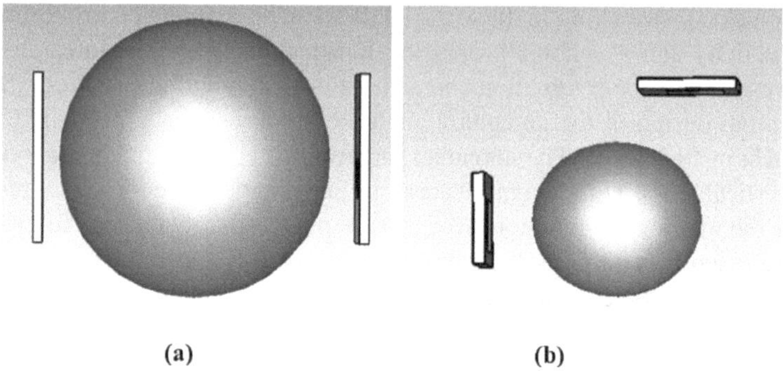

FIGURE 11.2 Tumor detection simulation setup (a) antenna positioned in front of each other (b) antenna positioned in perpendicular to each other (side by side).

Figure 11.2 shows how the antennas are set up to find tumors. Two antennas are positioned next to each other, facing one another. There's a small gap of 5 mm between the antennas and the breast model.

11.3 CONDUCTIVITY OF GRAPHENE

According to [92], graphene supports surface plasmon polariton waves in "transverse magnetic" modes, with an effective mode index (neff) described by the following equation:

$$\eta_{\mathrm{eff}}(\omega) = \sqrt{1 - \left(\frac{4\mu_o}{\varepsilon_0 \sigma \omega^2}\right)} \tag{11.1}$$

where $\sigma(\omega)$ represents the graphene conductivity. The total conductivity of graphene is given by the sum of the interband and intraband conductivity:

$$\sigma\,\mathrm{total}(\omega) = \sigma\,\mathrm{inter}(\omega) + \sigma\,\mathrm{intra}(\omega) \tag{11.2}$$

It is evaluated using the Kubo formula [17], ignoring the edge effect in graphene for a few micrometers. Interband conductivity is evaluated through the following equation:

$$\sigma_{\mathrm{inter}}(\omega) = \frac{e^2}{4h}\left(H\left(\frac{\omega}{2}\right) + i\frac{4\omega}{\pi}\int_0^{\infty} \frac{H(\varepsilon) - H\left(\frac{\omega}{2}\right)}{\omega^2 - 4\varepsilon^2}\partial\varepsilon\right) \tag{11.3}$$

$$H(\varepsilon) = \frac{\sinh\left(\frac{h\varepsilon}{K_B T}\right)}{\cosh\left(\frac{\mu_c}{K_B T}\right) + \cosh\left(\frac{h\varepsilon}{K_B T}\right)} \tag{11.4}$$

Here, intraband conductivity is evaluated through the following equation:

$$\sigma_{\text{intra}}(\omega) = \frac{2e^2 K_B T i}{\pi\hbar^2\left(\omega + i\tau^{-1}\right)} \ln\left(2\cosh\left(\frac{\mu_c}{2K_B T}\right)\right) \tag{11.5}$$

Here, K_B is the Boltzmann constant, $\hbar$ is the reduced Planck's constant, μ_c is the chemical potential, e is the electron charge, τ is the relaxation time, and T is the temperature."

Graphene is considered a very thin sheet of conducting material (thickness Δ). This material has a special property called "plasma frequency," which is given by [92,93]:

$$\omega_P = \sqrt{\frac{2e^2 K_B T}{\Delta\pi\varepsilon_0\hbar^2} \ln\left(2\cosh\left(\frac{\mu_c}{2K_B T}\right)\right)} \tag{11.6}$$

11.4 MWI THROUGH THE PROPOSED ANTENNA

This study found that healthy breast tissue has less water and a lower electrical property (permittivity) compared to cancerous tissue. Because of this difference, more of the signal scatters or reflects where the healthy and cancerous tissues meet. Figure 11.3 shows the antenna's scattered signal as the |S11| parameter. In free space, the antenna works well across a wide range of frequencies (3.1–11.5 GHz). This wide range is important because it allows the signal to reach different depths within the breast tissue. When the antenna is placed near healthy breast tissue (without a tumor), Figure 11.3 shows a clear change in the scattered-back signal. Similarly, placing a tumor in the tissue also changes the antenna's signal. Interestingly, only the frequencies between 7.0 and 7.5 GHz, and 10.0 and 10.5 GHz, aren't significantly affected by

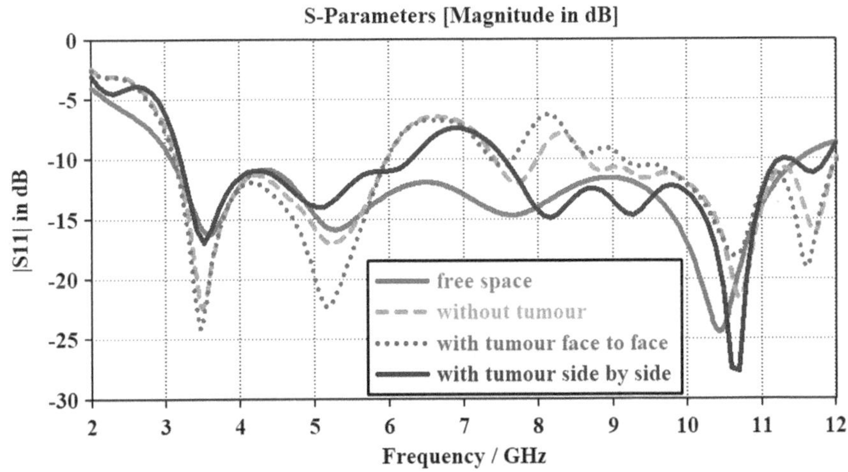

FIGURE 11.3 Backscattered |S11| of the proposed antenna.

the presence of a tumor. For almost all other frequencies, the scattered-back signal changes noticeably, which means this antenna design could be useful for detecting tumors.

The performance of two identical antennas is studied by locating them in side-by-side and front-to-front alignments (Figure 11.4). They focus on the transmission coefficient (|S21|) to understand how well the signal travels between the antennas across the entire bandwidth, considering the far-field effects. In both configurations, |S21| remains stable except for a dip at 5.0 GHz in the front-to-front setup. This behavior is similar to observations in other studies [21]. To explain the variations in power distribution within different tissue types, the authors have examined the electric field distribution (Figure 11.5). This graph reveals that power absorption by the tissue layers is not uniform, with some power being absorbed by a tumor. This uneven absorption is believed to cause deflections in the radiation patterns observed from different antenna orientations. This graph shows that the power from the antenna isn't absorbed equally by the different layers of tissue. Some of the power

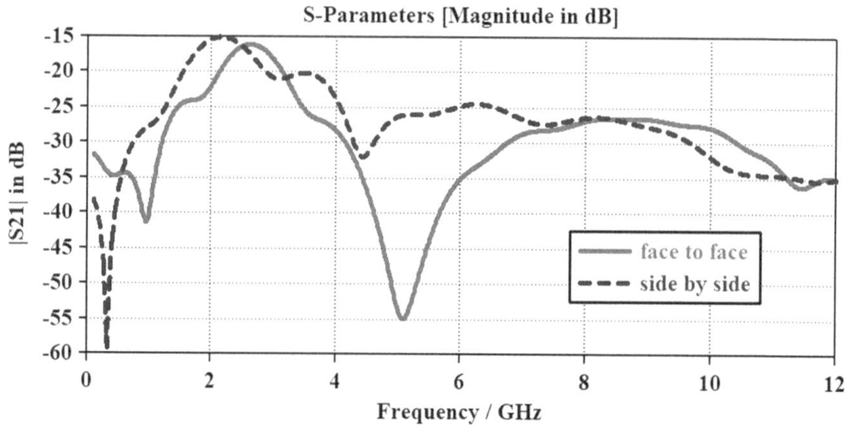

FIGURE 11.4 Transmission coefficient (|S21|) in side-by-side and face-to-face configuration.

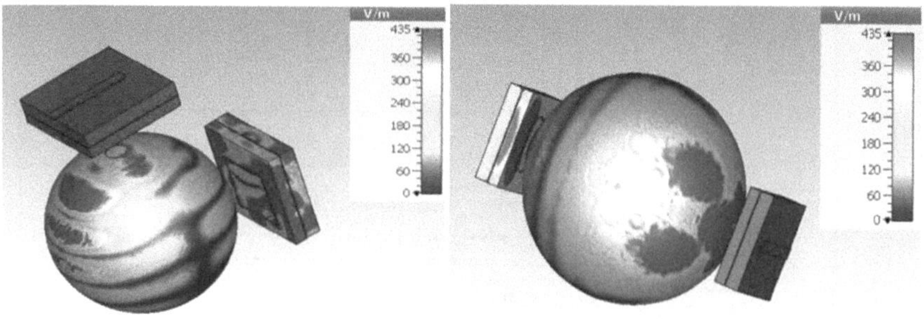

FIGURE 11.5 Electric field distribution for side-by-side and front-to-front configuration.

gets absorbed by the tumor itself. This uneven absorption is believed to be why the antenna's radiation patterns change depending on its position relative to the breast.

Despite the antenna's stable performance in the frequency domain (Figure 4), the authors investigate its time domain response for MWI applications (Figures 11.6 and 11.7). This figure shows the input signal alongside the received signals for both side by side and front-to-front arrangements, with and without a tumor present. In all cases, the received waveforms closely resemble the input signal, exhibiting only minor fluctuations, suggesting the antenna effectively transmits the signal even in the presence of a tumor.

Figure 11.8 depicts the group delay, which indicates how linearly the signal's phases travel across the channel at different frequencies. It's calculated as the negative rate of change of the phase response with frequency. The designed structure exhibits a linear phase distribution except for irregularities at 6 and 8 GHz in front-to-front

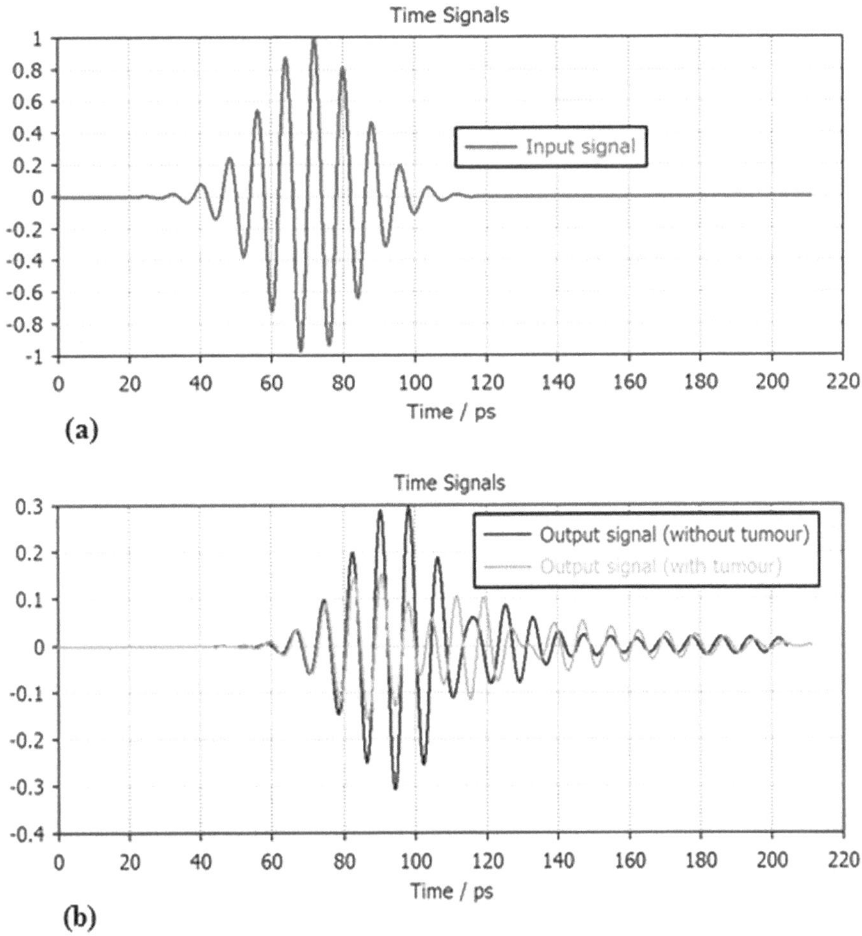

FIGURE 11.6 (a) Transmitted pulse (b) received pulse for front-to-front setup.

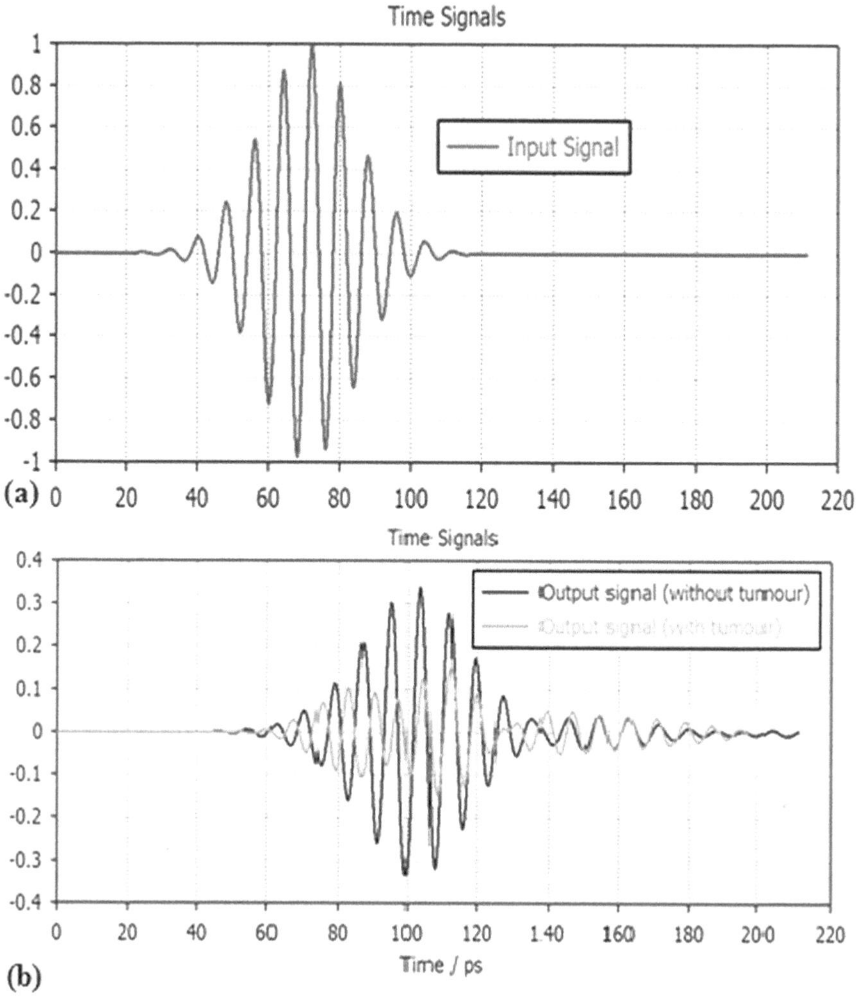

FIGURE 11.7 (a) Transmitted and (b) received pulse for side-by-side setup.

and side-by-side configurations, respectively. Notably, the group delay values (in picoseconds) for both setups fall within a similar range, making both configurations suitable for MWI applications.

As the antenna is designed for on-body use, it prioritizes minimal impact on healthy tissues during MWI procedures. Specific Absorption Rate (SAR) limits the maximum power the antenna can emit. Following safety standards, SAR is restricted to 1.6 W/Kg for 1 g of tissue. Figure 11.9 analyzes SAR values for breast tissue in the presence and absence of unhealthy cells. The results demonstrate sufficient variation in microwave power penetration into the tissue under both conditions. Furthermore, Figure 11.9a and b show that with an input power of 0.1 W, the simulated SAR on

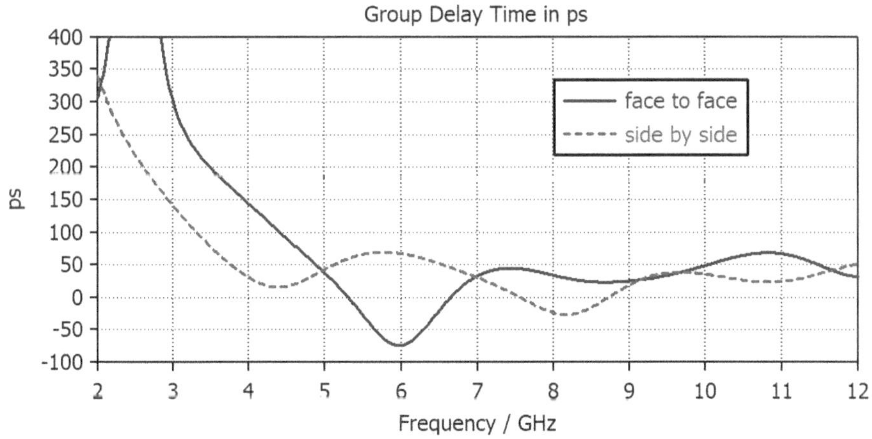

FIGURE 11.8 Group delay of the designed prototype.

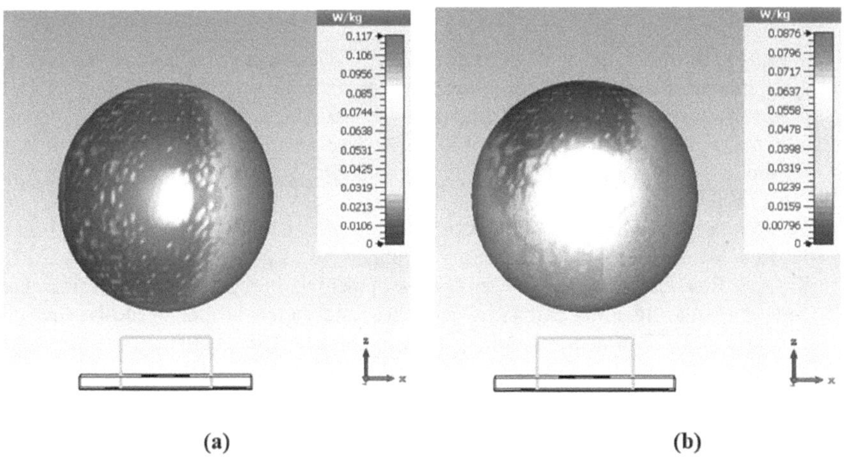

FIGURE 11.9 Simulated SAR plot at 5.0 GHz (a) healthy structure (b) unhealthy tissue structure.

1 g of tissue is 0.117 and 0.0876 W/Kg for healthy and unhealthy cells. Both values are well under the safe limits, ensuring the suitability of the proposed structure for operation in close proximity to the body and safe application in healthcare settings.

11.5 CONCLUSIONS

Carbon-based nanomaterials like graphene are attracting attention for their potential in antenna technologies at microwave frequencies due to their unique electrical and structural properties. Microwave radiation holds promise for applications like non-destructive imaging and communication. A key challenge involves efficiently controlling the amplitude of GHz waves, particularly for developing powerful and tunable frequency sources. Graphene, with its exceptional mobility arising from its massless Dirac fermions, has been identified as a promising candidate for ultra-broadband frequency sources. Researchers modified a standard MSA to attain a wide bandwidth (at least 10 dB below the peak response) suitable for UWB technologies. The design incorporates a defected ground structure with tapered slots. The antenna's performance for MWI was evaluated using a tissue phantom, demonstrating good coupling with the heterogeneous tissue. Backscattered signal variations were observed due to deviations in the electrical characteristics of the tissue, which can be used for image reconstruction. While the antenna exhibits stable performance, some dissimilarities in the |S11| parameter were observed, with an extreme deviation of 6 dB at 5.2 GHz (front to front) and 8 dB deviations at 8.1 and 10.8 GHz (side by side). The study also analyzed the non-uniform power distribution through electric field visualization and the linearity of the timing signals using group delay, received, and transmitted pulses. Finally, the study explored how power distribution varied within the tissue using electric field visualization and the consistency of timing signals.

REFERENCES

1. Awan, D., Bashir, S., Khan, S., Al-Bawri, S. S., & Dalarsson, M. (2024). UWB antenna with enhanced directivity for applications in microwave medical imaging. *Sensors (Basel, Switzerland)*, 24(4), 1315. https://doi.org/10.3390/s24041315
2. Abdulla, F. A. A., & Demirkol, A. (2024). A novel textile-based UWB patch antenna for breast cancer imaging. *Physical and Engineering Sciences in Medicine*. https://doi.org/10.1007/s13246-024-01409-w
3. Feng, Y., Spezia, M., Huang, S., Yuan, C., Zeng, Z., Zhang, L., Ji, X., Liu, W., Huang, B., Luo, W., Liu, B., Lei, Y., Du, S., Vuppalapati, A., Luu, H. H., Haydon, R. C., He, T.-C., & Ren, G. (2018). Breast cancer development and progression: risk factors, cancer stem cells, signaling pathways, genomics, and molecular pathogenesis. *Genes & Diseases*, 5(2), 77–106. https://doi.org/10.1016/j.gendis.2018.05.001
4. Arnold, M., Morgan, E., Rumgay, H., Mafra, A., Singh, D., Laversanne, M., Vignat, J., Gralow, J. R., Cardoso, F., Siesling, S., & Soerjomataram, I. (2022). Current and future burden of breast cancer: global statistics for 2020 and 2040. *Breast (Edinburgh, Scotland)*, 66, 15–23. https://doi.org/10.1016/j.breast.2022.08.010
5. Sung, H., Ferlay, J., Siegel, R. L., Laversanne, M., Soerjomataram, I., Jemal, A., & Bray, F. (2021). Global cancer statistics 2020: GLOBOCAN estimates of incidence and mortality worldwide for 36 cancers in 185 countries. *CA: A Cancer Journal for Clinicians*, 71(3), 209–249. https://doi.org/10.3322/caac.21660
6. Siegel, R. L., Miller, K. D., Wagle, N. S., & Jemal, A. (2023). Cancer statistics, 2023. *CA: A Cancer Journal for Clinicians*, 73(1), 17–48. https://doi.org/10.3322/caac.21763
7. Redig, A. J., & McAllister, S. S. (2013). Breast cancer as a systemic disease: a view of metastasis. *Journal of Internal Medicine*, 274(2), 113–126. https://doi.org/10.1111/joim.12084
8. Nover, A. B., Jagtap, S., Anjum, W., Yegingil, H., Shih, W. Y., Shih, W.-H., & Brooks, A. D. (2009). Modern breast cancer detection: a technological review. *International Journal of Biomedical Imaging*, 2009(1), 902326. https://doi.org/10.1155/2009/902326
9. Inderbir, A. K., Rehni, R., Kalra, G., Joshi, M., & Kumar, H. Y. (2007). Ion exchange resins: drug delivery and therapeutic applications. *Fabad Journal of Pharmaceutical Sciences*, 32(2), 91–100.

10. Hasan, A. M., Al-Waely, N. K., Ajobouri, H. K., Ibrahim, R. W., Jalab, H. A., & Meziane, F. (2023). A classifcation model of breast masses in DCE-MRI using kinetic curves features with Quantum-Raina's polynomial based fusion. *Biomed Signal Process Control*, 84.
11. Kumar, R., Ratnesh, R. K., Singh, J., Kumar, A., & Chandra, R. (2024). IoT-driven experimental framework for advancing electrical impedance tomography. *ECS Journal of Solid State Science and Technology. JSS*, 13(2), 027002. https://doi.org/10.1149/2162-8777/ad2331
12. Kumar, R., Ratnesh, R. K., Singh, J., Chandra, R., Singh, G., & Vishnoi, V. (2023). Recent prospects of medical imaging and sensing technologies based on electrical impedance data acquisition system. *Journal of the Electrochemical Society*, 170(11), 117507. https://doi.org/10.1149/1945-7111/ad050f
13. Kwon, S., & Lee, S. (2013). Instantaneous microwave imaging with time-domain measurements for breast cancer detection. *Electronics Letters*, 49(10), 639–641. https://doi.org/10.1049/el.2013.0248
14. Paulsen, K. D., Poplack, S. P., Li, D., Fanning, M. W., & Meaney, P. M. (2000). A clinical prototype for active microwave imaging of the breast. *IEEE Transactions on Microwave Theory and Techniques*, 48(11), 1841–1853. https://doi.org/10.1109/22.883861
15. Smith, K., Bourqui, J., Garrett, D., Zarnke, S., Owjimehr, M., D. Deutscher, Fung, T., & Fear, E. (2022). Microwave imaging of the breast: consistency of measurements over time. *IEEE Journal of Electromagnetics, RF and Microwaves in Medicine and Biology*, 6(1), 61–67. https://doi.org/10.1109/jerm.2021.3099014
16. Pastorino, M. (2004). Recent inversion procedures for microwave imaging in biomedical, subsurface detection and nondestructive evaluation applications. *Measurement: Journal of the International Measurement Confederation*, 36(3–4), 257–269. https://doi.org/10.1016/j.measurement.2004.09.006
17. Kumar, R., Kumar, S., & Sengupta, A. (2022). Optimization of bio-impedance techniques-based monitoring system for medical & industrial applications. *IETE Journal of Research*, 68(5), 3843–3854. https://doi.org/10.1080/03772063.2020.1780957
18. Benny, R., Anjit, T. A., & Mythili, P. (2020). An overview of microwave imaging f or breast tumor detection. *Progress in Electromagnetics Research B*, 87, 61–91.
19. Oloughlin, D., Halloran, O., Moloney, M., Glavin, B. M., Jones, M., & Elahi, E. (2018). Microwave breast imaging: clinical advances and remaining challenges. *IEEE Transactions on Biomedical Engineering*, 65(11), 2580–2590.
20. Nikolova, N. (2011). Microwave imaging for breast cancer. *IEEE Microwave Magazine*, 12(7), 78–94. https://doi.org/10.1109/mmm.2011.942702
21. Porter, E., Bahrami, H., Santorelli, A., Gosselin, B., Rusch, L. A., & Popovic, M. (2016). A wearable microwave antenna array for time-domain breast tumor screening. *IEEE Transactions on Medical Imaging*, 35(6), 1501–1509. https://doi.org/10.1109/TMI.2016.2518489
22. Abdullah, M. Z., Binajjaj, S. A., Zanoon, T. F., & Peyton, A. J. (2011). Highresolution imaging of dielectric profles by using a time-domain ultra wideband radar sensor. *Measurement*, 44(5), 859–870.
23. Fear, E. C., Meaney, P. M., & Stuchly, M. A. (2003). Microwaves for breast cancer detection? *IEEE Potentials*, 22(1), 12–18. https://doi.org/10.1109/mp.2003.1180933
24. Sachs, J. (2013). Handbook of Ultra-Wideband Short-Range Sensing: Theory, Sensors and Applications. New Jersey: John Wiley & Sons.
25. Islam, M., Mahmud, M., Islam, M. T., Kibria, S., & Samsuzzaman, M. (2019). A low cost and portable microwave imaging system for breast tumor detection using UWB directional antenna array. *Scientific Reports*, 9(1), 15491.

26. Holland, K. M., Jones, C., Vivolo-Kantor, A. M., Idaikkadar, N., Zwald, M., Hoots, B., Yard, E., D'Inverno, A., Swedo, E., Chen, M. S., Petrosky, E., Board, A., Martinez, P., Stone, D. M., Law, R., Coletta, M. A., Adjemian, J., Thomas, C., Puddy, R. W., … Houry, D. (2021). Trends in US emergency department visits for mental health, overdose, and violence outcomes before and during the COVID-19 pandemic. *JAMA Psychiatry (Chicago, Ill.)*, 78(4), 372–379. https://doi.org/10.1001/jamapsychiatry.2020.4402
27. Akintunde, T. Y., Akintunde, O. D., Musa, T. H., Sayibu, M., Tassang, A. E., Reed, L. M., & Chen, S. (2021). Expanding telemedicine to reduce the burden on the healthcare systems and poverty in Africa for a post-coronavirus disease 2019 (COVID-19) pandemic reformation. *Global Health Journal (Amsterdam, Netherlands)*, 5(3), 128–134. https://doi.org/10.1016/j.glohj.2021.07.006
28. Queyam, A. B., Kumar, R., Ratnesh, R. K., & Chauhan, R. K. (2024). LabVIEW-enabled synthetic signal for empowering fetal-maternal healthcare. *ECS Journal of Solid State Science and Technology*, 13(5). https://doi.org/10.1149/2162-8777/ad4dde
29. Kumar, R., & Singh, S. (2023). CPW fed conformal PIFA design for implantable IoMT devices with wideband performance. *IEEE Sensors Journal*, 1–1. https://doi.org/10.1109/jsen.2023.3329416
301. Bahrami, H., Porter, E., Santorelli, A., Gosselin, B., Popovic, M., & Rusch, L. A. (2014). Flexible sixteen monopole antenna array for microwave breast cancer detection. Annual International Conference of the IEEE Engineering in Medicine and Biology Society. IEEE Engineering in Medicine and Biology Society. Annual International Conference, 2014, Chicago, IL, USA, 3775–3778. https://doi.org/10.1109/EMBC.2014.6944445
31. Srinivasan, D., & Gopalakrishnan, M. (2019). Breast cancer detection using adaptable textile antenna design. *Journal of Medical Systems*, 43(6), 177. https://doi.org/10.1007/s10916-019-1314-5
32. Loss, C., Goncalves, R., Lopes, C., Salvado, R., & Pinho, P. (2016). Textile antenna for RF energy harvesting fully embedded in clothing. 2016 10th European Conference on Antennas and Propagation (EuCAP), Davos, Switzerland.
33. Saha, P., Mitra, D., & Parui, S. K. (2018). A circularly polarised implantable monopole antenna for biomedical applications. *Progress in Electromagnetics Research C. Pier C*, 85, 167–175. https://doi.org/10.2528/pierc18051807
34. Lin, X., Chen, Y., Gong, Z., Seet, B.-C., Huang, L., & Lu, Y. (2020). Ultrawideband textile antenna for wearable microwave medical imaging applications. *IEEE Transactions on Antennas and Propagation*, 68(6), 4238–4249. https://doi.org/10.1109/tap.2020.2970072
35. Sharma, N., & Bhatia, S. S. (2021). Design of printed UWB antenna with CPW and microstrip-line-fed for DCS/PCS/bluetooth/WLAN wireless applications. *International Journal of RF and Microwave Computer-Aided Engineering*, 31(1), e22488.
36. El Atrash, M., Abdalla, M. A., & Elhennawy, H. M. (2021). A fully-textile wideband AMC-backed antenna for wristband WiMAX and medical applications. *International Journal of Microwave and Wireless Technologies*, 13(6), 624–633. https://doi.org/10.1017/s1759078720001397
37. Elsheikh, D., & Eldamak, A. R. (2021). Microwave textile sensors for breast cancer detection. *2021 38th National Radio Science Conference (NRSC)*, Mansoura.
38. Chandra, R., Zhou, H., Balasingham, I., & Narayanan, R. M. (2015). On the opportunities and challenges in microwave medical sensing and imaging. *IEEE Transactions on Bio-Medical Engineering*, 62(7), 1667–1682. https://doi.org/10.1109/TBME.2015.2432137
39. Balanis, C. A. (2015). *Antenna Theory: Analysis and Design*. New Jersey: John Wiley & Sons.

40. Meaney, P. M., Fanning, M. W., Raynolds, T., Fox, C. J., Fang, Q., Kogel, C. A., Poplack, S. P., & Paulsen, K. D. (2007). Initial clinical experience with microwave breast imaging in women with normal mammography. *Academic Radiology*, 14(2), 207–218. https://doi.org/10.1016/j.acra.2006.10.016
41. Kang, K., Chu, X., Dilmaghani, R., & Ghavami, M. (2007). Low-complexity cole-cole expression for modelling human biological tissues in (FD) 2TD method. *Electronics Letters*, 43(3), 143–144.
42. Lazebnik, M., Okoniewski, M., Booske, J. H., & Hagness, S. C. (2007). Highly accurate Debye models for normal and malignant breast tissue dielectric properties at microwave frequencies. *IEEE Microwave and Wireless Components Letters: A Publication of the IEEE Microwave Theory and Techniques Society*, 17(12), 822–824. https://doi.org/10.1109/lmwc.2007.910465
43. Hasgall, P. A., Di Gennaro, F., Baumgartner, C., Neufeld, E., Lloyd, B., Gosselin, M. C., Payne, D., Klingenböck, A., & Kuster, N. (2022). IT'ISDatabase for thermal and electromagnetic parameters of biological tissues, Version 4.1. https://doi.org/10.13099/VIP21000-04-1
44. O'Halloran, M., Byrne, D., Conceição, R. C., Jones, E., & Glavin, M. (2016). Anatomy and dielectric properties of the breast and breast cancer. In *An Introduction to Microwave Imaging for Breast Cancer Detection*, 5-16.
45. Hossain, K., Sabapathy, T., Jusoh, M., Lee, S.-H., Rahman, K., & Kamarudin, M. R. (2022). Negative index metamaterial-based frequency-reconfgurable textile CPW antenna for microwave imaging of breast cancer. *Sensors*, 22(4), 1626.
46. Gupta, A., Kumar, V., Garg, D., Alsharif, M. H., & Jahid, A. (2023). Performance analysis of an aperture-coupled THz antenna for diagnosing breast cancer. *Micromachines*, 14(7). https://doi.org/10.3390/mi14071281
47. Gupta, A., Kansal, A., Chawla, P. (2021). An active and low-cost microwave imaging system design for detection of breast cancer using back scattered signal. *Current Medical Imaging*. https://doi.org/10.2174/1573405617666210129114536
48. Mahmud, M., Islam, M. T., Misran, N., Singh, M. J., & Mat, K. (2017). A negative index metamaterial to enhance the performance of miniaturized UWB antenna for microwave imaging applications. *Applied Sciences*, 7(11), 1149.
49. Islam, M., Islam, M. T., Samsuzzaman, M., Faruque, M. R., Misran, N., & Mansor, M. F. (2015). A miniaturized antenna with negative index metamaterial based on modified SRR and CLS unit cell for UWB microwave imaging applications. *Materials*, 8(2), 392–407.
50. Zhang, J., Fear, E. C., & Johnston, R. H. (2009). Cross-Vivaldi antenna for breast tumor detection. *Microwave and Optical Technology Letters*, 51(2), 275–280.
51. Abbak, M., & Çayören, M. (1127). Akduman, I. Microwave breast phantom measurements with a cavity-backed Vivaldi antenna. *IET Microwaves, Antennas & Propagation*, 8(3), 1127–1133.
52. Jafari, H. M., Deen, J. M., Hranilovic, S., & Nikolova, N. K. (2007). Co-polarised and cross-polarised antenna arrays for breast, cancer detection. *IET Microwaves Antennas & Propagation*, 1(5), 1055. https://doi.org/10.1049/iet-map:20060327
53. Mahmud, M. Z., Islam, M. T., Misran, N., Kibria, S., & Samsuzzaman, M. (2018). Microwave imaging for breast tumor detection using uniplanar AMC based CPW-fed microstrip antenna. *IEEE Access: Practical Innovations, Open Solutions*, 6, 44763–44775. https://doi.org/10.1109/access.2018.2859434
54. Mohammed, B. J., Abbosh, A. M., & Sharpe, P. (2013). Planar array of corrugated tapered slot antennas for ultrawideband biomedical microwave imaging system: planar array for microwave imaging system. *International Journal of RF and Microwave Computer-Aided Engineering*, 23(1), 59–66. https://doi.org/10.1002/mmce.20651

55. Molaei, A., Kaboli, M., Mirtaheri, S. A., & Abrishamian, M. S. (2014). Dielectric lens balanced antipodal Vivaldi antenna with low cross-polarisation for ultra-wideband applications. *IET Microwaves Antennas & Propagation*, 8(14), 1137–1142. https://doi.org/10.1049/iet-map.2014.0207
56. Abbak, Mehmet, Çayören, M., & Akduman, İ. (2014). Microwave breast phantom measurements with a cavity-backed Vivaldi antenna. *IET Microwaves Antennas & Propagation*, 8(13), 1127–1133. https://doi.org/10.1049/iet-map.2013.0484
57. Pandey, G. K., Verma, H., & Meshram, M. K. (2015). Compact antipodal Vivaldi antenna for UWB applications. *Electronics Letters*, 51(4), 308–310. https://doi.org/10.1049/el.2014.3540
58. Chiappe, M., & Gragnani, G. L. (2006). Vivaldi antennas for microwave imaging: theoretical analysis and design considerations. *IEEE Transactions on Instrumentation and Measurement*, 55(6), 1885–1891. https://doi.org/10.1109/tim.2006.884289
59. Kanj, H., & Popovic, M. (2005). Miniaturized microstrip-fed dark eyes antenna for near-field microwave sensing. *IEEE Antennas and Wireless Propagation Letters*, 4, 397–401.
60. Mahmud, M. Z., Islam, M. T., & Samsuzzaman, M. (2016). A high performance UWB antenna design for microwave imaging system. *Microwave and Optical Technology Letters*, 58(8), 1824–1831.
61. Kanj, H., & Popovic, M. (2008). A novel ultra-compact broadband antenna for microwave breast tumor detection. *Progress in Electromagnetics Research*, 86(9), 169–198.
62. Nilavalan, R., Craddock, I. J., Preece, A., Leendertz, J., & Benjamin, R. (2007). Wideband microstrip patch antenna design for breast cancer tumour detection. *IET Microwaves Antennas & Propagation*, 1(2), 277. https://doi.org/10.1049/iet-map:20050189
63. Janaswamy, R., & Schaubert, D. (1987). Analysis of the tapered slot antenna. *IEEE Transactions on Antennas and Propagation*, 35, 1058–1065.
64. Wu, B., Ji, Y., & Fang, G. (2009). Electronic measurement & instruments, ICEMI'09. 9th International Conference, Beijing, China, IEEE.
65. Nassar, I. T., & Weller, T. M. (2015). A novel method for improving antipodal Vivaldi antenna performance. *IEEE Transactions on Antennas and Propagation*, 63(7), 3321–3324. https://doi.org/10.1109/tap.2015.2429749
66. Ojaroudi, N., Ojaroudi, M., & Ghadimi, N. (2012). UWB omnidirectional square monopole antenna for use in circular cylindrical microwave imaging systems. *IEEE Antennas and Wireless Propagation Letters*, 11, 1350–1353. https://doi.org/10.1109/lawp.2012.2227137
67. Subramanian, S., Sundarambal, B., & Nirmal, D. (2018). Investigation on simulation-based specific absorption rate in ultra-wideband antenna for breast cancer detection. *IEEE Sensors Journal*, 18(24), 10002–10009. https://doi.org/10.1109/jsen.2018.2875621
68. Tiang, S. S., Sadoon, M., Zanoon, T. F., Ain, M. F., & Abdullah, M. Z. (2013). Radar sensing featuring biconical antenna and enhanced delay and sum algorithm for early stage breast cancer detection. *Progress in Electromagnetics Research B. Pier B*, 46, 299–316. https://doi.org/10.2528/pierb12102201
69. Moussakhani, K., Amineh, R. K., & Nikolova, N. K. (2011). High-efficiency TEM horn antenna for ultra-wide band microwave tissue imaging. *2011 IEEE International Symposium on Antennas and Propagation (APSURSI)*, Spokane, WA, USA.
70. Amineh, R. K., Trehan, A. & Nikolova, N. K. (2009). TEM horn antenna for ultra-wide band microwave breast imaging. *Progress in Electromagnetics Research*, 13(3), 59–74.
71. Nepote, M. S., Herrera, D. R., Tapia, D. F., Latif, S., & Pistorius, S. (2014). A comparison study between horn and vivaldi antennas for 1.5–6 GHz breast microwave radar imaging. *Proceedings of European Conference on Antennas and Propagation (EuCAP)*, The Hague, The Netherlands, 59–62.

72. Li, X., Hagness, S. C., Choi, M. K., & van der Weide, D. W. (2003). Numerical and experimental investigation of an ultrawideband ridged pyramidal horn antenna with curved launching plane for pulse radiation. *IEEE Antennas and Wireless Propagation Letters*, 2, 259–262. https://doi.org/10.1109/lawp.2003.820708
73. Yun, X., Fear, E. C., & Johnston, R. H. (2005). Compact antenna for radar-based breast cancer detection. *IEEE Transactions on Antennas and Propagation*, 53(8), 2374–2380. https://doi.org/10.1109/tap.2005.852308
74. Jalilvand, M., Vasanelli, C., Wu, C., Kowalewski, J., & Zwick, T. (2014). On the evaluation of a proposed bowtie antenna for microwave tomography. The 8th European Conference on Antennas and Propagation (EuCAP 2014), The Hague, The Netherlands.
75. See, C. H., Abd-Alhameed, R. A., Chung, S. W. J., Zhou, D., Al-Ahmad, H., & Excell, P. S. (2012). The design of a resistively loaded bowtie antenna for applications in breast cancer detection systems. *IEEE Transactions on Antennas and Propagation*, 60(5), 2526–2530. https://doi.org/10.1109/tap.2012.2189730
76. Lestari, A. A., Yarovoy, A. G., & Ligthart, L. P. (2004). RC-loaded bow-tie antenna for improved pulse radiation. *EEE Transactions on Antennas and Propagation*, 52(10), 2555–2563.
77. Shannon, C. J., Fear, E. C., & Okoniewski, M. (2005). Dielectric-filled slotline bowtie antenna for breast cancer detection. *Electronics Letters*, 41(7), 388–390.
78. Ambalg, A. P., & Sujata, S. K. (2021). Analysis of 8.71 GHz operated gold nanoparticles based multi-slotted patch antenna etched on epoxy dielectric material having broad-sided radiation characteristics. *Materials Today: Proceedings*, 43, 3810–3814. https://doi.org/10.1016/j.matpr.2020.10.997
79. Ambalgi, A. P., Kamalapurkar, S. S., Hunagund, P. V., & Patil, R. (2021). Design, modeling and experimental study of GHz patch antenna coated with conductive layer of nanomaterial for enhanced characteristics with defected ground structure in communication network applications. *SN Computer Science*, 2(3), 152.
80. Kashyap, P. A., Sarmah, K., Dakua, I., & Baruah, S. (2023). Gain and bandwidth enhancement of slotted microstrip antenna using metallic nanofilms for WLAN applications. *Journal of King Saud University. Science*, 35(1), 102374. https://doi.org/10.1016/j.jksus.2022.102374
81. Khan, M. A. H., Ali, S., Bae, J., & Lee, C. H. (2016). Inkjet printed transparent and bendable patch antenna based on polydimethylsiloxane and indium tin oxide nanoparticles. *Microwave and Optical Technology Letters*, 58(12), 2884–2887. https://doi.org/10.1002/mop.30171
82. Matyas, J., Slobodian, P., Munster, L., Olejnik, R., & Urbanek, P. (2017). Microstrip antenna from silver nanoparticles printed on a flexible polymer substrate. *Materials Today: Proceedings*, 4(4), 5030–5038. https://doi.org/10.1016/j.matpr.2017.04.110
83. Patil, R. R., Vani, R. M., & Hunagund, P. V. (2013). Design and simulation of nanotechnology based proximity coupled patch antenna at X-band. *International Journal of Advanced Research in Computer and Communication Engineering*, 2(9), 3344–3348.
84. Katsounaros, A., Cole, M. T., Tuncer, H. M., Milne, W. I., & Hao, Y. (2013). Near-field characterization of chemical vapor deposition graphene in the microwave regime. *Applied Physics Letters*, 102, 233104.
85. Huang, X., Leng, T., Zhu, M., Zhang, X., Chen, J., Chang, K., Aqeeli, M., Geim, A. K., Novoselov, K. S. & Hu, Z. (2015). Highly flexible and conductive printed graphene for wireless wearable communications applications. *Scientific Reports*, 5, 18298. https://doi.org/10.1038/srep18298
86. Leng, T., Huang, X., Chang, K., Chen, J., Abdalla, M. A., & Hu, Z. (2016). Graphene nanoflakes printed flexible meandered-line dipole antenna on paper substrate for low-cost RFID and sensing applications. *IEEE Antennas and Wireless Propagation Letters*, 15, 1565–1568. https://doi.org/10.1109/LAWP.2016.2518746

87. Akbari, M., Khan, M. W. A., Hasani, M., Björninen, T., Sydanheimo, L., Ukkonen, L., & Khan, M. W. A. (2015). Fabrication and characterization of graphene antenna for low-cost and environmentally friendly RFID tags. *IEEE Antennas and Wireless Propagation Letters*, 15, 1569–1572. https://doi.org/10.1109/LAWP.2015.2498944
88. Tang, D., Wang, Q., Wang, Z., Liu, Q., Zhang, B., He, D., Wu, Z., & Mu, S. (2018). Highly sensitive wearable sensor based on a flexible multi-layer graphene film antenna. *Science Bulletin*, 63, 574–579. https://doi.org/10.1016/j.scib.2018.03.015
89. Fan, C., Wu, B., Song, R., Zhao, Y., Zhang, Y., & He, D. (2019). Electromagnetic shielding and multi-beam radiation with high conductivity multilayer graphene film. *Carbon*, 155, 506–513. https://doi.org/10.1016/j.carbon.2019.06.062
90. Sa'don, S. N. H., Jamaluddin, M. H., Kamarudin, M. R., Ahmad, F., Yamada, Y., Kamardin, K., & Idris, I. H. (2019). Analysis of graphene antenna properties for 5G applications. *Sensors*, 19(22), 4835.
91. Castro, E. V., Novoselov, K. S., Morozov, S. V., Peres, N. M. R., Santos, J. M. B. L. dos, Nilsson, J., Guinea, F., Geim, A. K., & Neto, A. H. C. (2008). Electronic properties of a biased graphene bilayer. *In arXiv [cond-mat.mes-hall]*. https://arxiv.org/abs/0807.3348
92. Hanson, G. W. (2008). Dyadic green's functions for an anisotropic, non-local model of biased graphene. *IEEE Transactions on Antennas and Propagation*, 56(3), 747–757. https://doi.org/10.1109/tap.2008.917005
93. Mikhailov, S. A., & Ziegler, K. (2007). New electromagnetic mode in graphene. *Physical Review Letters*, 99(1), 016803. https://doi.org/10.1103/PhysRevLett.99.016803

12 Nanoscale Industry 4.0 Empowers Advanced Nanomaterials and Devices

Vineet Vishnoi, Shivam Chaudhary, Neeraj Joshi, and Prakhar Consul

12.1 INTRODUCTION TO INDUSTRY 4.0 AND NANOTECHNOLOGY

Introduction: Industry 4.0, often referred to as the fourth industrial revolution, represents a significant shift in the way industries operate, introducing a new level of automation and data exchange in manufacturing technologies. It incorporates advanced technologies such as the Internet of Things (IoT), artificial intelligence [1].

(AI), robotics, cloud computing, and big data analytics. Nanotechnology, on the other hand, is a field of research and innovation focused on building materials and devices at the scale of atoms and molecules, typically within the range of 1–100 nm. This technology leverages the unique properties of materials at the nanoscale to create highly specialized products with applications across medicine, electronics, energy, and materials science [2]. The framework of the chapter is given in Figure 12.1.

Industry 4.0: Industry 4.0 refers to the current trend of automation and data exchange in manufacturing technologies. It represents the fourth industrial revolution, following the previous revolutions brought by steam and water power, electricity and assembly lines, and computerization. Industry 4.0 is characterized by a fusion of technologies that blur the lines between the physical, digital, and biological spheres. The goal of Industry 4.0 is to create smart factories where machines and systems are interconnected and can communicate with each other, leading to more efficient, flexible, and customizable production processes [2,3]. This revolution aims to increase productivity, improve industrial processes, reduce waste, and enhance manufacturing flexibility. Different stages of industrial revolution from 1.0 to 5.0 are shown in Figure 12.2.

DOI: 10.1201/9781003634553-12

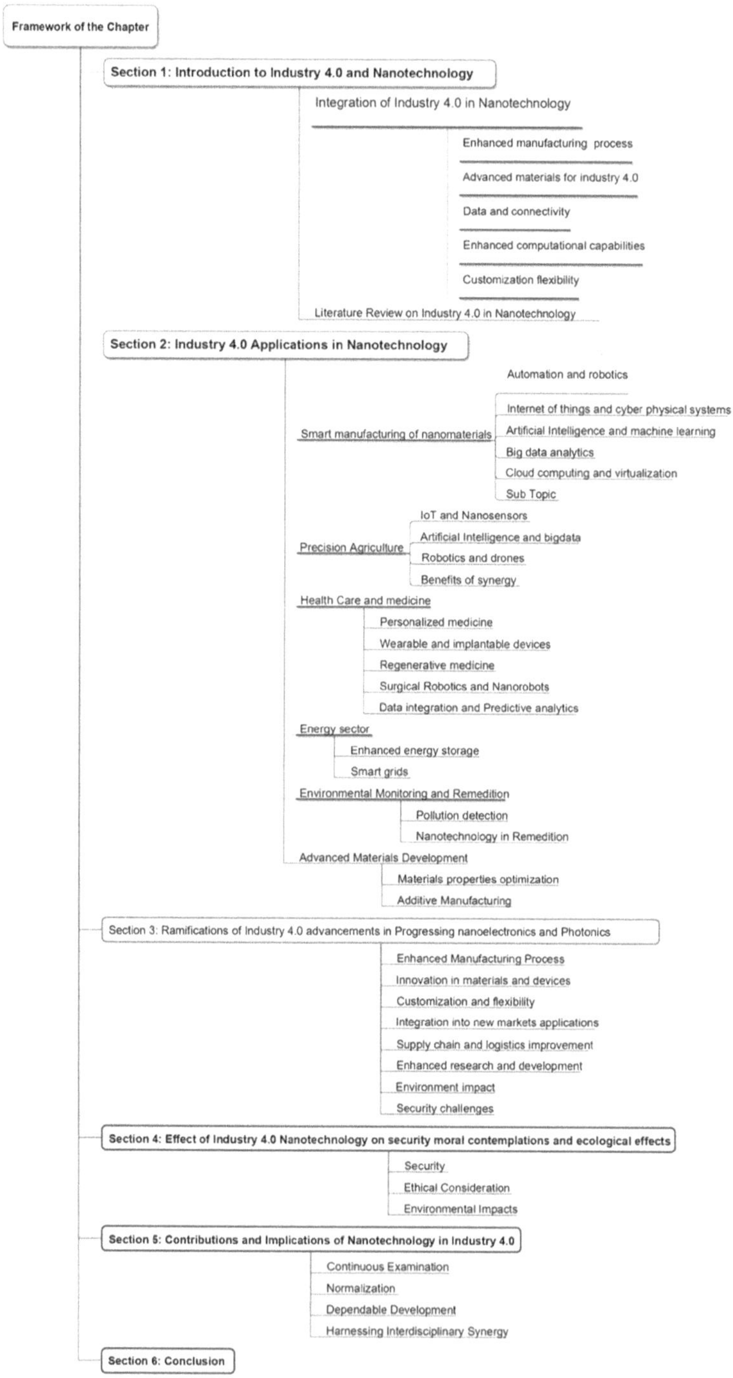

FIGURE 12.1 Framework of the chapter.

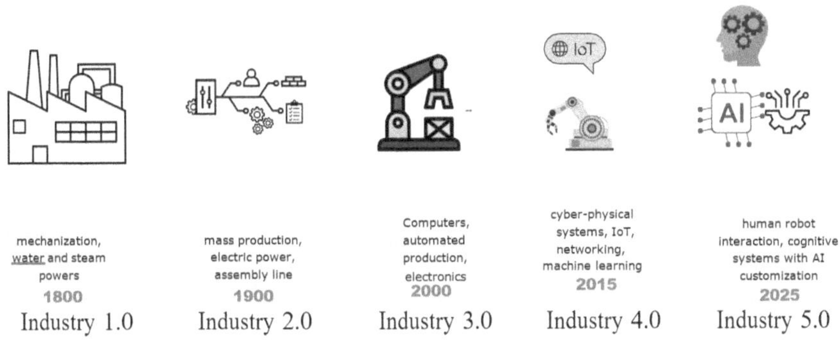

FIGURE 12.2 Industry 4.0.

Nanotechnology: Nanotechnology is the science, engineering, and application of materials and devices whose structures and components exhibit significantly improved physical, chemical, and biological properties, phenomena, and processes due to their nanoscale size. This field involves the manipulation and control of matter on an atomic, molecular, and supramolecular scale, generally within the range of 1–100 nm (a nanometer is one-billionth of a meter). The unique aspect of nanotechnology lies in the fact that at such small scales, materials begin to exhibit novel properties not observable in their bulk counterparts. These properties are a result of the increased surface area relative to volume, quantum effects, and the specific dynamics of how atoms interact at such scales. Nanotechnology spans multiple disciplines, including physics, chemistry, materials science, and biology, leading to applications across a broad range of fields [3,4,5]. Nanotechnology's interdisciplinary approach and the broad potential for impact make it a key area of research and development for future

Technological advances hold the promise of significant societal, environmental, and economic benefits. Different generations of nanotechnology are shown in Figure 12.3.

12.2 INTEGRATION OF INDUSTRY 4.0 IN NANOTECHNOLOGY

The integration of Industry 4.0 technologies with nanotechnology represents a frontier in both manufacturing and product innovation, bringing forth unprecedented opportunities for efficiency, customization, and new capabilities in various industries [6,7]. This integration is a two-way street: not only does nanotechnology benefit from Industry 4.0's advanced digital infrastructure, but Industry 4.0 also sees advancements through the incorporation of nanotechnological innovations [8,9,10]. A comparison of different costs in previous industry models, Industry 4.0, traditional technology, and nanotechnology is given in Table 12.1. Here is how this synergy unfolds:

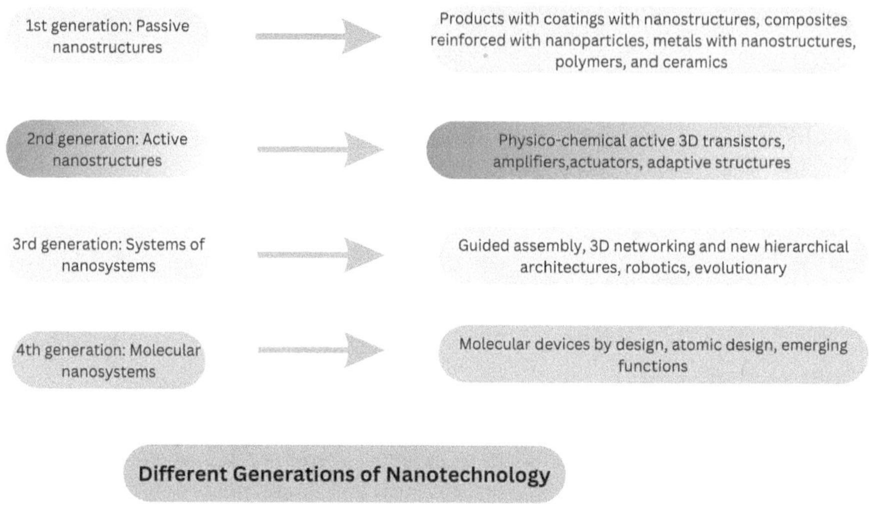

FIGURE 12.3 Nanotechnology generations.

TABLE 12.1
Comparison of Cost

Aspect	Previous Industry Models	Industry 4.0	Traditional Technology	Nanotechnology
Labor costs	High due to manual labor	Reduced due to automation and robotics	High due to manual and bulk production	Lower due to precise nanoscale processes
Production costs	Higher due to inefficiencies and downtime	Lower with optimized and continuous production	Higher due to less precise methods	Lower as nanoscale processes are refined
Material costs	Higher due to waste and inefficient use	Lower with precise control and reduced waste	Higher due to bulk usage	Lower due to efficient material use
Quality control	Costly with manual inspections	Enhanced with real-time monitoring and AI	Manual inspections increase costs	Improved with nanosensors and diagnostics
Development costs	High with longer R and D cycles	High initial setup, lower long-term costs	High with larger scale manufacturing	Initially high, decrease over time
Environmental costs	Higher due to waste and inefficiencies	Lower due to precision and reduced waste	Higher due to waste	Potentially lower with eco-friendly materials
Applications	Manual processes and limited automation	Smart manufacturing, IoT, autonomous robots	Bulk production and less customization	Medicine, electronics, energy, materials

12.2.1 Enhanced Manufacturing Processes

Precision and control: The precise control mechanisms of cyber-physical systems in Industry 4.0 enable the accurate manipulation of materials at the nanoscale, improving the quality and performance of nanomaterials and nanodevices.

Scalability: Industry 4.0's smart manufacturing processes, combined with additive manufacturing techniques, pave the way for scaling up the production of nanomaterials and nanostructures, making them more accessible for various applications.

12.2.2 Advanced Materials for Industry 4.0

Smart materials: Nanotechnology empowers the improvement of smart materials that can change their properties in light of external stimuli. These materials are fundamental to making versatile and responsive Industry 4.0 frameworks, for example, self-recuperating designs and sensors that can distinguish and respond to changes in their current circumstance.

Efficiency improvements: Nano-enhanced materials contribute to more productive energy storage, generation, and conversion frameworks, which are crucial for sustainable and energy-efficient Industry 4.0 activities.

12.2.3 Data and Connectivity

Sensors and IoT devices: Nanotechnology fundamentally improves the performance and capabilities of sensors and IoT devices. Nanosensors, being highly sensitive and requiring less power, can gather detailed data from manufacturing processes, products, and the environment, feeding into Industry 4.0's information analysis and dynamic systems.

Communication networks: Nanomaterials improve the performance of electronic parts utilized in specialized devices, supporting faster, more reliable data transmission that is essential for the interconnectedness of Industry 4.0 advancements.

12.2.4 Enhanced Computational Capabilities

Quantum computing: Nanotechnology is crucial in the advancement of quantum computing, which promises to increase computing power and efficiency dramatically. This can revolutionize data processing, simulation, and analysis in Industry 4.0, enabling more sophisticated modeling of complex systems.

12.2.5 Customization and Flexibility

Additive manufacturing (3D printing): The ability to control matter at the nanoscale enhances additive manufacturing processes, allowing for the production of products with novel properties (e.g., improved strength, lighter weight, or specific electrical or thermal conductivities) and the customization of products to individual specifications at a mass scale.

12.3 LITERATURE REVIEW ON INDUSTRY 4.0 IN NANOTECHNOLOGY

Tables 12.2 and 12.3 provide a comparative study of the different works done on Industry 4.0 in nanotechnology.

TABLE 12.2
Other Works in Industry 4.0 and Nanotechnology

Ref	Insight	Method Used	Results
[11]	Nanotechnology impacts cost structure in Industry 4.0 manufacturing systems.	Descriptive–analytic methods Comprehensive review of previous studies	Nanotechnology impacts cost structure and strategic cost management in production.
[12]	Industry 4.0 model includes advances in nanotechnology.	Industry 4.0 technologies	Illustrates advancements in manufacturing through modern technologies
[13]	Industry 4.0 focuses on AI, IoT, big data, robotics, and 3D printing.	Artificial intelligence, Internet of Things, big data, robotics, and three-dimensional printing	Defines Industry 4.0 as smart, connected production systems.
[14]	Nanomaterials replace silicon electronics in Industry 4.0 for added benefits.	Summarize recent state of nanomaterial-based electronic devices.	Nanomaterials explored as alternatives for MOS transistors to address limitations.
[15]	Industry 4.0 involves nanotechnology in disruptive technology convergence.	Described industry 4.0 disruption in major industrial sectors.	Identified 99 neologisms of Industry 4.0.
[16]	Nanotechnology is a growing technology shaping modern industry in Industry 4.0.	Industry 4.0 technologies: drones, 3D printing, AI, nanotechnology, robotics	Adoption of information technology blurs real–virtual world boundaries for growth.
[17]	Nanoscience contributes to industrial development in electronics, materials, and biomedicine.	Nanoscience in industry and its impact	Nanoscience and nanotechnology influence vast industrial panorama.
[18]	Industry 4.0 focuses on automation and data exchange in manufacturing.	Summarizes nanotechnology industry and its differences from manufacturing industries.	Provides further reading for additional information beyond main text.

(Continued)

TABLE 12.2 (*Continued*)
Other Works in Industry 4.0 and Nanotechnology

Ref	Insight	Method Used	Results
[19]	Nanotechnology enhances efficiency, design, and productivity in Industry 4.0.	Adoption of a broad-spectrum overview to present diverse applications of nanotechnology.	Nanotechnology has become a general-purpose technology with four generations of nanomaterials.
[20]	Nanotechnology innovations are beneficial across various industrial domains.	Discusses current nanotechnology examples successfully introduced in the market.	Challenges and roadmap for bringing nanotechnology to the market

TABLE 12.3
Other Works in Industry 4.0 and Nanotechnology

Ref	Insight	Method Used	Results
[21]	Industry 4.0 integrates physical, digital, and biological worlds.	It also highlights the policy frameworks required for implementing Industry 4.0.	The paper discusses the essential technologies behind Industry 4.0.
[22]	Industry 4.0 integrates nanotechnology for advanced manufacturing processes.	IoT, artificial intelligence	Describes the impact on modern industries
[23]	Industry 4.0 focuses on smart production using IoT technology.	Intelligent Web (Web 2.0), Internet of Things	Utilization of Intelligent Web and Internet of Things for operations
[24]	The paper presents a flexible big data platform that supports anomaly detection and can be applied to other use cases.	Sensor data combined with IT technology for production quality improvement and incident detection.	The paper discusses how sensor data and IT technology can improve production quality and incident detection in a die casting process.
[25]	The influence of industry 4.0 technologies on the environment	Described industry 4.0 disruption in major industrial sectors.	Enhanced flexibility, agility, and mass customization
[26]	Nanotechnology drives innovation for novel industrial applications in Industry 4.0.	Industry 4.0 technologies: drones, 3D printing, AI, nanotechnology, and robotics	Defines nanotechnology and advanced fabrication methods
[27]	Nanomaterials enhance efficiency and functionality in smart manufacturing processes.	Big data, additive manufacturing, augmented reality, and simulation technologies used	Industry 4.0 connects physical systems with cybersystems for efficiency.

(*Continued*)

TABLE 12.3 (*Continued*)
Other Works in Industry 4.0 and Nanotechnology

Ref	Insight	Method Used	Results
[28]	Nanotechnology can enhance Industry 4.0 through advanced materials and sensors.	Identifying and implementing Industry 4.0 scenarios	Cyber-physical systems, Internet of Things, cloud computing, and cognitive computing
[29]	AI is a key component of current technology that is revolutionizing the Industry 4.0.	Frequency distribution and pie charts were used for data analysis.	88.0%percent of respondents agree that AI-powered solutions enable smart factories.
[30]	Industry 4.0 focuses on ICTs, sensors, and data integration.	CAE tools with finite element method for design aid	Identifies ICTs implemented in industry for productivity and transformation

12.4 APPLICATIONS OF INDUSTRY 4.0 IN NANOTECHNOLOGY

Industry 4.0, characterized by the integration of digital technologies into manufacturing environments, has numerous applications in the field of nanotechnology. Figure 12.4 shows the different applications of Industry 4.0 in nanotechnology [1]. Furthermore, they have the potential to significantly impact sustainable development on both the social and environmental fronts. Organizations must consider how Industry 4.0 technologies contribute to sustainability [26–36].

12.4.1 Smart Manufacturing of Nanomaterials

Industry 4.0's integration into the smart manufacturing of nanomaterials represents a revolutionary step towards automation, efficiency, and precision in the nanotechnology sector. This convergence leverages digital technologies such as the IoT, AI,

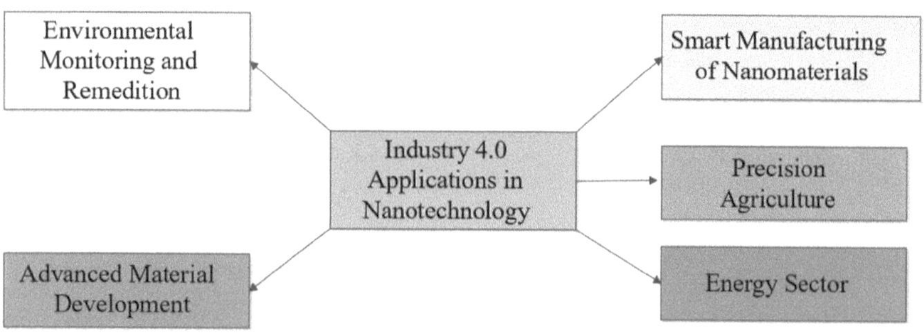

FIGURE 12.4 Industry 4.0 applications in nanotechnology.

machine learning (ML), big data analytics, and cyber-physical systems to enhance the production processes of nanomaterials. Here is a detailed look at how Industry 4.0 advancements change the smart manufacturing of nanomaterials [28].

12.4.1.1 Automation and Robotics

Precision handling and assembly: Robots and automated systems, guided by AI and precise control algorithms, can manipulate materials at the nanoscale with high precision. This capability is crucial for assembling nanodevices and structures, where the slightest error can lead to significant functional discrepancies [26].

High-throughput production: Automated systems enable the mass production of nanomaterials, ensuring consistency and quality while significantly reducing labor costs and human error.

12.4.1.2 IoT and Cyber-Physical Systems

Real-time monitoring: IoT devices and sensors integrated into manufacturing equipment collect real-time data on production processes. This includes parameters like temperature, pressure, and chemical concentrations critical for synthesizing nanomaterials [26].

Adaptive production systems: Cyber-physical systems interpret data from IoT devices to adjust manufacturing processes in real-time, ensuring optimal conditions for nanomaterial synthesis are maintained, thus improving yield and reducing waste.

12.4.1.3 AI and ML

Process optimization: AI and ML algorithms analyze vast amounts of data collected during the manufacturing process to identify patterns and insights. These insights can be used to predict optimal process parameters, reducing trial and error, and significantly speeding up the development of new nanomaterials [25].

Quality control: AI-driven image and pattern recognition tools can detect imperfections in nanomaterials and nanostructures at much earlier stages, ensuring only materials that meet strict quality standards proceed further in the production chain.

12.4.1.4 Big Data Analytics

Predictive analytics: The analysis of big data collected from manufacturing processes can predict equipment failures, material anomalies, and other potential issues before they become critical, reducing downtime and maintenance costs.

Supply chain optimization: Analytics can also streamline the supply chain, predicting demand for various nanomaterials and adjusting production schedules accordingly to meet market needs efficiently. Industry 4.0 is regarded as a new industrial stage where organizations can attain improved industrial performance through the integration of vertical and horizontal manufacturing processes and product connections [2].

12.4.1.5 Cloud Computing and Virtualization

Scalability and flexibility: Cloud computing provides scalable computational resources for data analysis, simulation, and modeling, which are essential for designing nanomaterials and planning production processes.

Collaborative platforms: Virtualization and cloud platforms enable collaboration across different geographical locations, allowing researchers and manufacturers to share insights, data, and simulations. This collaborative environment speeds up innovation and the optimization of nanomaterial manufacturing processes.

By integrating Industry 4.0 advances into the smart assembly of nanomaterials, organizations can accomplish unprecedented levels of efficiency, customization, and flexibility. This technological collaboration not only accelerates the pace of advancement in nanotechnology but also opens up additional opportunities for the utilization of nanomaterials across various industries, from electronics and healthcare to energy and environmental solutions [36].

12.4.2 Precision Agriculture

Nanotechnology, with its ability to manipulate materials at the molecular and atomic level, offers significant benefits for agriculture. It can lead to the development of more effective fertilizers and pesticides, nanosensors for soil and crop monitoring, and nanomaterials that improve water retention and reduce waste.

12.4.2.1 IoT and Nano Sensors

Soil and crop health monitoring: Nanosensors can distinguish soil moisture levels, pH balance, nutrient deficiencies, and the presence of pests or diseases at an early stage. When integrated with IoT devices, this data can be communicated in real-time to farmers and rural frameworks, enabling prompt and precise actions [31].

Water management: Nanotechnology can enhance water retention in soil, reducing water wastage. IoT-based irrigation systems can then use data from nanosensors to optimally manage water use, applying it only where and when needed.

12.4.2.2 AI and Big Data

Predictive analysis: AI algorithms can process the vast amounts of data collected from nanosensors and IoT devices to predict crop yields, pest outbreaks, and optimal planting times. This predictive capability allows for better planning and resource allocation.

Automated decision-making: AI can automate decision-making processes, from adjusting irrigation schedules and nutrient applications to optimizing the timing and methods of harvesting. Industry 4.0 is seen as a new industrial stage where digital solutions are provided by the convergence of numerous emerging technologies [3].

12.4.2.3 Robotics and Drones

Precision application of nanomaterials: Drones and robotic systems can precisely apply nano-enhanced fertilizers and pesticides, reducing the amount of chemicals needed and minimizing environmental impact.

Monitoring and mapping: Drones equipped with imaging technology can monitor crop health and growth at a granular level, supplementing data from nanosensors. This comprehensive view supports identifying issues such as nutrient deficiencies or water pressure [32].

12.4.2.4 Benefits of the Synergy

Increased crop yield and quality: Precise and timely interventions, informed by deep insights into soil and crop health, can significantly improve yields and the quality of produce.

Resource efficiency: The targeted application of water, fertilizers, and pesticides reduces waste and environmental impact. Nanotechnology can also contribute to developing more sustainable agricultural inputs.

Sustainability: Enhanced monitoring and management capabilities lead to more sustainable farming practices, preserving resources and minimizing chemical use. The use of energy and resources, emissions into the environment, and subsequently the impact on society are all greatly influenced by industrial and manufacturing systems [4].

12.4.3 Healthcare and Medicine

The application of Industry 4.0 and nanotechnology in healthcare and medicine represents a transformative convergence of advanced digital technologies and nanoscale innovations, leading to unprecedented improvements in medical diagnostics, treatment, and patient care. This synergy leverages the power of big data analytics, IoT, AI, robotics, and nanotechnology to create smarter, more efficient, and highly personalized healthcare solutions [32].

12.4.3.1 Personalized Medicine

Nanotechnology-enabled drug delivery: Tailored drug delivery systems using nanocarriers can precisely target diseased cells without affecting the surrounding healthy tissue, reducing side effects and improving the efficacy of treatments.

Smart diagnostics: Nanosensors integrated with AI algorithms can analyze biological markers at the nanoscale, enabling early detection of diseases with high precision.

12.4.3.2 Wearable and Implantable Devices

Continuous monitoring: Wearable devices equipped with nanosensors can monitor vital signs and biomarkers in real-time, transmitting data wirelessly for ongoing analysis. This continuous monitoring can lead to early detection of potential health issues.

Implantable nanodevices: Nanotechnology can enhance the functionality of implants, such as pacemakers or insulin pumps, making them more efficient and less invasive. Biosensors and nanoscale analytical tools have huge growth potential in the years to come [5].

12.4.3.3 Regenerative Medicine

Tissue engineering: Nanomaterials can mimic the structure of biological tissues, providing scaffolds for cell growth and tissue regeneration. This is crucial for healing wounds, replacing damaged tissues, and even regenerating organs.

Stem cell therapy: Nanotechnology can improve the delivery and differentiation of stem cells, enhancing their potential for regenerating damaged tissues or organs.

12.4.3.4 Surgical Robotics and Nanorobots

Precision surgery: Industry 4.0 advances like AI and robotics technology can direct surgeries with remarkable accuracy, decreasing human error. When integrated with nanotechnology, medical procedures become less invasive and recovery times shorter. The industry revolution is the dependent variable, while neurosciences and nanotechnology are regarded as independent. To calculate the significance of nanotechnology and neuroscience, utilize intelligent PLS software to generate several outcomes [6].

Nanorobots: These tiny robots could navigate through the body to perform tasks like clearing blood vessel plaques, precisely delivering medications, or performing minute repairs at the cellular level.

12.4.3.5 Data Integration and Predictive Analytics

Integrated health data: By combining patient information from different sources, including wearable sensors and electronic health records, healthcare providers can gain a comprehensive view of a patient's health status.

Predictive analytics: Machine learning and AI algorithms can analyze this integrated data to predict health trends, potential sickness episodes, or the likelihood of individual medical problems before they manifest.

12.4.4 Energy Sector

12.4.4.1 Enhanced Energy Storage

Nanotechnology contributes to the development of high-performance batteries and supercapacitors. Industry 4.0's data analysis and predictive maintenance can optimize the performance and lifespan [33].

12.4.4.2 Smart Grids

Nanotechnology enhances the efficiency and reliability of environmentally friendly power sources incorporated into smart frameworks, which are managed by cutting-edge Industry 4.0 innovations for better energy distribution and usage optimization. Nowadays, the design of products, including power generation systems, greatly benefits from the use of nanomaterials [7].

12.5.5 Environmental Monitoring and Remediation

12.5.5.1 Pollution Detection

Nanosensors, integrated with the IoT, can offer continuous, precise monitoring of air, water, and soil contamination levels. This information is pivotal for timely responses to environmental hazards and for informing policy decisions [31].

12.5.5.2 Nanotechnology in Remediation

AI models can anticipate the effectiveness of nanotechnology-based remediation methodologies (like nano-catalysts for breaking down pollutants), optimizing them for specific environmental conditions. Although nanotechnology is seen as a tool for improving human welfare and problem-solving, if applied carelessly, it can also present risks. When applied appropriately, nanotechnology can help achieve the 17 Sustainable Development Goals that must be accomplished by 2030 [8].

12.6.6 Advanced Materials Development

12.6.6.1 Material Properties Optimization

Using high-throughput computational models and simulations powered by AI, the development of new nanomaterials with desired properties for specific applications can be significantly accelerated.

12.6.6.2 Additive Manufacturing

3D printing at the nanoscale, guided by Industry 4.0 advancements, allows for the production of complex structures with nanoscale accuracy for use in electronics, aerospace, and medical implants.

12.5 RAMIFICATIONS OF INDUSTRY 4.0 ADVANCEMENTS IN PROGRESSING NANOELECTRONICS AND PHOTONICS

The implications of Industry 4.0 for nanoelectronics and photonics are profound and multifaceted, as illustrated in Figure 12.5.

12.5.1 Enhanced Manufacturing Processes

Industry 4.0 advancements can significantly improve the manufacturing processes of nanoelectronics and photonics. Automation and AI-driven frameworks can enhance production lines, reduce waste, and improve accuracy in the manufacture of tiny electronic and photonic parts. This leads to higher yields and better-quality products at lower costs.

12.5.2 Innovation in Materials and Devices

The integration of nanotechnology with Industry 4.0 tools can drive advancement in materials science and engineering, leading to the development of new nanoscale

FIGURE 12.5 Industry 4.0 implications in nanotechnology.

materials and devices. These advancements can transform sectors like energy (e.g., more efficient solar cells), healthcare services (e.g., better imaging frameworks [34]), and communications (e.g., faster and more reliable optical communication systems). The development of microelectromechanical system technology has led to the creation of precise and highly functional small device systems. These microscopic devices are so small that they are invisible to the human eye. Because of their small size, low power consumption, and high performance, they show excellent viability in miniaturizing sensors [9].

12.5.3 Customization and Flexibility

Industry 4.0 empowers greater customization and flexibility in manufacturing, which is especially valuable for the production of nanoelectronics and photonics where custom-made functionalities are often required. This could lead to more customized devices and photonics arrangements appropriate to specific applications or client needs [4].

12.5.4 Integration into New Markets and Applications

With Industry 4.0, nanoelectronics and photonics can be more easily incorporated into various applications, expanding their market reach. For instance, nano-empowered sensors can be coordinated into IoT devices for smarter home frameworks or into biomedical devices for more sensitive diagnostic devices [35].

12.5.5 Supply Chain and Logistics Improvement

Advanced data analytics and connectivity presented by Industry 4.0 can streamline supply chains and operations in the nanoelectronics and photonics sectors. Real-time tracking, predictive maintenance of equipment, and automated inventory management can further improve efficiency and reduce operational costs.

12.5.6 Enhanced Research and Development

ML and AI models can facilitate the research and development process in nanoelectronics and photonics by simulating and predicting the outcomes of experiments or by identifying promising materials and designs faster than traditional methods. This can shorten development cycles and bring innovations to market more quickly.

12.5.7 Environmental Impact

Industry 4.0's emphasis on sustainability can impact the production processes of nanoelectronics and photonics to be more environmentally friendly. Techniques like precision manufacturing help reduce material waste, and energy-efficient technologies can lower the carbon footprint of production facilities. A vast array of environmental problems can potentially have innovative solutions thanks to nanotechnology. These include enhanced techniques for remediation, environmental sensing, water treatment, pollution reduction, and increasing the affordability of alternative energy sources [10].

12.5.8 Security Challenges

As nanoelectronics and photonics increasingly become interconnected through digital architectural frameworks, security becomes a central issue. Protecting these frameworks from cyber threats is crucial, as weaknesses could lead to critical interruptions.

The advancements of Industry 4.0 are shaping the future of nanoelectronics and photonics in significant ways, improving capabilities, efficiency, and applications while also introducing new challenges and potential opportunities for development.

12.6 EFFECT OF INDUSTRY 4.0 ON NANOTECHNOLOGY, SECURITY, ETHICAL CONSIDERATIONS, AND ECOLOGICAL EFFECTS

The integration of Industry 4.0 and nanotechnology brings several transformative effects on security, ethical considerations, and environmental impacts. Each of these areas faces both significant challenges and opportunities:

12.6.1 Security

12.6.1.1 Cybersecurity Risks

As nanotechnology devices become more integrated into Industry 4.0 frameworks, the surface for cyberattacks expands altogether. Nanoscale devices can be integrated into basic frameworks, making them potential targets for hacking. Ensuring the security of these devices is essential to prevent data breaches, unauthorized access, and control.

12.6.1.2 Physical Security

Nanotechnology can enhance safety measures through cutting-edge materials for better security measures or through the development of new surveillance technologies. For instance, nanosensors can detect hazardous chemicals or biological threats, adding to public safety and health.

12.6.1.3 Data Integrity and Privacy

With the proliferation of nanotechnology-based sensors and devices collecting vast amounts of data, ensuring the integrity and privacy of this information becomes a critical concern. There is a requirement for vigorous encryption and information security measures to prevent misuse.

12.6.2 Ethical Considerations

12.6.2.1 Dual Use Dilemma

Nanotechnology, like other cutting-edge innovations, has both civilian and military applications. Moral problems emerge in regard to the development and utilization of such technologies, especially when it involves surveillance or weaponry that could infringe on personal freedoms or escalate military conflicts.

12.6.2.2 Health and Safety

The health implications of exposure to nanoparticles are not completely understood. There are concerns about the long-term impacts of nanoparticles on health, raising moral issues about worker safety in manufacturing environments and the broader public when such particles are released into the environment.

12.6.2.3 Equity and Access

There is a moral concern regarding the fair distribution of the advantages of nanotechnology. As with many cutting-edge innovations, there is a risk that these advantages could be disproportionately accessible to richer people and countries, fueling existing disparities.

12.6.3 Environmental Impacts

12.6.3.1 Pollution Reduction

Nanotechnology can contribute to reducing pollution by improving the efficiency of processes (utilizing less raw material and energy) and by enabling the development of cleaner production technologies. For instance, nanoparticles are utilized in pollution treatment technologies to clean hazardous wastes more efficiently than conventional techniques.

12.6.3.2 Resource Efficiency

Nanotechnology can enhance material and energy efficiency, which is critical for sustainable development. Nano-enhanced materials can be stronger, lighter, and more durable, reducing the need for frequent replacements and conserving resources.

12.6.3.3 Potential Risks

While nanotechnology offers ecological advantages, it also presents risks. Nanoparticles, because of their small size, can be released into the environment during production or disposal and may unknowingly affect ecosystems and human health. The environmental behavior of nanoparticles is not yet fully understood, raising concerns about bioaccumulation and toxicity.

By carefully managing these considerations, the potential of Industry 4.0 and nanotechnology can be realized in a way that maximizes benefits while minimizing adverse outcomes.

12.7 CONTRIBUTIONS AND IMPLICATIONS OF NANOTECHNOLOGY IN INDUSTRY 4.0

12.7.1 Continuous Examination

Nanotechnology enables continuous examination and real-time monitoring of manufacturing processes. By incorporating nanosensors and nanodevices into production lines, companies can achieve unprecedented levels of precision and control. These sensors can detect and measure variables at the nanoscale, providing data that was not previously accessible. This data is crucial for predictive maintenance, quality control, and efficiency optimization.

12.7.2 Normalization

Normalization refers to the standardization of processes and components to guarantee compatibility and interoperability across various frameworks and industries. Nanotechnology contributes to standardization by providing standardized nanomaterials and nanostructures that can be reliably produced and integrated into different applications. This normalization is essential for the widespread reception and increase of nanotechnology applications in the industry.

12.7.3 Dependable Development

The dependable development in the context of Industry 4.0 and nanotechnology involves the creation of robust, reliable, and safe nanotechnology applications. It focuses on developing technologies that are sustainable and have predictable performance over time. Dependable development also entails rigorous testing and compliance with environmental and safety regulations to ensure that nanotechnologies do not pose unforeseen risks.

12.7.4 Harnessing Interdisciplinary Synergy

The synergy between nanotechnology and other disciplines such as robotics, AI, and biotechnology is fueling innovative solutions and applications. For example, combining nanotechnology with AI can lead to smart nanomaterials that adapt their properties in response to environmental stimuli. These interdisciplinary approaches are leading to groundbreaking applications in sectors like healthcare, where nanotechnology is used for targeted drug delivery and diagnostics.

12.8 CONCLUSION

In conclusion, Industry 4.0 represents a transformative wave across various sectors, integrating cutting-edge technologies to revolutionize processes, products, and services. The role of nanotechnology within this system is especially effective, offering precise, efficient, and innovative solutions that were previously unattainable. As we have explored, the applications of nanotechnology within Industry 4.0 encompass consistent evaluation, standardization, and dependable development, each playing a crucial role in harnessing the full potential of this interdisciplinary synergy. Nanotechnology advancements are creating pivotal applications across various fields such as healthcare, electronics, energy, and environmental management. The combination of nanotechnology with AI, advanced robotics, and biotechnology is not only improving existing processes but also paving the way for innovative approaches and solutions that address some of the most pressing challenges facing society today. The journey of integrating nanotechnology within the Industry 4.0 framework is ongoing, and its full potential is yet to be fully realized. As industries continue to evolve and adopt these advanced technologies, the future looks promising, with the potential for significant economic, environmental, and societal benefits. This progress marks an exciting era of innovation and transformation, driven by the power of nanotechnology within Industry 4.0.

REFERENCES

1. Bai, C., P. Dallasega, G. Orzes, and J. Sarkis. "Industry 4.0 technologies assessment: a sustainability perspective." *International Journal of Production Economics* 229 (2020): 107776. https://doi.org/10.1016/j.ijpe.2020.107776

2. Dalenogare, L. S., G. B. Benitez, N. F. Ayala, and A. G. Frank. "The expected contribution of Industry 4.0 technologies for industrial performance." *International Journal of Production Economics* 204 (2018): 383–394. https://doi.org/10.1016/j.ijpe.2018.08.019
3. Frank, A. G., L. S. Dalenogare, and N. F. Ayala. "Industry 4.0 technologies: Implementation patterns in manufacturing companies." *International Journal of Production Economics* 210 (2019): 15–26. https://doi.org/10.1016/j.ijpe.2019.01.004
4. Franciosi, C., B. Iung, S. Miranda, and S. Riemma. "Maintenance for sustainability in the Industry 4.0 context: a scoping literature review." *IFAC-PapersOnLine* 51, no. 11 (2018): 903–908. https://doi.org/10.1016/j.ifacol.2018.08.459
5. Yadav, M. K., R. Kumar, R. K. Ratnesh, J. Singh, R. Chandra, A. Kumar, and A. K. Singh. "Revolutionizing technology with spintronics: devices and their transformative applications." *Materials Science and Engineering: B* 303 (2024): 117293. https://doi.org/10.1016/j.mseb.2024.117293
6. Perez, P. "The importance of neurosciences and nanotechnology in the industrial revolution." *Reviews of Adhesion and Adhesives* 10, no. 2 (2022). https://raajournal.com/menuscript/index.php/raajournal/article/view/283
7. Mouchou, R., T. Laseinde, T. -C. Jen, and K. Ukoba. "Developments in the application of nano materials for photovoltaic solar cell design, based on industry 4.0 integration scheme." In *Advances in Artificial Intelligence, Software and Systems Engineering: Proceedings of the AHFE 2021 Virtual Conferences July 25–29*, 2021, USA, pp. 510–521. Springer International Publishing, 2021. https://doi.org/10.1007/978-3-030-80624-8_64
8. Aithal, S., and P. S. Aithal. "Green and eco-friendly nanotechnology–concepts and industrial prospects." *International Journal of Management, Technology, and Social Sciences (IJMTS)* 6, no. 1 (2021): 1–31. https://ssrn.com/abstract=3778790
9. Kalovrektis, K., I. Dimos, A. Xanakis, and A. Kakarountas. "Nanotechnology as a tool for computational thinking skills using open hardware, embedded systems and repository platform, in Industry 4.0 era." In *2023 8th South-East Europe Design Automation, Computer Engineering, Computer Networks and Social Media Conference (SEEDA-CECNSM)*, pp. 1–6. IEEE, 2023. https://doi.org/10.1109/SEEDA-CECNSM61561.2023.10470788
10. Pathakoti, K., M. Manubolu, and H.-M. Hwang. "Nanotechnology applications for environmental industry." In *Handbook of Nanomaterials for Industrial Applications*, pp. 894–907. Elsevier, 2018. https://doi.org/10.1016/B978-0-12-813351-4.00050-X
11. Ali, A. M. Mohammad, S. M. Jabir, A. A. Kadhim, and A. Almagtome. "Nanotechnology practices and cost restructure for effective cost management under Industry 4.0 based manufacturing systems." *TEM Journal* 11, no. 3 (2022). https://www.ceeol.com/search/article-detail?id=1061864
12. Tsaramirsis, G., A. Kantaros, I. Al-Darraji, D. Piromalis, and C. Apostolopoulos, A. Pavlopoulou, M. Alrammal et al. "A modern approach towards an Industry 4.0 model: from driving technologies to management." *Journal of Sensors* 2022 (2022): 1–18. https://doi.org/10.1155/2022/5023011
13. Suleiman, Z., S. Shaikholla, D. Dikhanbayeva, E. Shehab, and A. Turkyilmaz. "Industry 4.0: Clustering of concepts and characteristics." *Cogent Engineering* 9, no. 1 (2022): 2034264. https://doi.org/10.1080/23311916.2022.2034264
14. Chethipuzha, M., A. R. Abraham, N. Kalarikkal, S. Thomas, and S. Sreeja. "Embracing nanotechnology concepts in the electronics industry." In *Fundamentals and Properties of Multifunctional Nanomaterials*, pp. 405–421. Elsevier, 2021. https://doi.org/10.1016/B978-0-12-822352-9.00004-3
15. Bongomin, O., A. Yemane, B. Kembabazi, C. Malanda, M. C. Mwape, N. S. Mpofu, and D. Tigalana. "Industry 4.0 disruption and its neologisms in major industrial sectors: A state of the art." *Journal of Engineering* 2020 (2020): 1–45. https://doi.org/10.1155/2020/8090521

16. Romanovs, A., I. Pichkalov, E. Sabanovic, and J. Skirelis. "Industry 4.0: methodologies, tools and applications." In *2019 Open Conference of Electrical, Electronic and Information Sciences (eStream)*, pp. 1–4. IEEE, 2019. https://doi.org/10.1109/eStream.2019.8732150
17. Malik, S., K. Muhammad, and Y. Waheed. "Nanotechnology: A revolution in modern industry." *Molecules* 28, no. 2 (2023): 661. https://doi.org/10.3390/molecules28020661
18. Ramsden, J. J. "The nanotechnology industry." *Nanotechnology Perceptions* 9, no. 2 (2013): 102–118. https://doi.org/10.4024/N06RA13A.NTP.09.02
19. Malik, S., K. Muhammad, and Y. Waheed. "Nanotechnology: A revolution in modern industry." *Molecules* 28, no. 2 (2023): 661. https://doi.org/10.3390/molecules28020661
20. Rambaran, T., and R. Schirhagl. "Nanotechnology from lab to industry – A look at current trends." *Nanoscale Advances* 4, no. 18 (2022): 3664–3675. https://doi.org/10.1039/D2NA00439A
21. Zalte, S., S. Deshmukh, P. Patil, M. Patil, and R. K. Kamat. *Industry 4.0: Design Principles, Technologies, and Applications, Handbook of Research on Technical, Privacy, and Security Challenges in a Modern World*, 2022. https://doi.org/10.4018/978-1-6684-5250-9.ch002
22. Nwulu, N., and U. Damisa, "Introduction to Industry 4.0 technologies." *Energy* 4: 1–26. https://doi.org/10.1063/9780735425163_001
23. Schöning, H. "Industry 4.0." *Information Technology* 60, no. 3 (2018): 121–123. https://doi.org/10.1515/ITIT-2018-0015
24. Nazmuz Sakib, S. M. "Comprehensive literature study on Industry 4.0." (2022). https://doi.org/10.33774/coe-2022-805gz
25. Luttge, R. Chapter 4 – Nanotechnology. In *Nano- and Microfabrication for Industrial and Biomedical Applications* (2016): 87–134. https://doi.org/10.1016/B978-0-323-37828-4.00004-2
26. Helmold, M., and B. Terry. Industry 4.0 and Artificial Intelligence (AI). In *Operations and Supply Management 4.0* (2021): 77–83. https://doi.org/10.1007/978-3-030-68696-3_5
27. Vinayendra Mani Tripathi. Role of Modern Technology in Industry 4.0. An Empirical Study in the Context of Artificial Intelligence. *Turkish Online Journal of Qualitative Inquiry* (2020). https://doi.org/10.52783/tojqi.v11i3.10010
28. Mon, A., H. Rene, D. M. Eduardo, Q. Matias, and C. Figuerola Evaluation of technological development for the definition of Industries 4.0. (2018). https://doi.org/10.1109/cacidi.2018.8584187
29. Duncan, T. V. Applications of nanotechnology in food packaging and food safety: Barrier materials, antimicrobials and sensors. *Journal of Colloid and Interface Science* 363, no. (1) (2011): 1–24. https://doi.org/10.1016/j.jcis.2011.07.017
30. Dalenogare, L. S., G. B. Benitez, N. F. Ayala, and A. G. Frank The expected contribution of Industry 4.0 technologies for industrial performance. *International Journal of Production Economics* 204, no. 1 (2018): 383–394. https://doi.org/10.1016/j.ijpe.2018.08.019
31. Kumar, R., R. K. Ratnesh, J. Singh, A. Kumar, and R. Chandra. "IoT-driven experimental framework for advancing electrical impedance tomography." *ECS Journal of Solid State Science and Technology* 13, no. 2 (2024): 027002. https://doi.org/10.1149/2162-8777/ad2331
32. Kumar, R., and R. Mahadeva. "An experimental measurement and control of human body stomach using electrical impedance tomography." *Journal of Circuits, Systems and Computers* 30, no. 6 (2021): 2150103. https://doi.org/10.1142/S0218126621501036.
33. Pandey, P. K., R. Kumar, V. Gupta, A. Kanungo, and S. Diwania. "An innovative hybrid controller-based combined grid-connected hybrid renewable energy system." *Electrical Engineering* (2024). https://doi.org/10.1007/s00202-024-02363-2

34. Kumar, R., and T. Shashank T. "A novel GUI-based image reconstruction algorithm of EIT imaging technique." *International Journal of Cognitive Informatics and Natural Intelligence (IJCINI)* 15, no. 3 (2021): 31–46. https://doi.org/10.4018/IJCINI.20210701.oa3
35. Kumar, R., M. Aljaidi, Singla, A., Gupta, A. M., Alhomoud, A. A., Alsuwaylimi, and S. M. Alenezi. "Development of a prototype global positioning system based stick for blind patients." *International Journal of Online & Biomedical Engineering* 20, no. 8 (2024): 21. https://doi.org/10.3991/ijoe.v20i08.49343
36. Queyam, A. B., R. Kumar, R. K. Ratnesh, and R. K. Chauhan. "LabVIEW-enabled synthetic signal for empowering fetal-maternal healthcare." *ECS Journal of Solid State Science and Technology* 13, no. 5 (2024). https://doi.org/10.1149/2162-8777/ad4dde

Index

Note: **Bold** page numbers refer to tables and *italic* page numbers refer to figures.

B

C

F

G

H

I

K

L

M

N

O

V

W

X

Kolibri 360 GmbH
Ferdinand-Jühlke-Straße 7
99095 Erfurt, Deutschland
produktsicherheit@kolibri360.de